James Shaw

Epitome of Mental Diseases

With the Present Methods of Certification of the Insane

James Shaw

Epitome of Mental Diseases
With the Present Methods of Certification of the Insane

ISBN/EAN: 9783337372743

Printed in Europe, USA, Canada, Australia, Japan

Cover: Foto ©berggeist007 / pixelio.de

More available books at **www.hansebooks.com**

EPITOME OF
MENTAL DISEASES,

WITH THE

PRESENT METHODS OF CERTIFICATION OF THE INSANE, AND
THE EXISTING REGULATIONS AS TO "SINGLE PATIENTS,"

FOR

PRACTITIONERS AND STUDENTS.

BY

JAMES SHAW, M.D., QU. UNIV., IREL.,

MASTER OF SURGERY;
MEMBER OF THE MEDICO-PSYCHOLOGICAL ASSOCIATION;

FORMERLY MEDICAL SUPERINTENDENT AND CO-LICENSEE, HAYDOCK LODGE ASYLUM,
LANCASHIRE;
ASSISTANT MEDICAL OFFICER, GROVE HALL ASYLUM, BOW, LONDON;
ASSISTANT MEDICAL OFFICER, NORFOLK COUNTY ASYLUM.

NEW YORK:
E. B. TREAT, 5, COOPER UNION.

1892.

PREFACE.

This work is intended to be a handy and practical book of reference for general practitioners, and to serve students as an introduction to the more comprehensive treatises and exhaustive monographs. Such a work is necessarily to a great extent a compilation, but it is not altogether so, some cases observed by me both in asylum and private practice having been brought to bear on the subject, especially in the chapters on Pathology and Treatment (VII and VIII). No new classification is offered, but a few of those already proposed are given. These range from the simple one of Esquirol, founded on the still simpler but time-honoured one of Pinel (Mania, Melancholia, Dementia) to the elaborate ones of Spitzka and Krafft-Ebing, the latter's recent classification being, I think, as nearly perfect as the present knowledge of encephalic pathology permits it to be. There is one fault, however,—the exclusion of Organic Dementia; it is difficult to account for this as the mental troubles certainly predominate in some cases.

A symptomatic grouping of the many forms (mostly etiological) is given at the end of Chapter III. to bring them more in line with the classification now commonly used.

With regard to some details: The name Originäre Paranoia (Congenital Paranoia, Chap. I., Originary or Primitive Paranoia) is applied by Krafft-Ebing to the small group of cases of Paranoia (Delusional Insanity), in which the disease develops at or before puberty, the patient having been peculiar mentally from early childhood. There are also frequently physical disturbances (pyrexia, etc.) and structural peculiarities (somatic stigmata), and there is always neurotic heredity. The full meaning of the term "Verbigeration" will be found in a combination of the definition given in the "Contents" table and the description in the text of Chap. II. Kahlbaum, who first described the symptom, distinguishes it from maniacal logorrhœa on the one hand and idiotic echolalia and babbling on the other. It may be mentioned that some authors believe cases of "Katatonia" to be merely instances of masturbatory or of adolescent insanity. To the causes of mental aberration given in Chap. IV., Influenza should be added. The recently observed alternation of asthma and insanity is also noteworthy. In Chap. V. the component parts of some of the senses have been separated, but merely for convenience of investigation; as it has been omitted there it may be mentioned here that the thermic sensibility is readily tested by applying to the skin several tubes containing water of various temperatures. The condition of the sense of weight or muscular resistance (part of the muscular sense) may be ascertained by placing in the hands coins of equal size but different weight (the patient's eyes being closed), or, the patient's eyes being open, small objects exactly alike in appearance but differing in weight may be placed in his hands or suspended from his feet. When investigating the muscular sense the patient's power of performing *active* movements without the aid of vision should be tested. He may be requested to place a limb in a given position, to touch his nose, etc. In addition to the test for *passive* movements given in the text, the leg may be lifted by the heel, and the patient asked to touch his great toe with his index finger. The localisation of focal cerebral disease, during life, by means of electricity, has not received notice. In Chap. VII. it will be seen that the weight of evidence is in favour of the view that the foci of the cortical areas of touch, of the muscular sense, and of voluntary motion are almost co-extensive, and are situated about the middle of the supero-lateral surface of the cerebrum, and merge posteriorly and inferiorly into the other sensory areas and anteriorly into the higher emotional and volitional regions. It is of interest to note that all the organs of sense are derived originally from the skin, that all knowledge is an attempt to reduce the other senses to terms of touch, and that, according to Mr. Herbert Spencer, Professor Bain, and Professor Wundt, the sensation of muscular tension is the primitive element in our intelligence. Dr. Hughlings Jackson and other authorities have formulated the opinion that every case of insanity presents motor and sensory symptoms; hence the utility of studying Organic Dementia in which the underlying lesions are well marked. In stating the hypothesis as to the genesis of sleep, it might be added that fatigue assists the first and second factors by diminishing the impressionability of the nervous elements, and by lessening cardiac action; on the other hand excessive fatigue perverts, or even prevents sleep by impairing the contractility of the cerebral arteries. Sensation and perception exclude each other, according to Mr. H. Spencer, with degrees of stringency which vary inversely: " if the sensations rise to extreme intensity consciousness becomes so absorbed in them, that only by great effort, if it all, can the thing causing them be thought about." The bearing of this on the hypothesis as to hypnosis will be seen when it is remembered that perceptions are almost entirely cortical, whilst sensations are sub-cortical. Mr. Spencer explains the causation of feelings of depression by low *nervous* pressure somewhat as follows:—The pleasurable channels are numerous but shallow, whilst the painful ones are few but deep; at low pressure many of the shallow channels are not permeated and the healthy equilibrium is disturbed, the painful feelings predominating. Sudden diminution of the *mechanical* pressure (as from hæmorrhage etc.) acts in a different way; the fibres convey impulses more readily, giving rise to convulsions. Diminished mechanical pressure may also constitute a factor in the pathogenesis of the Ideenjagd (idea-hunt) of the Germans, the inertia of the fibres being abnormally easy to overcome, though the encephalic nerve cells are weak. Some of the views of Professor Meynert, as to Treatment, Pathology, etc., given in this work, have not been quoted from books, but from his psychiatrical lectures, clinical and pathological demonstrations, and explanations of macroscopical and microscopical preparations during the session 1876-77. Several of the opinions and methods of M. M. Magnan, Luys, Charcot, and Voisin have been learnt from their lectures and cliniques during the session 1878-79.

In using the book for practical purposes, the chapters will be most advantageously taken in the following order, viz., first, Chap. II. (Index of Symptoms, etc.); secondly, Chap. V. (Diagnosis); thirdly, Chap. III. (Index of Diseases, etc.), and Chap. IV. (Etiology); the Prognosis (Chap. VI.) and Treatment (Chap. VIII.) being founded on the diagnosis thus formed; the section, "Morbid Anatomy of Symptoms" (Chap. VII.), will also be found useful in some cases.

Unless the case is a very plain one, it is advisable to see the patient at least twice before certifying, although the essential part of the certificate must be based on the facts observed at one of the visits only. One of the two medical certificates required for a private patient must now be signed by the patient's usual medical attendant, if possible. When it is necessary to certify, instructions for so doing will be found in the final chapter. These instructions are up to date so far as they relate to England and Wales, Ireland, and Scotland, and the State of New York. If the patient is to be kept as a single patient, the regulations as to forms to be sent to the Commissioners, entries to be made, etc., will be found in the same chapter.

I have to thank Dr. W. Z. Myles, Resident Medical Superintendent of the Kilkenny District Asylum, for most of the information as to the method of certification in Ireland.

Of the works in the accompanying "Reference List," I am especially indebted to the following, viz:—

Drs. Bucknill and D. Hack Tuke's "Psychological Medicine" (Churchill); Dr. Clouston's "Lectures on Mental Diseases" (Churchill); Dr. Savage's "Insanity and Allied Neuroses" (Cassell); Dr. Spitzka's "Insanity: Its Classification, Diagnosis, and Treatment"; Dr. Bra's "Manuel des Maladies Mentales"; Dr. Magnan's "Recherches sur les Centres Nerveux" (l'alcoolisme, paralysie générale, etc.); Dr. Voisin's "Traité de la Paralysie Générale des Aliénes"; Prof. Griesinger's "Die Pathologie und Therapie der Psychischen Krankheiten"; Prof. von Krafft-Ebing's "Lehrbuch der psychiatrie"; Prof. Meynert's "Psychiatrie. Klinik der Erkrankungen des Vorderhirns, begründet auf dessen Bau, Leistungen und Ernährung, Erste Hälfte"; Prof. Morselli's "Manuale di Semejotica delle Malattie Mentali," volumo primo.

Thanks are due to Messrs. Wright for the careful and painstaking way they have passed the work through the press, and for the preparation of the General Index.

JAMES SHAW.

63, *Kensington, Liverpool,*
January, 1892.

REFERENCE LIST.

(Bibliography, limited to books and journals referred to or consulted.)

1891, Lunacy Act.
,, "A Plea for the Scientific Study of Insanity" - Dr. J. BATTY TUKE.
,, "Medical Digest" - Dr. R. NEALE.

1890, "Lehrbuch der Psychiatrie" (4 Aufl.) - Prof. VON KRAFFT-EBING.
,, Lunacy Act.
,, "The Pulse" - Dr. BROADBENT.
,, "Sanity and Insanity" - Dr. C. MERCIER.

1889, "The Treatment of Epilepsy" - Dr. W. ALEXANDER.
,, "The Causation of Disease" - Dr. H. CAMPBELL.
,, "A Text Book of Mental Diseases" - Mr. BEVAN LEWIS.

1888, "The Descent of Man" - CHARLES DARWIN.

1887, "The Modern Treatment of Disease by the System of Massage" - Dr. STRETCH DOWSE.
,, "The Science of Thought" - Prof. MAX MÜLLER.

1886, "Clinical Manual" - Dr. FINLAYSON.
,, "Electricity in the Treatment of Disease" - Mr. TUNMER.
,, "Handbook of Diseases of the Ear" - Dr. U. PRITCHARD.
,, and 1880, "General Paralysis of the Insane" (2nd and 1st Editions) - Dr. JULIUS MICKLE.

1885, "The Blot upon the Brain" - Dr. W. W. IRELAND.
,, "Lectures on the Diagnosis of Diseases of the Brain" - Dr. GOWERS.
,, "Manuale di Semejotica delle Malattie Mentali" (Vol. I.) - Prof. MORSELLI.
,, "Nomenclature of Diseases" - R. C. P. LONDON.
,, "Responsibility in Mental Disease" (4th Ed.) - Dr. MAUDSLEY.

1884, "Insanity and Allied Neuroses" (1st Ed.) - Dr. SAVAGE.
,, "Psychiatrie. Klinik der Erkrankungen des Vorderhirns, begründet auf dessen Bau, Leistungen und Ernährung." Erste Hälfte - Prof. MEYNERT.
,, "Medical Electricity" (2nd Edition) - Dr. DE WATTEVILLE.
,, "Lectures on Mental Diseases" - Dr. W. H. O. SANKEY.
,, "Physiological and Pathological Chemistry" - Dr. T. CRANSTON CHARLES.

REFERENCE LIST.

1884, "A Treatise on the Chemical Constitution of the Brain" - - - Dr. THUDICHUM.
,, "A Handbook of Diseases of the Skin" Dr. LIVEING.

1883, "Insanity: Its Classification, Diagnosis, and Treatment" (1st Ed.) - Dr. SPITZKA.
,, "Lectures on Mental Diseases"(1st Ed.) Dr. CLOUSTON.
,, "Manuel des Maladies Mentales" - Dr. BRA.
,, "Diseases of Memory" (2nd Ed.) - Mons. TH. RIBOT.
,, "Mind and Body: The Theories of their Relation" (7th Ed.) - - - Prof. BAIN.

1882, "The Care and Treatment of the Insane in Private Dwellings" - - - Dr. WEATHERLY.
,, "A Dictionary of Medicine" - - Dr. QUAIN.
,, "Chapters in the History of the Insane in the British Isles" - - - Dr. D. HACK TUKE.

1881, "Illusions: A Psychological Study" - Mr. SULLY.
,, "A Treatise on the Diseases of the Nervous System" (1st Ed.) - - Dr. ROSS.
,, "The Factors of the Unsound Mind and the Plea of Insanity" - - Dr. GUY.

1880, "The Brain as an Organ of Mind" - Dr. CHARLTON BASTIAN.
,, "Leçons sur les Localisation Cérébrospinales" - - - - - - Prof. CHARCOT.
,, "Brain and Nerve Exhaustion, Neurasthenia" - - - - - - - Dr. STRETCH DOWSE.
,, "Physiological Chemistry of the Animal Body" (Vol. I.) - - - - Dr. GAMGEE.

1879, "A Manual of Psychological Medicine" (4th Ed.) - - - - - - } Drs. BUCKNILL AND D. HACK TUKE.
,, "Traité de la Paralysie Générale des Aliénes" - - - - - - Dr. AUG. VOISIN.
,, "Experimental Researches on the Regional Temperature of the Head" Dr. LOMBARD.
,, "Principles of Mental Physiology" - Dr. CARPENTER.
,, "Syphilis of the Brain and Spinal Cord" Dr. STRETCH DOWSE.
,, "A Manual of Medical Jurisprudence" (10th Ed.) - - - - - Dr. TAYLOR.
,, "Diseases of Modern Life" - - Dr. B. W. RICHARDSON.

1878, "Influence de l'Alcoolisme sur les Maladies Mentales" - - - Dr. MAGNAN.
,, "Le Cerveau et ses Fonctions" (3ième Ed.) - - - - - - - Dr. LUYS.

1877, "Archbold's Lunacy" (2nd Ed.) - } Mr. W. C. GLEN AND Mr. A. GLEN.

1876, "Die Pathologie und Therapie der Psychischen Krankheiten" (4te Auf.) - Prof. GRIESINGER.
,, "Recherches sur les Centres Nerveux" ("Troubles de l'Intelligence et des Sens dans l'Alcoolisme aigu et chronique," etc.) - - - - Dr. MAGNAN.

REFERENCE LIST.

1876, "The Functions of the Brain" (1st Ed.) — Prof. FERRIER.
,, "Skizzen über Umfang und wissenschaftliche Anordnung der Klinischen Psychiatrie" — Prof. MEYNERT.

1875, "Paralysis from Brain Disease" — Dr. CHARLTON BASTIAN.

1874, "Manual of Lunacy" — Dr. L. S. WINSLOW.

1873, "Body and Mind" — Dr. MAUDSLEY.
,, "Allegemeine Pathologie der Krankheiten des Nervensystems," I. Theil — Prof. HUGNENIN.

1872, "The Influence of the Mind upon the Body" — Dr. D. HACK TUKE.
,, "Der Bau der Gross-Hirnrinde und seine örtlichen Verschiedenheiten nebst einem pathologisch-anatomischen Corollarium" — Prof. MEYNERT.
,, "The Expression of the Emotions in Man and Animals" — CHARLES DARWIN.

1871, "Lectures on Insanity" (1st Ed.) — Dr. BLANDFORD.
,, "The Use of the Ophthalmoscope in Diseases of the Nervous System" — Dr. CLIFFORD ALLBUTT.

1870, "The Principles of Psychology" (2nd Ed.) — Mr. HERBERT SPENCER.

1868, "The Physiology and Pathology of Mind" (2nd Ed.) — Dr. MAUDSLEY.

The "Medical Annual."
"Braithwaite's Retrospect."
The "Liverpool Medico-Chirurgical Journal."
The "Journal of Mental Science."
"Brain."
"Die Allgemeine Zeitschrift. für Psychiatrie."
The "Journal of Nervous and Mental Diseases."
"L'Encéphale."
"Index Medicus."
The "Provincial Medical Journal."
The "Lancet."
The "British Medical Journal."
The "Medical Press and Circular."
"Le Progrès Médical."
"Le Mercredi Médical."
"Centralblatt für die Medicinischen Wissenschaften."

CONTENTS.

CHAPTER I.

DEFINITIONS OF INSANITY AND CLASSIFICATION
OF MENTAL DISEASES - - - - - PAGE 1

Definitions—Classifications—Esquirol's—Commissioners'—Morel's—Of International Congress of Alienists—Skae's—Krafft-Ebing's—Bra's—Clouston's—Spitzka's—Savage's grouping—Of London College of Physicians—Krafft-Ebing's Recent.

CHAPTER II.

INDEX OF SYMPTOMS SOMATIC, PHYSIOLOGICAL,
AND PSYCHICAL, WITH THE MENTAL DISEASES
IN WHICH THEY OCCUR - - - - PAGE 11

Abulia — Amnesia—Attention, defective — Bulimia — Consciousness impaired—Delusions—Emotional disturbances—Facial alterations —Grip or hand-grasp, feeble—Hallucinations—Illusions—Imperative conceptions or obsessions—Impulsive acts—Loquacity—Morbid propensities—Mutism—Noisiness—Optic nerve changes—Paralysis —Pupils, alterations of—Pulse changes—Reaction time—Restlessness—Retina, changes in—Sexual perversion—Somatic stigmata—Speech, abnormalities of—Temperature changes—Tongue changes —Tremor —Urine, alterations in — Verbigeration (a continued repetition of meaningless or disconnected sounds, words, or phrases. Kahlbaum)—Weeping—"Wet" in habits—Writing altered, etc., etc.

CHAPTER III.

INDEX OF MENTAL DISEASES WITH THEIR
SYNONYMS AND SYMPTOMS - - - - PAGE 49

Abdominal Disorders (insanity from)—Adolescent Insanity—Anæmic Insanity—Bright's Disease (insanity of)—Cataleptic Insanity—Choreic Insanity—Circular Insanity—Climacteric Insanity—Coarse Brain Disease (insanity from)—Confusional Insanity—Consecutive Insanity— Cyanosis from Bronchitis, Cardiac Disease and Asthma (insanity of)—Delirium, acute—Delusional Insanity (paranoia)—Dementia, Terminal—Deprivation of Senses (insanity from)—Diabetic Insanity — Epileptic Insanity —Exophthalmic Goitre (insanity with)—Folie à Deux—Folie du Doute—General Paralysis of the Insane—Gestational Insanity—Hypochondriasis—Hysterical Insanity—Idiocy (including Imbecility and Cretinism)—Impulsive Insanity—Katatonic Insanity—Lactational Insanity—Mania—Masturbational Insanity—Melancholia—Mental Deterioration, Primary—Metastatic Insanity—Moral Insanity—Myxœdema (insanity of)—Neurasthenia and Neurasthenic Insanity—Ovarian or Old Maid's Insanity—Oxaluria and Phosphaturia (insanity of)—Paralysis Agitans (insanity of)—Partial Emotional Aberration—Partial Exaltation or Amenomania—Pellagrous Insanity—Periodical Insanity—Phthisical Insanity—Podagrous or Gouty Insanity—Post-Connubial Insanity—Pubescent Insanity—Puerperal Insanity—Reasoning Insanity (Folie Raisonnante)—Rheumatic Insanity—Senile Insanity—Somnambulism (Pseudo-Insanity of)—Stupor, Anergic—Syphilitic Insanity—Toxic Insanity—Traumatic Insanity—Uterine or Amenorrhœal Insanity—Young Children, Delirium of.

Grouping of the foregoing forms of Mental Aberration according to one or two of the most prominent symptoms.

I. Mental Pain or Hindrance (Hampering) of Mental Action, or both—II. Emotional Exaltation, or Excitement (Mental or Motor), or both—III. Delusion, Hallucination, Fixed Idea (Obsession), Morbid Impulse, or Extraordinary Actions—IV. Acquired Mental Weakness—V. Acquired Mental Weakness with Delusions and Hallucinations—VI.—Acquired Mental Weakness with Paresis—VII. Acquired Mental Weakness with Paralysis—VIII. Stupor—IX. Delirium with Unconsciousness—X. Congenital, Mental or Moral Weakness.

CHAPTER IV.

ETIOLOGY - - - - - - - PAGE 134

A. General: Causes of Insanity in England and Wales from 1878 to 1887; Predisposing Causes; Exciting Causes—B. Special (Causes of the various forms of Insanity).

CHAPTER V.

DIAGNOSIS - - - - - - - PAGE 146

A Diagnosis of Insanity from other conditions (Eccentricity, Feigned Insanity, the Delirium of Fevers and Inflammations, Alcoholic or other Intoxication, Cerebral Meningitis, Aphasia)—Method of Examining a Patient—To investigate the Acuteness, etc., of the various Senses (touch, smell, taste, hearing, vision, and the muscular sense)—Description of Mental and Bodily Condition on Admission (to be made in "Case Book")—B. Differential Diagnosis of the Forms of Insanity—Alphabetical List of Mental Diseases with References to the preceding Paragraphs—Frequency of Principal Forms.

CHAPTER VI.

PROGNOSIS - - - - - - - PAGE 171

A. General Prognosis—I. As to Danger to Life—II. As to Recovery from Mental Derangement—B. Special Prognosis (including duration)—Most of the Forms in alphabetical order.

CHAPTER VII.

PATHOLOGICAL ANATOMY, PATHOLOGY, AND PATHOGENESIS - - - - - PAGE 189

Morbid Anatomy: A. General (encephalic)—B. Special (encephalic)—C. Morbid Anatomy (encephalic) of Symptoms (sensory, motor, emotional, etc.)—D. Lesions of Non-Nervous Organs, Tissues, etc.

Pathology and Pathogenesis: General Considerations—Cortical Cells and their Connections—Central Ganglia—Sleep, Hypothesis as to Physiology of—Cerebrum and Cerebellum—Hypothesis as to Functions of Cerebellum—Somnambulism—Hypnotism—Effects on Brain of Impressions from Abdominal and Thoracic Viscera—Effects of Cerebral Nutrition—Insanity in Children—Pathogenesis of Symptoms—Symptoms that are seldom absent—Climacteric Insanity.

CHAPTER VIII.

THERAPEUTICS AND HYGIENE - - - PAGE 242

Prophylaxis: Remedial Treatment.—I. The immediate relief of Urgent Symptoms, Insomnia, Excitement, Sitophobia, Suicidal or Homicidal Tendencies, etc.—II. Ultimate Care and Treatment A. Home Treatment—B. Private Care (Single Patients)—C. Asylum Treatment and Care: 1. Pauper Asylums; 2. Lunatic Hospitals; 3. Private Asylums. Treatment of some of the Forms of Insanity.

CHAPTER IX.

LEGAL REGULATIONS & FORENSIC PSYCHIATRY - PAGE 269

Certification of Insane Private Patients in England and Wales—Voluntary Boarders—Laws as to Keeping Single Patients in England and Wales—Chancery Patients—Uncertified Lunatics—Pauper Lunatics—Lunatics (not Paupers) not under proper Care and Control or Cruelly Treated or Neglected—Wandering Lunatics—Criminal Lunatics—Certification of the Insane in Scotland—Certification of the Insane in Ireland—Certification of the Insane in the State of New York—Certification of the Insane in the States of Connecticut, Pennsylvania, Massachusetts, and Illinois—Testamentary Capacity of the Insane—Evidence (Testimony) of the Insane—Legal Tests of Insanity and Legal Responsibility of the Insane.

GENERAL INDEX.

Epitome of Mental Diseases.

CHAPTER I.

Definitions of Insanity and Classifications of Mental Diseases.

DEFINITIONS.

FROM the physician's point of view insanity has to be considered as a disease of the brain or a disorder of the mind, quite apart from any consideration of responsibility whatever (Savage). A man must be considered as sane or insane in relation to himself (Savage). The statement that insanity is a perversion of the ego is absolutely true (Savage). The last two statements manifestly exclude congenital mental defects. Alteration of sentiments, disposition, and conduct in a morbid manner is sufficient to constitute an individual insane without the presence of a delusion (Griesinger). A disease of the brain (idiopathic or sympathetic) affecting the integrity of the mind whether marked by intellectual or emotional disorder, such affection not being the mere symptom or immediate result of fever or poison (Bucknill and Tuke).

Insanity may be (*a*,) Congenital, the mental powers or moral character being much below the normal average standard at the same age; or (*b*,) Acquired, the original character of the patient being morbidly altered (See "Legal Tests" of Insanity, Chap. IX.).

CLASSIFICATIONS.

ESQIROL'S CLASSIFICATION (FOUNDED ON PINEL'S).

(1,) Lypemania (Melancholy of the ancients); (2,) Monomania; (3,) Mania; (4,) Dementia; (5,) Imbecility or Idiocy.

THE PRINCIPAL FORMS OF INSANITY ARE CLASSIFIED BY THE COMMISSIONERS IN THEIR REPORT (1844) UNDER THE FOLLOWING HEADS:—

I.—Mania, which is thus divided: (1,) Acute Mania or Raving Madness; (2,) Ordinary Mania, or Chronic Madness of a less acute form; (3,) Periodical, or Intermittent Mania, with comparatively lucid intervals.

II.—Dementia, or decay and obliteration of the intellectual faculties.

III.—Melancholia.
IV.—Monomania.
V.—Moral Insanity.

(The three last mentioned forms are sometimes comprehended under the term Partial Insanity.)

VI.—Congenital Idiocy.
VII.—Congenital Imbecility.
VIII.—General Paralysis of the Insane.
IX.—Epilepsy.
X.—Delirium Tremens, which may perhaps be added to these heads, since it is mentioned as a form of insanity in the reports of some lunatic asylums ("Archbold's Lunacy," 2nd Ed., p. 651).

MOREL'S CLASSIFICATION (ABOUT 1860) PRINCIPALLY SOMATO.—ETIOLOGICAL.

Group I.—Hereditary Insanity: Class 1.—Those who are of congenitally nervous temperament; 2.—Those whose insanity is indicated by insane acts rather than insane conversation. Includes Prichard's Moral Insanity; 3.—Constitutes the transition state between Class 2, and idiots and imbeciles. The members of this class are marked by morbid impulses to incendiary acts, theft, etc.; 4.—Idiots and Imbeciles.

Group II.—Toxic Insanity: Class 1.—Caused by intoxicating substances as alcohol, opium, etc. Also poisonous ingredients as lead, mercury, etc.; 2.—Caused by insufficient or diseased food, as ergot of rye; 3.—Caused by Marsh Miasma or the geological constitution of the soil (*e.g.*, Cretinism).

Group III.—Insanity produced by the transformation of other cases: Class 1.—Hysterical Insanity; 2.—Epileptic ditto; 3.—Hypochondriacal ditto, consisting of three varieties.

Group IV.—Idiopathic Insanity: Class 1.—Progressive weakening or abolition of the intellectual faculties, resulting from chronic disease of the brain or its membranes; 2.—General Paralysis.

Group V.—Sympathetic Insanity.

Group VI.—Dementia "a terminative state" ("Bucknill and Tuke," pp. 40-41).

Classification proposed by the International Congress of Alienists (Paris 1867).

(1,) Simple Insanity comprehends the different varieties of Mania, Melancholia, and Monomania, Circular Insanity, and Mixed Insanity, Delusion of Persecution, Moral Insanity, and the Dementia following these different forms of insanity.

(2,) Epileptic Insanity means insanity with Epilepsy, whether the convulsive affection has preceded the insanity and has seemed to have been the cause, or has appeared during the course of the mental disease, only as a symptom or complication.

(3,) Paralytic Insanity or Dementia should be considered as a distinct morbid entity, and not at all as a complication, a termination of certain forms of Insanity. There should be comprehended, then, under the name of Paralytic Insane, all the insane who show in any degree whatever, the characteristic symptoms of this disease.

(4,) Senile Dementia is the slow and progressive enfeeblement of the intellectual and moral faculties consequent upon old age.

(5,) Organic Dementia embraces all the varieties of Dementia other than the preceding, and which are caused by organic lesions of the brain, nearly always local, and presenting, as almost constant symptoms, hemiplegic occurrences more or less prolonged.

(6,) Idiocy is characterised by the absence or arrest of the development of the intellectual and moral faculties, Imbecility and Weakness of Mind constituting two degrees or varieties.

(7,) Cretinism is characterised by a lesion of the intellectual faculties, more or less analogous to that observed in Idiocy, but with which is uniformly associated a characteristic vicious conformation of the body, an arrest of the development of the entirety of the organism.

Under the titles "Ill-defined Forms," "Other Forms," are to be set down all the varieties of Mental Alienation which it shall seem impossible to associate with any of the preceding typical forms ("Bucknill and Tuke," pp. 45-46).

SKAE'S CLASSIFICATION, ESSENTIALLY THOUGH NOT WHOLLY ETIOLOGICAL.

Moral and Intellectual Idiocy and Imbecility
Epileptic Insanity
Insanity of Masturbation
Insanity of Pubescence
Hysterical Mania
Amenorrhœal Mania
Post-connubial Mania
Puerperal Mania
Mania of Pregnancy
Mania of Lactation
Climacteric Mania
Ovario-Mania (Utero-Mania)
Post-febrile Mania

Mania of Oxaluria and Phosphaturia
Senile Mania
Phthisical Mania
Metastatic Mania
Traumatic Mania
Syphilitic Mania
Delirium Tremens
Dipsomania
Mania of Alcoholism
General Paralysis with Insanity
Epidemic Mania

Idiopathic Insanity } Sthenic / Asthenic

KRAFFT-EBING'S CLASSIFICATION ("LEHRBUCH DER PSYCHIATRIE," 1879. FROM SPITZKA'S "INSANITY," pp. 116-117).

Group A.—Mental affections of the developed brain.
I.—*Psychoneuroses*:—
(1,) Primary curable conditions
(*a*,) Melancholia ; (a,) Melancholia passiva ; (β,) Melancholia Attonita.
(*b*,) Mania : (a,) Maniacal Exaltation ; (β,) Maniacal Frenzy.
(*c*,) Stupor.
(2,) Secondary incurable states.
(*a*,) Secondary monomania (Secundäre Verrücktheit).
(*b*,) Terminal dementia ; (a,) Dementia agitata ; (β,) Dementia Apathetica.
II.—*Psychical Degenerative States*:—
(*a*,) Constitutional affective insanity (folie raisonnante)
(*b*,) Moral insanity.
(*c*,) Primary monomania (primäre Verrücktheit): (a,) With delusions ; (aa,) Of a persecutory tinge ; (ββ,) Of an ambitious tinge ; (β,) With imperative conceptions.
(*d*,) Insanities transformed from the constitutional neuroses ; (a,) Epileptic ; (β,) Hysterical ; (γ,) Hypochondriacal.
(*e*,) Periodical insanity.
III.—*Brain Diseases with Predominating Mental Symptoms*:—
(*a*,) Dementia paralytica.
(*b*,) Lues cerebralis.

(*c*,) Chronic alcoholism.
(*d*,) Senile dementia.
(*e*,) Acute delirium.
Group B.—Mental Results of Arrested Brain Development, Idiocy and Cretinism.

BRA'S CLASSIFICATION, 1883 ("MANUEL DES MALADIES MENTALES.") BASED ON ETIOLOGY. FOUNDED ON THE CLASSIFICATIONS OF BALL AND MOREL :—

(1,) *Vesanic* (without actually well determined lesions). General delirium ; mania, melancholia, folie circulaire. Partial delirium ; delirium of persecution, religious insanity, mania of suspicion, dipsomania.

(2,) *Neuropathic*.—Hysterical ; epileptic ; choreic ; cataleptic, etc.

(3,) *Diathetic*.—Gouty ; tuberculous ; syphilitic.
(4,) *Sympathetic*.—Genital ; puerperal.
(5,) *Toxic*.—Alcoholic ; saturnine.
(6,) *Organic*.—Acute delirium ; general paralysis ; dementia.
(7,) *Congenital* or *Morphological*.—Idiocy ; imbecility ; cretinism.

CLOUSTON'S SYMPTOMATOLOGICAL CLASSIFICATION, 1883, ("CLINICAL LECTURES ON MENTAL DISEASES," pp. 19-20).

(1,) States of mental depression (melancholia, psychalgia) :—
(*a*,) Simple melancholia ; (*b*,) Hypochondriacal melancholia ; (*c*,) Delusional Melancholia ; (*d*,) Excited melancholia ; (*e*,) Resistive (obstinate) melancholia ; (*f*,) Convulsive melancholia ; (*g*,) Organic melancholia ; (*h*,) Suicidal and homicidal melancholia.

(2,) States of mental exaltation (mania, psychlampsia) :—(*a*,) Simple mania ; (*b*,) Acute mania ; (*c*,) Delusional mania ; (*d*,) Chronic mania ; (*e*,) Ephemeral mania (mania transitoria) ; (*f*,) Homicidal mania.

(3,) States of regularly alternating mental conditions (folie circulaire, psychorythm, folie à double forme, circular insanity, periodic mania, recurrent mania, katatonia).

(4,) States of fixed and limited delusion (monomania, monopsychosis) :—(*a*,) Monomania of pride and grandeur ; (*b*,) Monomania of unseen agency ; (*c*,) Monomania of suspicion.

(5,) States of mental enfeeblement (dementia and amentia, psychoparesis, congenital imbecility, idiocy) :—(*a*,) Secondary (ordinary) dementia (following mania and melancholia) ; (*b*,) Primary enfeeblement (imbecility, idiocy, cretinism, the result of deficient brain development, or of brain disease in very early life);

(*c*,) Senile dementia ; (*d*,) Organic dementia (the result of gross organic brain disease).

(6,) States of mental stupor (stupor, psychocoma):—(*a*,) Melancholic stupor, "melancholia attonita"; (*b*,) Anergic stupor, primary dementia, "dementia attonita"; (*c*,) Secondary stupor (transitory after acute mania).

(7,) States of defective inhibition (psychokinesia hyperkinesia, impulsive insanity, volitional insanity, uncontrollable impulse; insanity without delusion):—(*a*,) General impulsiveness; (*b*,) Epileptiform impulse; (*c*,) Animal, sexual, and organic impulse; (*d*,) Homicidal impulse; (*e*,) Suicidal impulse; (*f*,) Destructive impulse; (*g*,) Dipsomania; (*h*,) Kleptomania; (*i*,) Pyromania; (*k*,) Moral insanity.

(8,) The insane diathesis (psychoneurosis, neurosis insana, neurosis spasmodica).

CLOUSTON'S CLINICAL CLASSIFICATION (*op. cit.* p. 21), INCLUDING THE PATHOLOGICAL VARIETIES OF MENTAL DISEASE.

(1,) General paralysis ; (2,) Paralytic insanity (organic dementia); (3,) Traumatic insanity ; (4,) Epileptic insanity ; (5,) Syphilitic insanity ; (6,) Alcoholic (and toxic) insanity ; (7,) Rheumatic and choreic insanity ; (8,) Gouty (podagrous) insanity ; (9,) Phthisical insanity : (10,) Uterine insanity ; (11,) Ovarian insanity ; (12,) Hysterical insanity ; (13,) Masturbational insanity ; (14,) Puerperal insanity ; (15,) Lactational insanity ; (16,) Insanity of pregnancy ; (17,) Insanity of puberty and adolescence ; (18,) Climacteric insanity ; (19,) Senile insanity. With a number of more rare and less important varieties, viz :—(1,) Anæmic insanity ; (2,) Diabetic insanity ; (3,) Insanity from Bright's disease ; (4,) The insanity of oxaluria and phosphaturia; (5,) The insanity of cyanosis from bronchitis, cardiac disease and asthma ; (6,) Metastatic insanity ; (7,) Post-febrile insanity ; (8,) Insanity from deprivation of the senses ; (9,) The insanity of myxœdema ; (10,) The insanity of exophthalmic goitre ; (11,) The delirium of young children ; (12,) The insanity of lead poisoning ; (13,) Post-connubial insanity ; (14,) The pseudo-insanity of somnambulism.

SPITZKA'S CLASSIFICATION ("INSANITY, ITS CLASSIFICATION," Etc., p. 126) 1883.

GROUP FIRST, PURE INSANITIES.

Sub-Group (A,)—Simple insanity, not essentially the manifestation of a constitutional neurotic condition.

First Class.—Not associated with demonstrable active organic changes of the brain.

I. Division.—Attacking the individual irrespective of the physiological periods.
(α,) Order of primary origin.
Sub-Order (A,)—Characterised by a fundamental emotional disturbance :—
Genus (1,) Of a pleasurable and expansive character—*Simple Mania*.
Genus (2,) Of a painful character—*Simple Melancholia*.
Genus (3,) Of a pathetic character—*Katatonia*.
Genus (4,) Of an explosive transitory kind—*Transitory Frenzy*.
Sub-Order (B,)—Not characterised by a fundamental emotional disturbance :—
Genus (5,) With simple impairment or abolition of mental energy—*Stuporous Insanity*.
Genus (6,) With confusional delirium—*Primary Confusional Insanity*.
Genus (7,) With uncomplicated progressive mental impairment—*Primary Deterioration*.
(β,) Order : Of secondary origin :—
Genus (8,) *Secondary Confusional Insanity*.
Genus (9,) *Terminal Dementia*.
II. Division.—Attacking the individual in essential connection with the developmental or involutional periods (a single order).
Genus (10,) With senile involution—*Senile Dementia*.
Genus (11,) With the period of puberty—*Insanity of Pubescence* (Hebephrenia).
Second Class.—Associated with demonstrable active organic changes of the brain (orders coincide with genera).
Genus (12,) Which are diffuse in distribution, primarily vasomotor in origin, chronic in course and destructive in their results—*Paretic Dementia*.
Genus (13,) Having the specific luetic character—*Syphilitic Dementia*.
Genus (14,) Of the kind ordinarily encountered by the neurologist, such as encephalomalacia, hæmorrhage, neoplasms, meningitis, parasites, etc.—*Dementia from Coarse Brain Disease*.
Genus (15,) Which are primarily congestive in character and furibund in development—*Delirium Grave* (Acute Delirium, *Manie grave*).
Sub-Group (B,)—Constitutional insanity, essentially the expression of a continuous neurotic condition.
Third Class.--Dependent on the great neuroses (orders and genera coincide).
I. Division.—The toxic neuroses :—
Genus (16,) Due to alcoholic abuse—*Alcoholic Insanity*.

(Analogous forms, such as those due to abuse of opium, the bromides, and chloral, need not be enumerated here owing to their rarity.)

II. Division.—The natural neuroses :—
Genus (17,) The hysterical neurosis—*Hysterical Insanity.*
Genus (18,) The epileptic neurosis—*Epileptic Insanity.*
Fourth Class.—Independent of the great neuroses (representing a single order).
Genus (19,) In periodical exacerbations—*Periodical Insanity.*
Order.—Arrested development :—
Genus (20,) *Idiocy* and *Imbecility.*
Genus (21,) *Cretinism.*
Genus (22,) Manifesting itself in primary dissociation of the mental elements, or in a failure of the logical inhibitory power, or of both—*Monomania.*

GROUP SECOND, COMPLICATING INSANITIES.—These may be divided into the following main orders, which, as a general thing, are at the same time genera : *traumatic, choreic, post-febrile, rheumatic, gouty, phthisical, sympathetic, pellagrous.*

SAVAGE MAKES USE OF THE FOLLOWING GROUPS ("INSANITY AND ALLIED NEUROSES," p. 12, 1884).

Hysteria—mania ; hypochondriasis—melancholia ; Dementia, general and partial, primary and secondary ; states of mental weakness—chronic mania and melancholia ; recurrent insanity : delusional insanity ; general paralysis of the insane ; paralytic insanity—epileptic insanity ; puerperal insanity—post-connubial, puerperal, lactation ; toxic insanity—alcohol, lead, opium, chloral, gout, etc. ; visceral insanity—renal, cardiac, pulmonary ; insanity with syphilis—myxœdema—Graves' disease—asthma—diabetes : idiocy in its various forms.

CLASSIFICATION ADOPTED BY THE COMMITTEE OF THE LONDON COLLEGE OF PHYSICIANS, 1885 ("NOMENCLATURE OF DISEASES," pp. 29-31).

They divide Mental Diseases into Hypochondriasis and Insanity, and sub-divide Insanity into :—Mania : melancholia ; dementia; including acquired imbecility ; idiocy, synonym, congenital imbecility ; general paralysis of the insane ; puerperal insanity : epileptic insanity ; insanity of puberty ; climacteric insanity : senile insanity ; toxic insanity, from alcohol, gout, lead, etc., variety, delirium tremens ; traumatic insanity ; insanity associated with obvious morbid change or changes in the brain ; consecutive insanity, from fevers, visceral inflammations, etc.

NOTE.—So-called cases of monomania should be named according as the prevailing symptoms are those of mania, melancholia, or dementia; and distinct hereditary tendency should be mentioned.

KRAFFT-EBING'S RECENT CLASSIFICATION (LEHRBUCH DER PSYCHIATRIE, 4th Ed., 1890, pp. 325-326).

A.—Psychical affections of the developed brain.

I.—Diseases without any discernible lesion—functional psychoses.

AA.—Psychoneuroses, that is to say, diseased conditions of the normally constituted and previously healthy brain.

(1,) Melancholia (inhibiting neurosis of the psychical organ): (*a*,) Melancholia simplex; (*b*,) Melancholia cum stupore.

(2,) Mania (discharging neurosis): (*a*,) Maniacal exaltation; (*b*,) Maniacal Frenzy (*tobsucht*).

(3,) Stupor or acute and curable dementia (neurosis of exhaustion).

(4,) Hallucinatory delirium, hallucinatory psychoneurosis (Wahnsinn).

Chronic mania or *chronic confusional insanity* (secundäre Verrücktheit) and *secondary dementia* are terminal incurable states of the above four conditions. There are two clinical varieties of secondary dementia, viz., *agitated* and *apathetic*.

BB.—Psychical degenerative states, that is to say, affections of the morbidly constituted or weakened brain.

(1,) Constitutional affective insanity ("folie raisonnante").

(2,) Paranoia: (*a*,) Congenital form; (*b*,) Acquired form.

(*a*,) Paranoia persecutoria (primary and predominating delusions as to injury of the patient by others): (*aa*,) Typical form; (*ββ*,) Paranoia querulans.

(*β*,) Paranoia expansiva (primary and predominating delusions of advanced interests of the personality): (*aa*,) Paranoia inventtoria or paranoia reformatoria; (*ββ*,) Paranoia religiosa; (*γγ*,) Paranoia erotica.

(3,) Periodical insanity.

(4,) Insanity proceeding from the constitutional neuroses: (*a*,) Neurasthenic insanity; (*b*,) Epileptic insanity; (*c*,) Hysterical insanity; (*d*,) Hypochondriacal insanity.

II.—Diseases in which lesions are constantly found—Brain diseases with predominating mental troubles—Organic psychoses.

(1,) Delirium acutum (transudative hyperæmia passing into periencephalitis diffusa acuta).

(2,) Chronic paralysis or dementia paralytica (periencephalomeningitis diffusa chronica).

(3,) Lues cerebralis.

(4,) Dementia senilis (primary cerebral atrophy).

The intoxications constitute a transition group between I. and II :—(1,) Alcoholismus chronicus ; (2,) Morphinismus.

B.—States of arrested psychical development—Idiocy (with eventual bodily degeneration—Cretinism) : (*a*,) with predominating intellectual defect—(congenital imbecility and idiocy) ; (*b*,) With predominating ethical defect (congenital moral imbecility and idiocy).

CHAPTER II.

INDEX OF SYMPTOMS SOMATIC AND PSYCHICAL, WITH THE MENTAL DISEASES IN WHICH THEY OCCUR.

Abruptness of Outbreak.—In the attacks of periodical insanity (usually).

Absent-Mindedness.—Primary mental deterioration (simple primary dementia).

Abulia.—Weakness or want of will. In simple melancholia; prodromal period of mania; prodromal period of general paralysis; alcoholic insanity; periodical melancholia; forms ending in general mental enfeeblement; monomania with overwhelming hallucinations and delusion (Spitzka). In chronic hysterical insanity.

Activity, Unusual and Useless.—In the prodromal period of general paralysis.

Acts, Extraordinary.—In moral insanity.

Acts, Monotonous.—In katatonic insanity (katatonia).

Affective Sensibility, Perverted or Paralysed.—Natural affection altered, diminished, or lost. In general paralysis; chronic alcoholic insanity; puerperal insanity; lactational insanity; gestational insanity; advanced consecutive insanity (*post*-febrile); acute mania; simple melancholia; climacteric insanity; prodromal period of general paralysis; second stage of masturbational insanity; diabetic insanity; moral insanity; periodical insanity; pubescent insanity (in boys at beginning of attack); chronic mania.

Ageustia.—Loss of the sense of taste. In confirmed general paralysis.

Aggressiveness.—In acute mania; chronic mania; some cases of agitated terminal dementia; epileptic insanity; traumatic insanity; periodical mania; monomania; sometimes in general paralysis.

Agitation, Excessive.—In agitated melancholia; delirium tremens; maniacal form of saturnine insanity; acute mania; in some cases of depressive general paralysis; in acute hysterical insanity; constant in expansive syphilitic insanity; some cases of delirium of young children.

Alcoholic Excesses, Tendency to.—Toxic insanity (alcoholic form); impulsive insanity (dipsomaniacal form); traumatic insanity; periodical insanity; circular insanity; climacteric insanity; general paralysis (prodromal period and first stage).

Alternating Exaltation and Depression, with or without Lucid Interval.—Circular insanity (folie circulaire, folie alternante, and folie à double forme); occasionally in anæmic insanity.

Ambitious Ideas.—Ambitious delusional insanity (ambitious monomania); religious delusional insanity (religious monomania, theomania); general paralysis.

Amenorrhœa.—In prodromal period of general paralysis. In most forms of insanity at some period.

Amnesia, marked (apart from Unconsciousness).—Loss of memory.

(1,) For recent events. In prodromal period and first stage of general paralysis; primary mental deterioration (simple primary dementia); senile insanity; some cases of chronic mania and terminal dementia; some cases of organic dementia; some cases of chronic alcoholic insanity.

(2,) For long past and recent events. Advanced terminal dementia; advanced chronic alcoholic insanity; epileptic insanity; third stage of general paralysis; simple depressive syphilitic insanity (advanced); advanced senile insanity; organic dementia (some cases); organic melancholia; some cases of acute mania; melancholy form of delirious saturnine insanity; advanced consecutive insanity (*post*-febrile form); choreic insanity; insanity of myxœdema; most severe form of cataleptic insanity; delirium of young children; sometimes after traumatisms.

Anæmia.—In anæmic insanity; early chronic hysterical insanity.

Anæsthesia.—See "Cutaneous Anæsthesia."

Annoyance at Trifles.—In incipient climacteric insanity.

Anorexia.—Loss of appetite. Alcoholic pseudo-general paralysis. Many cases of melancholia (See "Refusal of Food").

Anosmia.—Absence or loss of the sense of smell. In some cases of general paralysis at first, and in all at last; some cases of organic dementia; some cases of idiocy.

Answers Questions Irrelevantly.—In mania; primary mental deterioration (simple primary dementia); terminal dementia (secondary dementia).

Anxiety.—Alcoholic insanity; excited melancholia; hypochondriacal melancholia; puerperal insanity.

Apathy.—Anergic stupor (acute dementia, acute primary dementia); mild gestational insanity (mild insanity of pregnancy); chronic alcoholic insanity.

Aphasia.—Loss of speech: here used in a general and comprehensive sense. In organic dementia; general paralysis; senile insanity.

Apoplectiform Attacks.—Insanity from coarse brain disease; often precede the mental troubles of syphilitic pseudo-general paralysis; general paralysis; chronic alcoholic insanity.

Apparent Unconsciousness.—Anergic stupor (acute primary dementia); stuporous melancholia (melancholia attonita, melancolie avec stupeur); acute epileptic insanity during the attack; acute delirium (acute delirious mania, delirium grave); epileptic and general paralytic stupor; melancholic (most common form) true rheumatic insanity.

Apprehensiveness.—In melancholia. Insanity complicated with heart disease; climacteric insanity; lactational insanity; gestational insanity; advanced consecutive insanity; incipient general paralysis (some cases); incipient mania (some cases). It is a frequent prodroma of acute delirium, acute mania, and general paralysis. Folie du doute; partial emotional aberration; neurasthenic insanity.

Arcus Senilis.—In senile insanity; insanity from coarse brain disease (organic dementia and organic melancholia).

Arteries Atheromatous.—In senile insanity; organic dementia; organic melancholia.

Ataxia.—In general paralysis, increasing as the disease progresses, not affected by shutting eyes, as in tabes dorsalis; in the third stage of general paralysis—in some cases the patient can neither stand nor walk; in severe and fatal choreic insanity; in some cases of acute delirium (acute delirious mania).

Atony.—In katatonia (katatonic insanity).

Attention, Power of, Defective.—In primary mental deterioration (simple primary dementia); terminal or secondary dementia; imbecility; general paralysis; mania; melancholia; organic dementia; organic melancholia; senile insanity; chronic epileptic insanity; simple depressive syphilitic insanity; choreic insanity; lost in most cases of delirium of young children. One of the prodromata of general paralysis.

Attitude Immobile.—In simple melancholia.

Attitude Insinuating.—In sexual perversion; erotomania.

Attitude Listless.—In atonic simple melancholia.

Attitude Suggestive of Auditory Hallucinations.—In delusional insanity (monomania), especially the persecutory and ambitious forms; melancholia; mania; alcoholic insanity; general paralysis; puerperal insanity; consecutive insanity.

Attitude Suggestive of Delusions of Grandeur.—In general paralysis; ambitious delusional insanity (ambitious monomania); acute mania.

Automatic Ideas and Words.—Ideas arise that the patient knows to be false; he says things he knows to be wrong but "cannot help it." In impulsive insanity; pubescent insanity; climacteric insanity; senile insanity.

Babbling and Chattering.—Excessive and incoherent in delirium tremens; puerperal mania; consecutive insanity.

Back, Weakness in.—In masturbational insanity.

Barometric and Seasonal Conditions, much Influenced by.—Periodical insanity; epileptic insanity.

Bed, Refusal to Leave.—Katatonia (katatonic insanity).

Bedsores, Liability to.—In third stage of general paralysis; in insanity from coarse brain disease on paralysed side.

Biliousness.—In adolescent insanity.

Blindness.—In insanity from coarse brain disease when caused by tumours; in some cases of general paralysis.

Blood, Alteration of.—Coagulates with difficulty or not at all in the third stage of general paralysis, and in other cachectic states of the insane (Voisin); hæmoglobin diminished in general paralysis (Bevan Lewis), and in pubescent and adolescent insanity, with notable stupor; red corpuscles and hæmoglobin diminished in anæmic insanity.

Bodily Functions, Active Disturbance of the.—In acute delirium (acute delirious mania); melancholia; anergic stupor (acute dementia, acute primary dementia, stuporous insanity); katatonia; frenzy (of transitory mania, of alcoholic insanity, or of melancholia); initial and terminal periods of general paralysis; senile dementia; persecutory delusional insanity (persecutory monomania) (Spitzka).

Bodily Symmetry, Want of.—In idiocy; monomania (delusional insanity); moral imbecility.

Brutishness.—In chronic alcoholic insanity; traumatic insanity.

Bulimia.—Ravenous appetite. (See "Voracity.")

Cachexia.—In third stage of general paralysis; often very pronounced in syphilitic pseudo-general paralysis.

Calculation, Power of, Defective.—Terminal dementia (secondary dementia); primary mental deterioration (simple primary dementia); imbecility; general paralysis; organic dementia; organic melancholia; senile insanity.

Catalepsy.—In cataleptic insanity; katatonia (katatonic insanity).

Catamenia, Disorders of.—See "Menstruation," "Amenorrhœa," etc.

Cephalalgia.—See "Headache."

Change of Place, Desire for.—Expansive intellectual petit mal.

Character, Change of.—More or less in all acquired insanity.

Marked in prodromal period of general paralysis; in simple depressive syphilitic insanity; in most cases of incipient rheumatic insanity, and often coincident with diminution of articular pains; in moral insanity; in these forms it is often the most prominent and sometimes almost the only symptom.

Childishness, in Actions.—In chronic epileptic insanity (epileptic dementia); idiocy; imbecility.

Choreic Movements.—In choreic insanity; sometimes in maniacal true rheumatic insanity.

Coma, or Somnolence.—In comatose saturnine insanity, patient answers questions but falls again into state of somnolence or torpor; in convulsive form of saturnine insanity after convulsive attack; at termination of insanity from coarse brain disease when caused by tumour; fatal coma is a frequent termination of acute delirium (acute delirious mania); sometimes in fatal cases of rheumatic insanity (comatose form).

Complexion, Pale and Pasty.—In masturbational insanity.

Concentration of the Intellectual Operations round one Idea or set of Ideas.—In folie du doute.

Concentration of Thoughts and Feelings on Patient's own Health, Organs, etc.—In hypochondriacal melancholia (hypochondriacal insanity).

Conduct and Disposition, Change in.—One of the first symptoms in almost all forms of insanity, and one of the most noticeable symptoms in incipient general paralysis, in simple mania, and in simple melancholia.

Conduct, Extraordinary.—In moral insanity.

Confusion, Feeling of.—In traumatic insanity.

Confusion of Ideas.—In katatonia; in the tumour form of organic insanity (insanity from coarse brain disease); in insanity of cyanosis from bronchitis, cardiac disease, and asthma.

Connect Everything with Self, Disposition to.—In hypochondriacal melancholia.

Consciousness Abolished.—In the most severe form of cataleptic insanity (Bra); in most cases of delirium of young children (Clouston).

Consciousness Confused.—In primary or acute confusional insanity (Spitzka).

Consciousness Markedly and Demonstrably Impaired.—In epileptic insanity; transitory frenzy (transitory or ephemeral mania); anergic stupor (acute dementia, acute primary dementia); melancholic frenzy; alcoholic frenzy; acute delirium (acute delirious mania, delirium grave); frenzy of mania; frenzy of general paralysis (paretic dementia); cataleptic phases of katatonia (Spitzka).

Consciousness, Temporary Losses of.—In the prodromal period of general paralysis.

Constantly making the same Gesture.—In idiocy; acute mania; chronic mania; terminal dementia; hysterical insanity; some cases of each.

Constipation.— In idiocy; melancholia; organic dementia: organic melancholia; climacteric insanity; prodromal period of general paralysis; chronic hysterical insanity; obstinate in acute delirium (acute delirious mania).

Continuity, Want of, in Thought and Action.—In incipient masturbational insanity.

Controlled and Overpowered, Sense of Being.—In simple melancholia.

Convulsions.—(1,) General: In epileptic insanity; in epileptic idiocy, and sometimes in other forms of idiocy and in some cases of imbecility; general paralysis; syphilitic insanity; organic dementia; katatonia; saturnine insanity, convulsive saturnine seizures resemble epileptic seizures but they have no aura; saturnine pseudo-general paralysis; some cases of rheumatic insanity, acute and sub-acute; some cases of traumatic insanity; accidentally in other forms. (2,) Local or Jacksonian: In general paralysis; organic dementia.

Countenance Expressionless.—In apathetic terminal dementia; acute delirium (acute delirious mania); confirmed general paralysis; anergic stupor (acute dementia).

Countenance, Expressive of Distrust, Indifference and Inertia, Inquietude, or Self-Effacement.—In simple melancholia.

Countenance Expressive of Wretchedness and Misery.—In stuporous melancholia.

Courage, Failure of.—In climacteric insanity; masturbational insanity.

Cramps.—In acute delirium (acute delirious mania); katatonia.

Cruelty.—In congenital moral insanity (moral imbecility).

Cursatory Impulses.—A prodroma of the attacks in acute epileptic insanity.

Cutaneous Anæsthesia. — In confirmed general paralysis; hysterical insanity; organic dementia and organic melancholia (insanity from coarse brain disease); alcoholic pseudo-general paralysis; traumatic insanity.

Dejection.—In hypochondriacal melancholia (hypochondriacal insanity).

Delirium.—In acute delirium; delirium tremens; delirium of young children; furious and more or less continuous in the maniacal form of saturnine insanity; it is present sometimes in pubescent insanity; in pneumonic consecutive insanity; sometimes in insanity from coarse brain disease, tumour form; in insanity of cyanosis from bronchitis, cardiac disease, and asthma; rarely in uterine or amenorrhœal insanity.

Delirium with Remissions.—In insanity of Bright's disease.

Delusions.—Faulty ideas growing out of a perversion or weakening of the logical apparatus (Spitzka).

False notions and ideas (independently of false inductions) which have no immediate reference to the senses (Bucknill and Tuke).

Legal definition of a delusion :—A faulty belief, out of which the subject cannot be reasoned by adequate methods for the time being.

Delusions are divided by Spitzka into *genuine* and *spurious*.

(*A*,) GENUINE.—Those created by the patient himself. Genuine delusions are divided into :—

(*a*,) *Systematised*.—Systematised delusions are distinct and fixed and present circumstances are incorporated in a pseudo-logical chain (Spitzka). The patient can give reasons for his belief. Systematised delusions are :—

(α,) *Expansive*.—Expansive systematised delusions may be : (1,) Ambitious. In ambitious delusional insanity (monomania of pride or grandeur, delusional insanity with exalted delusions); (2,) Erotic. Voluptuous delusions occur more frequently in females than in males. In erotic delusional insanity (erotomania); (3,) Religious. In religious delusional insanity (religious monomania, religious mania, theomania).

(β,) *Depressive*.—Depressive systematised delusions may be : (1,) Hypochondriacal. In hypochondriacal melancholia; (2,) Persecutory. In persecutory delusional insanity (monomania of suspicion, delusional insanity with delusions of persecution).

(*b*,) *Unsystematised*.—Unsystematised delusions are vague and changeable, and the logical power of the patient is in whole or in part, in abeyance with regard to them (Spitzka). They are met with in the acute insanities and chronic deteriorations (Spitzka). They may be :—

(α,) *Expansive*. — The delusions of grandeur of general paralysis are examples; they are due to destruction of the logical associating force (Spitzka). Expansive unsystematised delusions are found in chronic mania, delusional mania, acute mania, periodical mania, hysterical insanity, imbecility; fragmentary in agitated terminal dementia.

(β,) *Depressive*.—In delusional melancholia (having committed the "unpardonable sin," being very poor, etc.); religious melancholia (eternal damnation, having led wicked lives, etc.); senile melancholia (delusions of suspicion of a possible nature, such as stealing, etc., by members of the household); senile dementia; secondary or terminal dementia; periodical melancholia; lactational insanity (delusions of suspicion); gestational insanity (delusions of suspicion, poison); chronic melancholia, included by Spitzka in chronic confusional insanity; chronic alcoholic insanity; katatonia; severe climacteric insanity; general delusional state in depressive syphilitic insanity with incoherence; some cases of advanced traumatic insanity; some cases of imbecility; some cases of true melancholia due to emotional disturbance (Spitzka); some cases of general paralysis (depressive unsystematised delusions of persecution); of poverty in diabetic insanity, and some cases of depressive general paralysis; fragmentary in some cases of agitated dementia.

(γ,) *Mingled.*—Expansive and depressive mingled: in primary or acute confusional insanity (delusions of identity are very common in this disorder); senile dementia; epileptic insanity; katatonia; pubescent insanity (insanity of pubescence) (Spitzka); as to surrounding persons in insanity of Bright's disease.

(B,) SPURIOUS.—Spurious delusions are those adopted from others (Spitzka). These delusions are very frequently persecutory (unseen agency, the telephone, electricity, etc.). In folie à deux.

For examples illustrating the difference between a delusion, a hallucination, and an illusion, see " Illusions."

Depression.—In melancholia; puerperal melancholia; lactational insanity; gestational insanity; senile melancholia; periodical melancholia; climacteric insanity; most cases of sub-acute or true rheumatic insanity; anæmic insanity, occasionally alternating with an acutely maniacal condition; phthisical insanity; depressive confirmed general paralysis (not common); mild cataleptic insanity; insanity from deprivation of the senses; insanity of oxaluria and phosphaturia; insanity from abdominal disorders, especially from hepatic derangements; diabetic insanity; post-connubial insanity; two thirds of cases of uterine insanity; chloral insanity; katatonia; great in pellagrous insanity; pubescent insanity, in early stages and afterwards alternating with delirium, etc.; sometimes at commencement of adolescent insanity; commencement of organic dementia; commencing saturnine insanity; traumatic insanity, at first; sometimes in incipient general paralysis; early puerperal mania; often prodromal of general paralysis; a frequent prodroma of acute mania; sometimes prodromal of the outbursts of periodical mania; at end of acute delirium; advanced consecutive insanity; sometimes after acute symptoms of acute mania pass off; at end of puerperal mania sometimes; in some cases of syphilitic depression with incoherence; some cases of hysterical insanity; some cases of choreic insanity; some cases of insanity of myxœdema; some cases of delirium of young children.

Despondency, Intense Religious.—In religious melancholia

Destructiveness.—In destructive mania (a form of impulsive insanity); acute mania; sometimes in chronic mania; agitated terminal dementia; epileptic insanity during the paroxysms; persecutory delusional insanity (monomania of suspicion) at times; maniacal cases of rheumatic insanity; insanity with exophthalmic goitre.

Diarrhœa.—Succeeds constipation near the termination of acute delirium (acute delirious mania); in advanced general paralysis.

Diminutives, Tendency to use.—In katatonia (katatonic insanity).

Dirty in Habits: (*a*,) *From Inattention, Indifference or Perverseness*.—In terminal or secondary dementia, especially the apathetic form; third stage of general paralysis; idiocy; puerperal insanity; acute mania; chronic mania; insanity with exophthalmic goitre. (*b*,) *From Unconsciousness*.—In anergic stupor (acute primary dementia); during the paroxysms of epileptic insanity. (*c*,) *From Paralysis of Sphincter*.—In organic dementia; third stage of general paralysis; saturnine pseudo-general paralysis.

Discontentedness.—In some cases of simple melancholia.

Disposition, Change in.—See "Conduct and Disposition, Change in."

Distortion of Surrounding Objects and Persons.—One of the prodromata of acute delirium (acute delirious mania).

Dread, Unfounded.—Especially on awaking from sleep early in the morning in podagrous or gouty insanity.

Dreaminess.—In incipient folie du doute.

Dress and Undress, Inability to.—In acute delirium (acute delirious mania); anergic stupor (acute primary dementia); second and third stages of general paralysis; epileptic insanity during the paroxysmal excitement; organic dementia; organic melancholia; stuporous melancholia.

Dressing Fantastically.—In feigned insanity; acute mania; chronic mania; periodical mania; maniacal phase of folie circulaire.

Dressing Negligently.—In melancholia.

Dress Suggestive of Exalted Ideas.—In ambitious delusional insanity (monomania of pride or grandeur); general paralysis.

Drink, Intense Craving for.—In alcoholic insanity; dipsomania, periodically; often the most prominent, and sometimes the only, symptom of climacteric insanity; periodical mania; folie circulaire; many cases of traumatic insanity.

Drowsiness.—In cretinism; after meals in incipient general paralysis.

Dulness and Indifference.—In choreic insanity.

Dynamometric Indications.—See "Grip Feeble."

Dysmenorrhœa and Irregular Menstruation.—In incipient chronic hysterical insanity and the prodromal period of the same disease.

Dyspepsia.—One of the prodromata of chronic hysterical insanity; in primary mental deterioration; prodromal of general paralysis. (See "Gastric Embarrassment.")

Echolalia.—"The thoughtless repetition of words and phrases spoken by others, the subject not associating any mental conception with them" (Spitzka); in imbecility, insanity of puberty, dementia (Spitzka).

Egotism, Increased.—Ambitious delusional insanity; general paralysis, often very much in the prodromal period and first stage; pubescent insanity; adolescent insanity; early masturbational insanity; senile dementia; senile melancholia; chronic hysterical insanity; partial exaltation.

Electric Sensibility, Abolished or Altered.—In confirmed general paralysis.

Emaciation.—In acute delirium; acute mania; acute melancholia; epileptic insanity during paroxysmal excitement; masturbational insanity; phthisical insanity; climacteric insanity; chronic hysterical insanity; melancholy phase of folie circulaire; insanity from coarse brain disease; diabetic insanity; early lactational insanity; some cases of general paralysis.

Emotional.—In organic dementia; advanced general paralysis.

Emotional Disturbance (marked).—May be present with or without intellectual motive:—

(*A*,) WITHOUT INTELLECTUAL MOTIVE.—May be angry, expansive, or depressed:—

(*a*,) *Angry*.—(1,) *Simply*: In maniacal furor; furor of general paralysis (paretic dementia); periodical mania; (2,) *Angry and Treacherous*: In epileptic, alcoholic, and general paralytic furor; (3,) *Angry and Anxious*: In melancholic frenzy (paroxysms of agitated melancholia); transitory frenzy (transitory mania); katatonia.

(*b*,) *Expansive, Good-humoured, or Pleasurable.*—In simple mania; periodical mania; general paralysis (paretic dementia).

(*c*,) *Depressed, Sad and Anxious.*—In simple melancholia; periodical melancholia; alcoholic insanity; epileptic insanity; early general paralysis; katatonia; insanity of puberty.

(*B*,) WITH INTELLECTUAL MOTIVE.—May be angry and expansive, or depressed:—

(*a*,) *Angry and Expansive,*—In episodical delirium of monomania (delusional insanity).

(*b*,) *Depressed.*—In prodromal period of mania; monomania with depression; primary mental deterioration (Spitzka). (See "Depression," "Exaltation.")

Emotions, Blunted.—In sub-lucid intervals of long-standing periodical insanity.

Energy, Diminished.—In sub-lucid intervals of long-standing periodical insanity.

Enjoyment of Life much Diminished or Lost.—In organic melancholia; simple melancholia; prodromal period of erotic and religious delusional insanity.

Epileptic Seizures.—In epileptic insanity; early general paralysis; insanity from absinthe; alcoholic insanity; saturnine pseudo-general paralysis; cerebral syphilis; some cases of traumatic insanity.

Epileptiform Attacks.—In general paralysis; organic insanity (insanity from coarse brain disease); senile dementia; alcoholic insanity; alcoholic pseudo-general paralysis; precede comatose saturnine insanity; often precede psychical troubles in syphilitic pseudo-general paralysis; in many cases of katatonia.

Erotic Ideas or Tendency.—In erotomania (erotic delusional insanity); ovarian or old maid's insanity, occurring in single females aged 35 to 43; in puerperal insanity; early general paralysis; mania; epileptic insanity; opium or morphia insanity; often in consecutive insanity (post-febrile); prodromal of religious delusional insanity, also a symptom of the fully developed disease; some cases of chronic hysterical insanity.

Exacerbations at Menstrual Periods.—In chronic hysterical insanity; periodical insanity.

Exaggeration, Ideas of.—In confirmed general paralysis (expansive or most common form).

Exaggeration of Trifles.—In incipient delusional insanity (incipient monomania).

Exaggeration, Proneness to.—In expansive syphilitic insanity.

Exaltation.—Ambitious delusional insanity (monomania of pride or grandeur); partial exaltation or amenomania; theomania; erotomania; most cases of general paralysis; mania; periodical mania; maniacal phase of folie circulaire; adolescent insanity; third or maniacal stage of masturbational insanity; expansive syphilitic insanity; about one-third of the cases of uterine insanity; occasionally in anæmic insanity alternating with a melancholic condition.

Exaltation and Depression Alternating Momentarily.—In acute rheumatic insanity.

Excess in Eating (Bulimia) and Drinking.—One of the prodromata of general paralysis. (See "Voracity.")

Excess, Sexual.—One of the prodromata of general paralysis.

Excitement, Motor and Mental.—May be constant, occasional, or periodical: (*a*,) *Constant.*—In acute delirium; acute mania; agitated melancholia; delirium tremens; puerperal mania; insanity with exophthalmic goitre; severe choreic insanity; (*b*,) *Occasional.*—In delusional insanity (monomania); epileptic insanity; general paralysis; phthisical insanity; depressive syphilitic insanity, with incoherence; idiocy; mania, simple, delusional, and chronic; terminal or secondary dementia; katatonia. (*c*,) *Periodical.*—In periodical mania; folie circulaire; followed by prostration in épilepsie larvée.

Excretions, Deficient.—In stuporous melancholia.

Exertion, Mental and Bodily, Feeling of Incapacity for.—In traumatic insanity.

Exhaustion, Nervous.—In acute delirium; severe delirium tremens; primary mental deterioration; consecutive insanity.

Extravagance (as to Expenditure, etc.).—In moral insanity.

Extremities Cold and Bluish.—Often in idiocy.

Eye, Averted.—In masturbational insanity.

Eyes, Downcast.—In simple melancholia.

Eyes, Fixed.—In melancholic true rheumatic insanity.

Eyes, Glistening.—In acute mania.

Eyes, Hollow.—In melancholic true rheumatic insanity.

Eyes, Injected.—In acute mania; acute delirium; prodromal period of general paralysis.

Face, Flushed.—In acute mania.

Face, Haggard.—In masturbational insanity.

Face, Sudden Redness of.—In prodromal period of general paralysis.

Facial Circulation Defective.—In general paralysis.

Facial Expression, Denoting Terror.—In delirium tremens; agitated or anxious melancholia; some cases of stuporous melancholia.

Facial Expression, Dull and Apathetic.—In primary mental deterioration (simple primary dementia); organic dementia.

Facial Expression, Dull and Self-absorbed.—In puerperal mania at commencement.

Facial Expression, Gloomy.—In melancholia.

Facial Expression, Indicating Elation.—In ambitious delusional insanity (monomania of pride); general paralysis; hysterical mania; partial exaltation or amenomania.

Facial Expression, Vacant.—In anergic stupor (acute dementia, acute primary dementia); apathetic terminal dementia; long-continual agitated terminal dementia, especially when the patient is asked a question; idiocy; primary mental deterioration; general paralysis; some cases of stuporous melancholia. (See "Countenance, Expressive of, etc.")

Facial Muscles, Chorea-like Movements of.—In katatonia.

False and Malevolent Assertions, Making.—In moral insanity.

Fatigued, Easily.—In hypochondriacal melancholia; one of the prodromata of general paralysis.

Fatuity.—In advanced terminal dementia; epileptic dementia; advanced senile dementia; alcoholic dementia; advanced primary mental deterioration (simple primary dementia).

Features Contracted.—In stuporous melancholia; true rheumatic insanity.

Fed Forcibly, Requiring to be.—In some cases of melancholia; some cases of persecutory monomania; some cases of puerperal insanity; some cases of terminal dementia.

Feverishness.—See "Temperature."

Fickleness.—In simple mania.

Fly or Hide, Tendency to.—In some cases of persecutory delusional insanity (persecutory monomania) in the early stages before the delusions have become systematised and fixed.

Food, Refusal of.—See "Refusal of Food."

Force, Want of, in Thought and Action.—In incipient masturbational insanity.

Forgetfulness.—In primary mental deterioration (simple primary dementia); a prodroma of general paralysis; in organic dementia; simple depressive syphilitic insanity.

Formication.—Prodromal of an attack of acute epileptic insanity; prodromal of general paralysis.

Gaining Flesh.—In mania becoming chronic, the mental symptoms not improving; in some cases of general paralysis and in the remissions of that disease; terminal or secondary dementia; during convalescence from mania, melancholia, and other mental diseases.

Gait, Unsteady, Staggering, or Halting.—In confirmed general paralysis; organic dementia; organic melancholia; acute delirium; alcoholic insanity; syphilitic insanity; idiocy.

Gastric Embarrassment.—In delirium tremens; alcoholic pseudo-general paralysis; insanity from opium; primary mental deterioration; prodromal of general paralysis. (See "Dyspepsia.")

Giddiness (Vertigo).—In epileptic insanity; organic dementia; organic melancholia; traumatic insanity; a prodroma of general paralysis, mania, melancholia, and lactational insanity; in incipient acute rheumatic insanity; some cases of senile dementia; sometimes in commencing saturnine insanity; pneumonic consecutive insanity; sometimes precedes an attack of periodical mania.

Giving Away Property.—In general paralysis; acute mania; ambitious delusional insanity (monomania of pride or ambition); some cases of simple mania.

Glance, Unusually Vivacious.—In prodromal period of general paralysis.

Gloominess.—In epilepsy; in folie du doute at the commencement; melancholia.

Grinding Teeth.—In confirmed general paralysis; acute delirious mania (acute delirium, delirium grave).

Grip, Feeble.—The grip (as tested by the dynamometer) of almost all insane persons, except sufferers from early epileptic insanity, is weaker than that of sane persons of somewhat similar age and physique. It is very weak in general paralysis and melancholia, and in organic dementia the hand of the apparently sound side is much less powerful than that of a healthy person: (*a*,) *Of both Hands equally.*—In general paralysis; melancholia; (*b*,) *Of one Hand more than the other.*—In organic dementia; organic melancholia; some cases of general paralysis.

Groaning.—In melancholia; senile melancholia.

Gustatory Hallucinations.—See "Hallucinations."

Gyratory Impulses.—Prodromal of acute epileptic insanity.

Habits, Change of.—In all forms of acquired insanity. Prominent in simple mania and moral insanity; prodromal of mania, melancholia, and general paralysis in many cases.

Hæmatoma Auris or Othæmatoma.—A swelling of the external ear containing blood. Sanguineous or serous cyst of auricle (U. Pritchard). These othæmatomata are not peculiar to the insane, and are not necessarily indicative of incurability, but they occur much more frequently in chronic incurable cases of insanity than in recent and curable ones, and are rare in the sane compared with the insane.

Opinions vary as to the part played by violence in the causation of hæmatoma auris.

In general paralysis; mania, especially chronic mania; melancholia; organic dementia; terminal or secondary dementia; epileptic insanity.

Hæmorrhages, Mucous.—Liable to occur towards the termination of general paralysis.

Hallucinations.—A hallucination is a perception without an object (Ball). Hallucinations are most frequently met with in monomania (delusional insanity) and melancholia, but are not uncommon in mania (Bucknill & Tuke). For other forms in which they occur see the hallucinations of the several senses, vision, hearing, etc. They may be *simple* or *compound*:—

(A,) SIMPLE.—Simple hallucinations are those affecting only one sense. The hallucinations are painful and terrifying in choreic insanity and true rheumatic insanity. They are of a depressive character in katatonia. The hallucinations and delusions are mixed and contradictory, and the former often preponderate from the first in primary or acute confusional insanity. Hallucinations may be auditory, visual, gustatory, olfactory, tactual or cutaneous, sexual, or visceral or internal.

(1,) *Auditory.*—Hallucinations of hearing are more frequent than those of any of the other senses (Bucknill & Tuke, Savage, Bra). They may range from mere buzzings, thumpings, and ringing of bells to "voices." Auditory hallucinations are very common in delusional insanity, especially the persecutory and ambitious forms; according to Griesinger they are specially frequent in connection with diseases of the abdomen and genital organs; insanity from abdominal disorders (disease of bladder, etc.); mania, acute, chronic, and delusional; melancholia, acute and delusional: delirium tremens; chronic alcoholic insanity; puerperal insanity; climacteric insanity; general paralysis; early stage of terminal dementia; post-febrile consecutive insanity; pneumonic consecutive insanity; hummings, whistlings, sound of bells prodromal of general paralysis; choreic insanity; of a disagreeable character in melancholic uterine insanity; insanity from deprivation of senses; insanity of myxœdema; lactational insanity.

(2,) *Visual.*—Next in frequency to those of hearing. Vary from blurs, clouds, or haloes, to flashes of light, bright colour perceptions, faces and figures of persons at some occupation, animals, etc. (Spitzka). Especially frequent in acute delirium and delirium tremens; in the latter disease they are painful, mobile, nocturnal, in the former terrifying; occur also in alcoholic pseudo-general paralysis; monomania, especially the religious form; chronic alcoholic insanity, painful, mobile, nocturnal; early stages of terminal dementia; acute and chronic mania; puerperal insanity; lactational insanity; general paralysis; painful and terrifying in choreic insanity, acute and subacute rheumatic insanity, and in melancholic form of delirious saturnine insanity; worse at night in insanity of cyanosis from bronchitis, etc.; occur in pneumonic consecutive insanity; saturnine pseudo-general paralysis; sometimes in post-febrile consecutive insanity; insanity of myxœdema; insanity of abdominal disorders; frightful in some cases of delirium of young children; they occur in epileptic insanity and hysterical insanity.

(3,) *Gustatory.*—Hallucinations of taste, though much less often met with than those of sight, come next in frequency. They are most frequently observed in persecutory delusional insanity (suspicions of poisoning), religious delusional insanity, hypochondriacal melancholia, delusional melancholia, chronic alcoholic insanity; they occur in general paralysis, insanity of puberty (where they point to masturbation according to Spitzka), acute mania, acute melancholia, and in conditions of weak mindedness, ovarian disease, and phthisis (Savage); occur in puerperal mania and choreic insanity; acute epileptic insanity.

(4,) *Olfactory.*—Hallucinations of smell appear chiefly to belong to early stages of insanity (Griesinger). Like those of taste they rarely occur uncomplicated with those of other senses (Bucknill & Tuke). They usually coexist with those of taste (Spitzka). They are common in hypochondriacal melancholia, and in general paralysis, and are almost characteristic of masturbatory insanity (Krafft-Ebing, Spitzka). They are occasionally pleasant in the excitement of mania and general paralysis and in religious monomania, but in many cases of mental depression especially those associated with ovarian and uterine troubles the smells are of an unpleasant kind (Savage). Unpleasant in early puerperal insanity; disagreeable in chronic alcoholic insanity; occur in choreic insanity; lactational insanity; acute epileptic insanity.

(5,) *Tactual, Tactile, Cutaneous, or of Common Sensibility.*—Patients complain of feeling electric shocks, of being magnetised, of having chemical substances applied to the skin, of vermin crawling on the skin, etc. In climacteric insanity (Savage). According to Savage ("Insanity," page 74), "Where we have ovarian troubles we may expect to find hallucinations of smell and touch." Occur occasionally in insanity of puberty, and in melancholia; also in persecutory delusional insanity; chronic alcoholic insanity; delirium tremens; choreic insanity.

(6,) *Sexual* (Bra).—In erotomania; puerperal insanity; general paralysis; chronic alcoholic insanity.

(7,) *Visceral or Internal.*—Difficult or impossible to distinguish from visceral or internal *illusions* (q. v.).

(B,) COMPOUND HALLUCINATIONS.—Hallucinations affecting two or more senses. Hallucinations of several senses are more common than those of one; sometimes, though rarely, all the senses are involved (Bucknill & Tuke). Lactational insanity, smell, sight, and hearing; acute epileptic insanity, sweet tastes and strong odours (Bra); mania; alcoholic insanity; general paralysis; masturbational insanity; puerperal insanity; phthisical insanity; monomania; choreic insanity; melancholia.

For examples illustrating difference between "Delusions," "Hallucinations," and "Illusions," see "Illusions."

Happiness, Feeling of.—In partial exaltation; confirmed general paralysis.

Headache.—Intense in organic dementia from tumours; severe in puerperal mania; occipital in katatonia (Kahlbaum); with uneasy sensations at top of head in early lactational insanity; at beginning and end of primary confusional insanity; pulsatory or grinding in traumatic insanity; frontal in early rheumatic insanity (often); commencing saturnine insanity; prodromal of general paralysis, acute delirium, acute mania, melancholia, apyretic delirium tremens, and sometimes periodical mania.

Head, Flashes of Heat to.—Prodromal of general paralysis. In neurasthenia.

Head, Pains in.—Masturbational insanity.

Head Proportionately Large.—In hydrocephalic idiocy; cretinism.

Head, Sensation of Electric Currents in.—Prodromal of general paralysis.

Head, Sensation of Fulness or Bursting of.—In acute mania.

Head, Sensation of Pressure or Fulness in.—In primary mental deterioration. Neurasthenia and neurasthenic insanity.

Head Very Small.—In microcephalic idiocy.

Hearing Impaired.—In insanity from myxœdema.

Hearing Voices.—See "Hallucinations," Auditory.

Heart, Palpitation of.—In masturbational insanity.

Heat Sense, Abolished or Altered.—In confirmed general paralysis.

Hemianæsthesia.—In organic insanity; hysterical insanity; general paralysis.

Hemiplegia.—See "Paralysis."

Homicidal Impulse.—See "Kill, Impulse to," under "Impulsive Acts."

Hopelessness.—In climacteric insanity.

Hyperacousia.—In first stage of general paralysis; in acute mania at commencement; precursory of apyretic delirium tremens.

Hyperæsthesia (Cutaneous).—Temporary in early general paralysis; hysterical insanity; some cases of senile dementia; precursory of apyretic delirium tremens and of alcoholic pseudo-general paralysis; traumatic insanity.

Hyperbulia.—Wilfulness. In mania; general paralysis; periodical insanity; expansive monomania (ambitious delusional insanity).

Hypochondriasis.—Prodromal of general paralysis; occurs in climacteric insanity; chronic hysterical insanity; during lucid intervals of traumatic insanity; second stage of masturbational insanity; insanity of oxaluria and phosphaturia; precedes hypochondriacal melancholia. In neurasthenia and neurasthenic insanity.

Hysterical.—In sub-lucid intervals of periodical insanity.

Hysterical Convulsions.—In many cases of katatonia (Spitzka).

Hysterical or Hystero-Epileptic Fits.—Occurring in middle-aged men suggest general paralysis (Savage); are often replaced by the maniacal attacks of acute hysterical insanity.

Ideas, Mobile and Futile.—In choreic insanity.

Ideas, Multiple, Mobile, Absurd, and Contradictory.—In the first stage of confirmed general paralysis.

Ideas, Paucity of.—In melancholia.

Identity, Mistakes of.—In delusional insanity; senile insanity; puerperal insanity; lactational insanity; mania; general paralysis; agitated terminal dementia; chronic confusional insanity; some cases of imbecility.

Ill-being, Sense of.—Profound in simple melancholia. In neurasthenia and neurasthenic insanity.

Illusions.—An illusion is a transformation of a peripheral sensation (Griesinger). "In scientific works treating of the pathology of the subject, the word ("illusion") is confined to what are specially known as illusions of the senses, that is to say to false or illusory perceptions. And there is very good reason for this limitation since such illusions of the senses are the most palpable and striking symptoms of mental disease. In addition to this it must be allowed that, to the ordinary reader, the term first of all calls up this same idea of a deception of the senses" (Sully, "Illusions," pp. 4-5). "The slight, scarcely noticeable illusions of normal life lead up to the most startling hallucinations of abnormal life. From the two poles of the higher centres of attention and imagination on the one side, and the lower regions of nervous action involved in sensation on the other side, issue forces which may, under certain circumstances, develop into full hallucinatory percepts. Thus closely is health attached to morbid mental life" (Sully, *op. cit.*, pp. 120-121). If a patient believes a perfect stranger really present to be some intimate friend, or that an inarticulate shout is a word of reproach or a name, he is the subject of an illusion. But if he hears a voice where no sound has been emitted, or sees a person where no one is present, he is the subject of a hallucination. If he believes that taking food will cause him to lose his soul, or that he has many millions of pounds hidden somewhere under a stone, or that he is God, or that he has committed the unpardonable sin, he is labouring under a delusion. A sane person may suffer from illusions and hallucinations; but if he begins to think they are real and acts in accordance with that idea, he becomes insane. A person with a decided delusion can hardly be called sane. Illusions are common in semi-insane, weakly eccentric people, and render them a nuisance to themselves and all about them. Illusions are not so common in monomania and melancholia as hallucinations, but are more frequent than the latter in mania. They occur more frequently than either hallucinations or delusions in periodical insanity. Illusions may affect one or all of the senses: (1,) *Visual.*—The most frequent form of illusion. Illusions of sight generally relate to persons (Spitzka). Visual illusions should be distinguished from optical illusions; the latter being phenomena of refraction and reflection are entirely objective, and may be perceived by many persons simultaneously; the mirage of the desert is an example. Visual illusions occur in periodical insanity; early stages of toxic insanity; early on in acute mania; persecutory delusional insanity (monomania of suspicion); general paralysis. (2,) *Auditory.*—In toxic insanity; persecutory delusional insanity; periodical insanity; mania. (3,) *Gustatory.*—Often arise from furred tongue. So-called hallucinations of taste are often really illusions. They frequently disappear with the relief of dyspepsia (Spitzka). In hypochondriacal melancholia (hypochondriacal insanity); persecutory delusional insanity. (4,) *Olfactory.*—So-called hallucinations of smell are often really illusions, as there is some slight odour which is much exaggerated and altered in the consciousness of the patient. Melancholics (especially masturbatory neurasthenics) sometimes fancy they emit a horrible odour themselves. (5,) *Tactile, Cutaneous, or of General Sensibility.*—Touch often suffers from illusions (Bucknill and Tuke). (6,) *Visceral.*—So-called visceral hallucinations are often illusions arising from cancer of the stomach, etc. In hypochondriacal melancholia. (7,) *Sexual* (Griesinger).—In erotic delusional insanity (monomania of suspicion).

Imagination Weakened.—In chronic alcoholic insanity.

Immoral, Grossly and Openly.—In general paralysis; impulsive insanity; moral insanity; simple mania.

Impatience.—In chronic hysterical insanity.

Imperative Conceptions.—("Fixed Ideas," "Zwangsvorstellungen"); Reflections and suspicions which tyrannise the patient's thoughts and sometimes his acts as markedly as the most firmly rooted organised insane idea. They differ from the delusion in that the patient is able to reason himself out of them, and to recognise their absurdity at times. They arise suddenly without any obvious connection with previous thoughts; they appear like spontaneous explosions of some uncontrolled segment of the nervous system. They sometimes arise from suggestions quite inadequate to produce such impressions in a healthy state. Agoraphobia (fear of open places), clanstrophobia (fear of narrow quarters), mysophobia (fear of defilement), are examples of imperative conceptions which remain *in statu quo* for years. Imperative conceptions are more common in females than in males, in youthful and imbecile than in aged and strong-minded persons, and under such conditions as pregnancy, menstruation, and the convalescence from fevers; it may, therefore, be assumed that there is a morbid impressionability of the nervous system in these cases. The imperative conception often leads to the imperative act (morbid impulse) (Spitzka, "Insanity," pp. 35–36).

They may be (1,) *Continuous.*—In monomania, imbecility, folie du doute, and partial emotional aberration ; (2,) *Periodical.*— In periodical insanity ; or (3,) *Episodical.*—In melancholia, hysterical insanity, general paralysis, monomania, imbecility (Spitzka); partial emotional aberration ; senile insanity ; some cases of chronic mania.

Improvidence and Absurdity of Actions.—In confirmed general paralysis. Simple mania.

Impulsive Acts (Morbid Impulses).—The result of defective inhibition. These acts may be performed (1,) *Consciously.*—In mania ; imbecility ; moral insanity; dementia ; (2,) *Unconsciously.* In epileptic insanity ; in somnambulism ; under the influence of hypnotism ; (3,) *Sometimes Consciously, Sometimes Unconsciously.* In impulsive insanity.

There are many forms of morbid impulse, to destroy, to kill, to steal, etc., etc. (*a,*) *Destroy, Impulse to.*—Destructive mania. In mania ; imbecility ;. moral insanity ; dementia ; (*b,*) *Drink, Impulse to.*—Dipsomania (a form of impulsive insanity). In periodical mania ; circular insanity ; climacteric insanity ; gestational insanity ; (*c,*) *Exhume and Eat Dead Bodies* (Clouston), *or Defile Dead Bodies* (Spitzka), *Impulse to.*—Necrophilism ; (*d,*) *Kill, Impulse to.*—Homicidal mania (a form of impulsive insanity). In puerperal insanity ; epileptic insanity ; traumatic insanity ; alcoholic insanity, especially mania *a potu* ; religious delusional insanity (religious monomania); occasionally in severe gestational insanity ; imbecility, some cases ; sometimes in masturbational insanity ; (*e,*) *Sexual Impulse.*—Satyriasis in the male, nymphomania in the female (forms of impulsive insanity). In periodical mania ; acute epileptic insanity ; (*f,*) *Steal, Impulse to.*—Kleptomania (a form of impulsive insanity). In periodical mania ; gestational insanity ; imbecility ; chronic mania ; incipient general paralysis ; (*g,*) *Suicidal Impulse, without Depression.*—Suicidal mania (a form of impulsive insanity). In puerperal mania ; acute epileptic insanity ; religious monomania ; (*h*), *The Incendiary Impulse.*—Pyromania (a form of impulsive insanity). In periodical insanity ; acute epileptic insanity ; religious monomania ; pubescent insanity ; some cases of expansive syphilitic insanity ; (*i,*) *Wander from Home and throw off the Restraints of Society, Impulse to.*—Planomania (Clouston); (*j,*) *Wild Beast, Impulse to act like.*—Lycanthropia.

Incoherence.—In pubescent insanity; acute mania ; general paralysis; imbecility; acute or primary confusional insanity ; absolute in acute delirium (acute delirious mania); agitated terminal dementia; alcoholic insanity; epileptic insanity; puerperal insanity; senile insanity; maniacal saturnine insanity; transitory mania; syphilitic depression with incoherence; may be absolute in maniacal true rheumatic insanity; insanity with exophthalmic goitre; sometimes in chronic mania.

Inconsistencies in Speech and Acts.—Prodromal of general paralysis.

Indecision.—In organic melancholia; chronic hysterical insanity; partial dementia.

Indifference.—In some cases of simple melancholia.

Inertia.—In primary mental deterioration (simple, primary dementia); prodromal of anergic stupor (acute dementia); in simple melancholia; organic melancholia; climacteric insanity; insanity of oxaluria and phosphaturia.

Infantile Convulsions, Succeeding.—Eclampsic idiocy.

Inhibition, Loss of Power of.—In impulsive insanity; in many cases of epileptic insanity, imbecility, climacteric insanity, melancholia, puerperal insanity, gestational insanity, pubescent insanity, periodical insanity.

Injuring Husband and Children.—In puerperal insanity; climacteric insanity.

Injuring Self.—Self-injury apart from suicidal tendency and as a result of excitement, terror, delusions, or unconsciousness occurs in acute mania; pubescent insanity; agitated melancholia; puerperal insanity; monomania; epileptic insanity; delirium tremens.

Inquietude.—In depressive intellectual *petit mal*; prodromal of acute delirium (acute delirious mania); premonitory of acute mania; in early delusional insanity (monomania); in neurasthenia and neurasthenic insanity.

Insomnia.—In primary mental deterioration (simple primary dementia), prodromal period, and fully-developed disease; climacteric insanity; senile insanity; pubescent insanity; adolescent insanity; puerperal insanity; lactational insanity; acute delirium; delirium tremens; general paralysis; at night with drowsiness after meals prodromal of general paralysis; epileptic insanity; prodromal of acute epileptic insanity; in melancholia; severe simple mania, acute mania (usually also a premonitory symptom), chronic mania (often), transitory mania; phthisical insanity; organic dementia, especially in cases with neurotic heredity; complete in expansive syphilitic insanity; incipient true rheumatic insanity; neurasthenia and neurasthenic insanity; insanity from opium; chloral insanity; incipient saturnine insanity; post-febrile consecutive insanity; traumatic insanity; insanity of cyanosis from bronchitis, etc.; insanity of myxoedema.

Instability.—Prodromal of general paralysis; in syphilitic depression with incoherence.

Instructed, Incapable of being.—In congenital moral insanity (moral imbecility).

Intellectual Faculties Absent.—In some cases of idiocy; some cases of cretinism; advanced terminal dementia.

Interest, Loss of, in Surroundings.—In the tumour form of insanity from coarse brain disease; monomania; lucid or sub-lucid intervals of long-standing periodical insanity.

Intestinal Disturbances.—Often at menstrual periods in chronic hysterical insanity.

Introspection Shallow and Conceited.—In incipient masturbational insanity.

Irascibility.—In epileptic insanity; imbecility; simple melancholia; early chronic hysterical insanity; choreic insanity; incipient true rheumatic insanity; traumatic insanity; partial exaltation; sub-lucid intervals of long-standing periodical insanity.

Irritability.—In epileptic insanity; alcoholic insanity; imbecility; persecutory monomania; phthisical insanity; early puerperal insanity; insanity from coarse brain disease, tumour form; chronic hysterical insanity; excessive in early choreic insanity; traumatic insanity; second stage of masturbational insanity; neurasthenic neurosis and psychosis; lactational insanity; senile melancholia; simple melancholia; hypochondriacal melancholia; climacteric insanity; incipient rheumatic insanity; partial exaltation; lucid intervals of periodical insanity; insanity of oxaluria and phosphaturia; insanity from deprivation of senses. It is a symptom of brain exhaustion.

Jaws, Champing of.—In the second stage of confirmed general paralysis.

Jealousy, Insane.—In jealous monomania; chronic alcoholic insanity.

Judgment, Defective.—In imbecility; simple mania; prodromal of general paralysis.

Kill, Impulse to.—See "Impulsive Acts."

Language, Blasphemous or Obscene, or Both.—Liable to occur in many forms, but most frequently and markedly in puerperal insanity, climacteric insanity, epileptic insanity, chronic mania, imbecility, and general paralysis.

Lassitude.—One of the prodromata of acute delirium and general paralysis; a prodroma of primary mental deterioration. In neurasthenia.

Laughter, Bursts of.—In expansive intellectual *petit mal*; acute mania; expansive syphilitic insanity.

Laziness.—In epilepsy; simple melancholia.

Lethargy.—In lactational insanity.
Limited Speech or Action, or Both.—In all forms of melancholia.
Listlessness.—In advanced post-febrile consecutive insanity; apathetic terminal dementia.
Localised Paralysis or Paresis.—See "Paralysis."
Lochia Altered, Diminished, or Suppressed.—Often in puerperal insanity.
Locomotion, Positive Disturbances of.—In general paralysis; syphilitic dementia; acute delirium; organic dementia; epileptic insanity; alcoholic insanity.
Loquacity.—In acute mania; maniacal phase of circular insanity; puerperal mania; delirium tremens; mania *a potu*; early expansive general paralysis; senile mania; epileptic insanity during paroxysmal excitement; acute delirium; unceasing in expansive syphilitic insanity; simple mania; acute rheumatic insanity; maniacal form of sub-acute or true rheumatic insanity; partial exaltation; chronic mania; agitated terminal dementia; adolescent insanity; imbecility.
Loss of Ability to Write.—In organic dementia; organic melancholia; advanced general paralysis; terminal dementia; temporarily in delirium tremens.
Low Company, Disposition to Keep.—In moral insanity.
Malaise.—Profound in depressive intellectual petit mal; often prodromal of melancholia; neurasthenic neurosis and psychosis.
Manner, Fierce.—In some cases of acute mania.
Manner, Foolish.—In simple mania.
Manner, Jolly.—In some cases of acute mania.
Manual Inability, Leading to Awkwardness in Manipulations.—In confirmed general paralysis.
Marasmus.—In incipient chronic hysterical insanity.
Masturbation.—In masturbational insanity; pubescent insanity; adolescent insanity; epileptic insanity; nymphomania; satyriasis; acute mania; chronic mania; early general paralysis.
Memory Defective.—See "Amnesia."
Memory, Loss of.—See "Amnesia."
Mendacity.—In moral insanity, acquired and congenital; simple mania; gestational insanity; maniacal phase of circular insanity.
Menstruation Irregular or Profuse, or Both.—Towards climacteric; prodromal of climacteric insanity.
Mental Enfeeblement.—In sub-acute, or true rheumatic insanity organic melancholia; katatonia. See "Mental Weakness," etc.
Mental Weakness (Limited).—In monomania (Spitzka).
Mental Weakness Prominently Developed.—See "Mental Enfeeblement." (*a,*) *Involving the mental faculties generally.*—In idiocy; imbecility; primary deterioration; dementia, whether terminal,

alcoholic, epileptic, or organic; acute delirium. (*b*,) *With focal lacunæ.*—In general paralysis; syphilitic dementia; chronic alcoholic insanity; chronic mania; primary confusional insanity; pubescent insanity (Spitzka).

Metastasis.—In metastatic insanity; diabetic insanity; rheumatic insanity; gouty insanity.

Migraine.—Frequently at menstrual periods in chronic hysterical insanity.

Misanthropic.—In some cases of imbecility.

Mischievousness and Mockery, Tendency to.—In maniacal phase of circular insanity.

Moaning.—In melancholia; senile melancholia.

Mobility of Character.—In epilepsy.

Mobility of Ideas.—In incipient delusional insanity (monomania); expansive confirmed general paralysis; choreic insanity; second stage of masturbational insanity.

Monoparaplegia.—See "Paralysis."

Monoplegia.—See "Paralysis."

Monotony of Speech or Action or Both.—In all forms of melancholia.

Monotony of Thoughts and Movements.—In melancholia.

Mood, Quickly Changing.—In alcoholic insanity; second stage of masturbational insanity.

Moral Deterioration.—In acquired moral insanity; simple mania; early general paralysis; senile dementia; advanced chronic alcoholic insanity; folie circulaire (circular insanity); periodical insanity; one of the prodromata of general paralysis.

Moral Perverseness.—In traumatic insanity.

Moral Perversion.—In periodical insanity. (See "Moral Deterioration" and "Moral Perverseness.")

Morbid Propensities.—These are perversions of the two main instinctive tendencies of the human race: the desire for food and the sexual appetite. Anthropophagy and sexual perversions are the most important morbid propensities from a medico-legal point of view. (Spitzka, "Insanity," p. 38.) In pubescent insanity; periodical insanity.

Morbid Sensations; Formication, etc.—In some cases of hypochondriacal melancholia.

Morbid Sensations of Lightness, Heaviness, etc.—Prodromal of general paralysis.

Moroseness.—In epilepsy; some cases of chronic hysterical insanity in early stage; traumatic insanity.

Motiveless Actions.—In simple mania; depressive syphilitic insanity, with incoherence; expansive syphilitic insanity.

Motiveless Laughter.—In expansive syphilitic insanity; alternating with weeping in choreic insanity.

Motor Restlessness.—See "Restlessness, Motor."
Movement, Aversion to.—In simple melancholia.
Muscular Contractures.—In insanity from coarse brain disease; third stage of general paralysis, but only temporary in uncomplicated cases of general paralysis.
Muscular Relaxation.—In masturbational insanity; neurasthenia and neurasthenic insanity.
Muscular Resistance, Diminished.—In anergic stupor (acute dementia, acute primary dementia).
Muscular Sense, Abolished, Altered, or Hallucinated.—In confirmed general paralysis.
Muscular Weakness, not amounting to Paralysis.—In prodromal period and beginning of first stage of general paralysis; delirium tremens; acute delirium; primary mental deterioration; neurasthenia and neurasthenic insanity; climacteric insanity; some cases of traumatic insanity; most marked in extensor muscles in apathetic terminal dementia, and causes wrist-drop.
Mutism.—In anergic stupor; stuperous melancholia; most cases of true rheumatic insanity; periodical melancholia; melancholic phase of folie circulaire; severe form of cataleptic insanity; cataleptic phase of katatonia; some cases of uterine insanity; a few cases of monomania; some cases of partial exaltation; some cases of advanced general paralysis.
Muttering Isolated Words.—In the third stage of general paralysis; terminal dementia; idiocy.
Naso-Labial Fold or Folds Flattened or Effaced.—In general paralysis; organic dementia; organic melancholia.
Natural Affection, Loss of.—See "Affective Sensibility Perverted or Paralysed."
Nervous.—In lucid or sub-lucid intervals of periodical insanity.
Neuralgia, General (and Mobile) or Local.—Prodromal of general paralysis; saturnine pseudo-general paralysis; prodromal of outburst of periodical mania, in some cases.
Neuroses Affecting Eye.—In podagrous or gouty insanity.
Never Speaking.—See "Mutism" and "Speech Congenitally Absent or Permanently Lost."
Nocturnal Exacerbations.—In senile insanity; alcoholic insanity.
Noisiness.—In acute mania; exalted phase of folie circulaire; puerperal mania; acute delirium; mania a potu; delirium tremens; agitated melancholia; paroxysmal excitement of epileptic insanity; senile mania; general paralysis; organic dementia, generally in cases with neurotic heredity; severe choreic insanity; maniacal form of true rheumatic insanity.
Nosophobia.—The fear that serious illness is impending. In neurasthenia and neurasthenic insanity.

Nystagmus.—Convulsive oscillation or rolling of the eyeball. In idiocy and imbecility. Occasionally in acquired insanity and then, according to Griesinger, it is generally symptomatic of the transition from the acute to the chronic stage.

Obscenity.—In puerperal mania; gestational insanity; severe simple mania.

Obstinacy.—In resistive melancholia; persecutory delusional insanity.

Occupation, Change of.—One of the prodomata of general paralysis.

Oddness and Peculiarity from Birth.—In congenital moral insanity (moral imbecility).

Odour Exhaled.—Alcoholic in delirium tremens; breath malodorous in acute delirium, masturbational insanity, acute epileptic insanity, chronic mania, terminal dementia, and in all cases where food is refused. The skin often emits a repulsive, acrid, ammoniacal odour in the third stage of general paralysis. So-called "mousey" odour in room during or after attacks of maniacal excitement. Bromides impart halcine odour to breath.

Œdema.—In anergic stupor (acute dementia).

Olfactory Hallucinations.—See "Hallucinations."

Optic Nerve Changes, Atrophy, etc.—Prodromal of general paralysis in some cases; atrophy in general paralysis in many cases (Clifford Allbutt, Wiglesworth. etc.). According to Aug. Voisin, tortuosity of the retinal arteries is more frequent in the third stage of general paralysis than optic nerve atrophy. According to Clifford Allbutt, the disc is pale from ischæmia during an attack of maniacal excitement, but is obscured by the general redness of the fundus after the subsidence of the attack.

Optic Neuritis.—In insanity from coarse brain disease when caused by tumour.

Originating Power, Want of.—In insanity of oxaluria and phosphaturia; neurasthenia and neurasthenic insanity.

Othæmatoma.—See "Hæmatoma Auris."

Overwhelmed, Apparently.—In melancholic true rheumatic insanity (the most common form of true rheumatic insanity).

Own Words and Acts, Paying Extreme Attention to.—In folie du doute.

Painful Ideas.—In incipient delusional insanity.

Painful Morbid Sensations: Heat, Cold, Pressure, etc.—Prodromal of general paralysis; in neurasthenia and neurasthenic insanity.

Pain, Mental.—In organic melancholia; simple melancholia.

Pain, Mental, Outward Signs of.—In melancholia; senile melancholia; persecutory monomania.

Pains.—In general paralysis (prodromal period and first stage); alcoholic insanity; saturnine insanity; syphilitic insanity.

Pallor.—In anæmic insanity; lactational insanity; face pale and earthy with red malar prominences in acute delirium; chronic hysterical insanity; early masturbational insanity.

Palpitation.—In masturbational insanity; early lactational insanity; prodromal period of general paralysis; in neurasthenia and neurasthenic insanity.

Paraplegia.—See "Paralysis and Paresis."

Paralysis and Paresis.—Motor paralysis may be incomplete in degree or partial in extent; example of latter, ptosis. Paresis, a term used by some authors to indicate a milder degree of paralysis. *Hemiplegia* = paralysis of the whole of one side; *monoparaplegia* or hemiparaplegia or crural monoplegia = paralysis of one leg; *facial monoplegia* = paralysis of one side of face; *brachial monoplegia* = paralysis of one arm; *brachio-crural monoplegia* = paralysis of arm and leg on same side; *paraplegia* = paralysis of both lower extremities; *lingual monoplegia* = paralysis of one side of tongue; *diplegia* = paralysis of both arms, etc.

All forms, but most commonly hemiplegia, in insanity from coarse brain disease; all forms in general paralysis; transient hemiplegia in senile melancholia; paraplegia and hemiplegia in senile dementia; incomplete paralysis of arms and legs in severe delirium tremens; incomplete partial paralyses commencing at distal extremities of limbs in chronic alcoholic insanity; paraplegia in alcoholic dementia of females; commencing at distal ends of limbs in alcoholic pseudo-general paralysis; affecting flexor muscles of arms and respecting supinator longus in saturnine pseudo-general paralysis; pronounced and partial (*e. g.* ptosis) in syphilitic pseudo-general paralysis; in idiocy (mostly in paralytic idiocy) may take form of either hemiplegia or paraplegia; in imbecility; partial and especially affecting the muscles of the eyeball in traumatic insanity; in hysterical insanity; pneumonic consecutive insanity; in the third stage of general paralysis the paralysis is real and most frequently unilateral.

Parsimony.—In moral insanity (some cases).

Passive Suffering.—In simple melancholia.

Pathos.—In katatonia.

Peevishness.—In hypochondriachal melancholia.

Penuriousness.—In senile dementia.

Periodicity.—In periodical insanity; adolescent insanity: circular insanity.

Persecution, Ideas of.—Persecutory monomania; true rheumatic insanity; traumatic insanity; opium or morphia insanity; post-febrile consecutive insanity; katatonia; saturnine pseudo-general paralysis; syphilitic depression with incoherence;

some cases of depressive general paralysis; some cases of neurasthenic insanity; prodroma of acute delirium.

Perspiration, Profuse.—In severe delirium tremens; acute rheumatic insanity.

Photopsia.—In traumatic insanity.

Picking Fingers.—In agitated melancholia.

Place, Incorrect Ideas of.—In terminal (secondary), alcoholic, epileptic, and syphilitic dementia; general paralysis; some cases of imbecility.

Poisoning, Ideas of.—In persecutory or suspicious monomania; phthisical insanity; alcoholic insanity; opium insanity; saturnine pseudo-general paralysis; acute delirium; katatonia.

Premature Grayness.—In primary mental deterioration (simple primary dementia); chronic mania; terminal dementia.

Presentiments of Evil.—Prodromal of general paralysis, acute delirium, and acute mania; neurasthenia.

Pressure or Fulness in Head, Sensation of.—In primary mental deterioration (simple primary dementia); neurasthenia.

Propensities, Morbid.—See "Morbid Propensities."

Prostration.—Extreme in depressive intellectual petit mal; in most cases of true rheumatic insanity; during remissions of insanity of Bright's disease.

Ptosis.—In syphilitic insanity; general paralysis; saturnine insanity.

Pulling Out Hair.—In agitated melancholia.

Punctiliousness.—At commencement of folie du doute (dubious or doubting insanity).

Pupils, Dilated.—In hysterical insanity; anergic stupor (acute dementia); stuporous melancholia (melancholia attonita); general paralysis in the later stages; some cases of neurasthenic insanity.

Pupils, Dilated and Immobile.—In stuporous melancholia.

Pupils Extremely Contracted.—Prodromal of general paralysis.

Pupils, Fixed.—Prodromal of general paralysis.

Pupils, Irregular.—In general paralysis; syphilitic insanity and syphilitic pseudo-general paralysis.

Pupils, Mobile.—In epileptic insanity.

Pupils Unequal.—In general paralysis; simple primary deterioration; alcoholic pseudo-general paralysis; saturnine pseudo-general paralysis; sometimes unequally contracted in severe delirium tremens; occasionally unequal in mania (acute and chronic); hysterical insanity; liable to occur in all forms of insanity as a result of aneurism, bromism, cervical wounds, tumours, dental caries, etc.

Pulse.—Small and frequent in anergic stupor; compressible with dicrotism in general paralysis; according to Broadbent ("The Pulse," p. 275) general paralysis in its early stages usually has the arteries contracted and a pulse of tension, later the pulse becomes weak and toneless. High ascent, flat top, interrupted wavy descent, characteristics of sphygmogram of general paralysis (Voisin). Frequent (may reach 150), small, and irregular in acute delirium. Frequent (sometimes 120 or more), weak, and thready in puerperal mania. Weak in phthisical insanity. Small and often accelerated in severe delirium tremens; accelerated in commencing saturnine insanity and in the fully developed psychosis. Slow and small in simple melancholia; according to Broadbent high arterial tension is usual in melancholia, and where extremely low tension is present the prognosis is bad. In mania the pulse is "singularly little affected" (Broadbent). Small and frequent in acute rheumatic insanity; frequent in sub-acute or true rheumatic insanity. Insomnia is sometimes associated with high arterial tension, sometimes with low. In the former case a mercurial aperient lowers the tension and procures sleep, and in the latter, change of air and vascular tonics—iron, acids, strychnine, and digitalis are appropriate measures (Broadbent).

Quarrelsomeness.—In traumatic insanity.

Quietude.—In insanity from deprivation of senses.

Reaction Time.—The reaction to acoustic and optic stimuli is abnormally slow in acute and sub-acute mania, melancholia, alcoholic insanity, and epileptic insanity (Lewis).

Reaction to Alcohol or Drugs Increased.—In traumatic insanity; prodromal of general paralysis.

Reflexes Exaggerated.—In insanity from coarse brain disease; acute delirium; many cases of general paralysis (rather more than a third according to Lewis); some cases of neurasthenic insanity.

Refusal of Food.—In melancholia in all its forms; puerperal insanity; chronic alcoholic insanity; delirium tremens; mania a potu; persecutory delusional insanity; phthisical insanity; alcoholic pseudo-general paralysis; insanity from opium or morphia; post-febrile consecutive insanity; prodromal of acute epileptic insanity; some cases of chronic hysterical insanity; most cases of sub-acute or true rheumatic insanity; diabetic insanity; katatonia.

Relatives and Friends, Unreasonable Antipathy to.—To husband in severe gestational insanity. (See "Affective Sensibility, Perverted or Paralysed.")

Religious Tinge.—In some cases of hysterical insanity.

Remorse.—In climacteric insanity.

Repetition (Constant) of Same Words to Self.—In folie du doute.

Repetition (Frequent) of One Word or Phrase.—In chronic mania; idiocy; terminal dementia.

Reproductive Faculties Absent.—In some cases of idiocy; some cases of cretinism.

Repugnance to Husband.—In puerperal insanity; often in climacteric insanity; severe gestational insanity. (See "Affective Sensibility, Perverted or Paralysed.')

Resistance to Movement, Feeding, etc.—In stuporous melancholia (melancholia attonita, mélancolie avec stupeur).

Respirations, Affected.—Slow and irregular in simple melancholia; accelerated in puerperal mania, in some cases 56 or 60; early lactational insanity; acute rheumatic insanity.

Restlessness, Mental.—In agitated terminal dementia; simple mania; chronic mania; hypochondriacal melancholia; climacteric insanity; prodromal of general paralysis.

Restlessness, Motor.—In acute delirium (acute delirious mania); acute mania; most cases of sub-acute, chronic, and delusional mania; severe simple mania; periodical insanity (maniacal form); circular insanity (maniacal phase); extreme in agitated melancholia; agitated terminal dementia; senile insanity, especially at night; pubescent insanity; adolescent insanity; delirium tremens; some cases of climacteric insanity; incipient true rheumatic insanity and maniacal form of developed true rheumatic insanity; third or maniacal stage of masturbational insanity; extreme in insanity of Bright's disease; insanity of myxœdema (some cases); organic dementia, worst cases, those with neurotic heredity (Clouston); one of the prodromata of general paralysis.

Retention of Urine.—From indifference or from diminished sensibility, atony, or paralysis of bladder. In melancholia; acute delirium; insanity from coarse brain disease.

Retina, Changes in.—Anæmia of retina frequent in melancholia (Allbutt). In mania if there is decided excitement retinal congestion is often met with (Monti). Retinal arteries very frequently tortuous in third stage of general paralysis (Voisin). Albuminuric retinitis in insanity of Bright's disease. Choroido-retinitis in some cases of syphilitic insanity.

Rhyming Speech.—In katatonia; pubescent insanity; epileptic mental states; sometimes in any episodical excitement (Spitzka).

Rhythmical Movements.—In acute mania; apathetic terminal dementia.

Rigidity, Cataleptic.—In cataleptic insanity; katatonia; melancholic forms and stages of pubescent insanity.

Rigidity of Muscles, of Trunk, or Extremities.—In confirmed general paralysis.

Sallowness.—In phthisical insanity; stuporous melancholia; saturnine insanity.

Satisfaction, Ideas of.—In expansive general paralysis; expansive intellectual petit-mal; partial exaltation.

Satyriasis.—A form of impulsive insanity and a symptom. Often a symptom of senile insanity, the patients committing criminal assaults on young girls and even infants.

Screaming.—Violent in some cases of delirium of young children.

Secretions, Diminished.—In simple melancholia.

Self-Abasement.—In some cases of simple melancholia.

Self-Absorption.—In some cases of simple melancholia; most cases of true rheumatic insanity.

Self-Abuse (Onanism).—In masturbational insanity; satyriasis; nymphomania; pubescent insanity; adolescent insanity; chronic mania; erotomania; idiocy; expansive general paralysis; epileptic insanity.

Self-Accusations of Hypocrisy, Impiety, etc.—In religious melancholia.

Self-Confidence, Loss of.—In folie du doute.

Self-Control, Loss of.—In hypochondriacal melancholia; climacteric insanity; one of the prodromata of general paralysis.

Self-Esteem, Sometimes Exaggerated, Sometimes Diminished.—In katatonia.

Self-Feeling, Morbid.—In incipient masturbational insanity.

Self-Injury.—See "Injuring Self."

Self-Interrogation.—In folie du doute.

Self-Mutilation.—In religious monomania.

Self-Reproaches of Masturbation, etc.—In the initial depression of katatonia.

Senses, Absence of Two or More.—In idiocy by deprivation; insanity by deprivation.

Senses (Special) Enfeebled.—In simple melancholia.

Senses (Special) Supernaturally Acute.—In early stages of acute mania.

Sensibility, Extreme.—In hypochondriacal melancholia.

Sensibility (General) Enfeebled or Impaired.—In simple melancholia; anergic stupor (acute dementia); general paralysis, especially in second and third stages.

Sentences Uncompleted (in Talking).—In acute or primary confusional insanity. Early general paralysis.

Sentiments, Alteration of.—In all forms of acquired insanity.

Seriousness Increased.—In early chronic hysterical insanity.

Sexual Appetite Diminished.—In simple melancholia; confirmed general paralysis; in neurasthenia except episodically (Krafft-Ebing).

Sexual Appetite Lost.—In the second stage of confirmed general paralysis it is permanently lost.

Sexual Excitement.—In satyriasis ; nymphomania ; erotomania; religious monomania ; folie circulaire (maniacal phase) ; periodical mania ; simple mania ; prodroma of general paralysis.

Sexual Perversion.—In impulsive insanity ; periodical insanity ; moral insanity ; simple mania ; early general paralysis ; pubescent insanity to the extent of dislike and avoidance of opposite sex ; folie circulaire (some cases) ; masturbational insanity, avoidance of opposite sex.

Shutting Eyes.—In acute mania.

Sight, Impaired.—In masturbational insanity ; insanity from myxœdema.

Similarity of Repeated Attacks in same Patient.—In acute epileptic insanity ; periodical insanity.

Sin, Preoccupation with Ideas About.—In religious delusional insanity (theomania, religious mania) ; religious melancholia ; depressive general paralysis (some cases).

Sitophobia.—Fear to take food, resulting in refusal to take it. In simple melancholia.

Skin, Changes in.—Dull in tint, and either dry or greasy in third stage of general paralysis ; cold and clammy in masturbational insanity ; hard and dry in insanity from myxœdema ; muddy and more than normally pigmented in climacteric insanity, increasing as the disease progresses.

Skin Irritation.—In climacteric insanity, causing picking and scratching. One of the prodromata of general paralysis. In insanity from coarse brain disease arising from softening of the inferior surface of the temporo-sphenoidal lobe.

Sleeping Badly and Sleeplessness.—See "Insomnia." Sleep absent, insufficient, or unrefreshing in neurasthenia and neurasthenic insanity (Krafft-Ebing).

Slovenly and Untidy.—In melancholia ; acute and chronic mania ; general paralysis.

Slow Pulse.—In melancholia. (See "Pulse.")

Slowness of Ideation.—In simple depressive syphilitic insanity.

Smell, Hallucinations of.—Disagreeable in masturbational insanity, and with false taste perceptions almost characteristic (Spitzka). (See "Hallucinations.")

Solitary.—In the second stage and the demented or fourth and last stage of masturbational insanity. Desire for solitude in simple melancholia.

Somatic Stigmata.—Such as cranial malformations, unilateral atrophy, dental irregularities, vaulted palate, cleft palate, club foot, etc., etc. In idiocy, cretinism, imbecility, monomania, epileptic insanity ; periodical insanity ; hysterical insanity ; exceptionally and then non-essential in all other forms (Spitzka).

Sombreness.—In early chronic hysterical insanity.

Sores on Extremities, Tendency to the Formation of.—In diabetic insanity.

Speech, Abnormalities of.—These may be (*a*,) *Acquired.*—In second stage of general paralysis slow, drawling, and hesitant from the cerebral lesions; stammering, stuttering, and tremulous from the bulbar; it becomes unintelligible, and in some cases ends in mutism (Voisin). The articulation is thick in organic dementia, and the defect is not greatest at the end of a sentence as in general paralysis; timid, hesitant, and interrupted in simple melancholia; articulation affected in primary mental deterioration; articulation often embarrassed in acute delirium; affected in alcoholic insanity; embarrassed in alcoholic pseudo-general paralysis; stammering in saturnine pseudo-general paralysis, and at first often unintelligible; jerky in true rheumatic insanity; slow and muffled in insanity of myxœdema; (*b*,) *Congenital.*—In idiocy, imbecility, monomania.

Speech Congenitally Absent or Permanently Lost.—In some cases of idiocy; some cases of cretinism; some cases of advanced terminal dementia; some cases of general paralysis in third stage. (See "Mutism.")

Speech, Senile.—In senile insanity. For description of "senile speech," see symptoms of senile melancholia ("Senile Insanity," Chapter III).

Spermatorrhœa.—Sometimes in masturbational insanity; in neurasthenia and neurasthenic insanity.

Spontaneity Impaired.—In climacteric insanity.

Squander, Disposition to.—In general paralysis; simple mania; moral insanity; incipient acute mania; imbecility.

State, Restless and Unsettled.—In incipient masturbational insanity.

Statements Inconsistent and Contradictory.—In primary mental deterioration; terminal dementia; organic dementia; general paralysis; mania.

Stripping Naked in Public.—In erotomania; organic dementia; general paralysis; nymphomania.

"Stunnings."—In early general paralysis; saturnine pseudo-general paralysis.

Stupidity.—In post-connubial insanity.

Stupor.—In acute primary dementia (acute dementia, anergic stupor); with retention of memory in melancholia attonita (mélancolie avec stupeur, stuporous melancholia); moderately severe cataleptic insanity; most cases of true rheumatic insanity; uterine insanity in young women of nervous heredity; sometimes in anæmic insanity; sometimes in epileptic insanity; sometimes in general paralysis; transitory after acute mania (secondary stupor, Clouston).

Subsultus Tendinum.—In acute delirium towards end; severe pyretic delirium tremens.

Suicidal Tendency.—In suicidal form (suicidal mania) of impulsive insanity; melancholia, especially suicidal melancholia; alcoholic insanity; puerperal insanity; lactational insanity; severe gestational insanity; climacteric insanity, but in this disease not usually intense; melancholic form of periodical insanity; monomania, especially the religious form (theomania), sometimes in the persecutory form; sometimes in chronic hysterical insanity; post-connubial insanity; frequently in pellagrous insanity; some cases of acute rheumatic insanity; sometimes in insanity of cyanosis from bronchitis, etc.; severe chloral insanity; rarely in depressive syphilitic insanity; epileptic insanity.

Suppression of Catamenia.—See "Amenorrhœa."

Surrounding Persons, Delusions as to.—In insanity of Bright's disease.

Suspiciousness.—In persecutory or suspicious monomania; lactational insanity; gestational insanity; phthisical insanity, suspicious of being poisoned; senile insanity, suspicious of being defrauded or robbed; primary mental deterioration; semi-insanity with illusions; delirium tremens; chronic alcoholic insanity; hypochondriacal melancholia; true rheumatic insanity; traumatic insanity; chronic morphismus; some cases of imbecility; insanity from myxœdema; incipient acute mania; premonitory of melancholia; in nearly all forms in which there are auditory hallucinations.

Symptoms Influenced by Physiological Periods.—In periodical insanity.

Taciturn.—In simple melancholia; some cases of persecutory monomania.

Tactile Hallucinations.—See "Hallucinations."

Talkativeness.—See "Loquacity."

Talking Incessantly even when Alone.—In acute mania.

Talking Irrelevantly.—In acute or primary confusional insanity.

Talking to Self.—In mania; monomania; alcoholic insanity; imbecility; general paralysis.

Taste, Hallucinations of.—See "Hallucinations."

Temperature.—(*a*,) *Subnormal.*—In anergic stupor (acute primary dementia; primary mental deterioration (simple primary dementia); melancholia; generally in third stage of general paralysis; in phthisical insanity until the lungs become affected (Clouston); in insanity from myxœdema; in apathetic terminal dementia. (*b*,) *Supranormal.*—Oscillates between 102° and 104° in acute delirium (acute delirious mania); in mild pyretic delirium tremens from 100°·2 to 101°·4, in severe from 102° to 105°·8 (Magnan); puerperal mania in many cases, exceeding 100° in some and 103° in a few (Clouston); evening rise in phthisical insanity after lungs have become affected; periodically in first stage of general paralysis; in complicating delirium tremens it follows the course of the complicating disease; post-febrile consecutive insanity (some cases); in some cases of acute mania before the maniacal outburst it may rise to 100°; in a few cases of lactational insanity, but it rarely rises over 100°; rheumatic insanity is often announced by a matutinal rise; in acute rheumatic insanity (acute cerebral rheumatism) it may reach 109°, 111°, or even 112° in the rectum; in true rheumatic insanity it is normal or nearly so.

Temperature, Regional, of Head.—In the quiescent mental state the anterior region of the head has the highest, and the middle region the lowest average temperature; high temperatures occur most frequently in the right anterior and left posterior regions.

The greatest elevations of temperature found at the surface of the head during intellectual work *of all kinds* are usually met with in the space lying over the tract of the brain-surface formed by the posterior portions of the first and second frontal, the ascending frontal, and the ascending parietal (anterior part) convolutions.

During emotional activity the rise of temperature is, in the greater number of cases, in all three regions, higher on the left side than on the right; the highest rises being over the same tract as in intellectual work (Lombard).

In headache due to overwork the temperature of the scalp over the region affected is higher than that of the surrounding parts and this can be easily ascertained by means of an ordinary mercurial surface thermometer.

Temper, Change of.—One of the prodomata of general paralysis.

Temper, Unequal.—Simple depressive syphilitic insanity.

Terror.—In acute delirium (acute delirious mania).

Threatening to Commit Suicide.—In general paralysis; monomania; agitated melancholia; alcoholic insanity; sometimes, though not by any means always, in suicidal mania and suicidal melancholia.

Threatening Violence to Others.—In general paralysis; mania; periodical mania; maniacal phase of circular insanity; persecutory monomania; imbecility; epileptic insanity.

Thyroid Gland Affected.—Enlarged in many cases of cretinism and in most of insanity of exophthalmic goitre. Diminished in size in insanity of myxœdema. Affected in sexual neurasthenia (Krafft-Ebing).

Time, Ideas of Incorrect.—In terminal dementia; alcoholic, epileptic, and syphilitic dementia; some cases of imbecility.

Timidity.—At commencement of folie du doute.

Tinnitus Aurium.—In traumatic insanity.

Tongue.—Furred in acute mania and in melancholia; black and dry in acute delirium; displaying fibrillary tremor in general paralysis and in alcoholic, saturnine, and syphilitic pseudo-general paralysis; tremulous and often protruded to one side in insanity from coarse brain disease; tremulous in chronic alcoholic insanity and delirium tremens; incapable of being protruded in advanced general paralysis; sometimes fissured in syphilitic insanity, and bitten at the sides in epileptic insanity.

Torpidity.—In mild cases of organic dementia.

Torpor, Mental.—Often towards end of cyanosis from bronchitis, etc.

Tremor.—May be (*a*,) *General*.—In severe delirium tremens, especially severe pyretic delirium tremens; acute rheumatic insanity; general and massive in alcoholic pseudo-general paralysis; general paralysis. (*b*,) *Localised*.—Hands alone, arms alone, arms and legs, legs and face in mild apyretic and complicating delirium tremens; ataxic of upper and lower limbs in confirmed general paralysis. Facial in acute delirium, acute mania, chronic alcoholic insanity, saturnine pseudo-general paralysis, sometimes in syphilitic pseudo-general paralysis; transitory mania, episodical delirium of monomania. Of head and neck in organic dementia (where there is multiple sclerosis), insanity from paralysis agitans, episodical excitement in mania and monomania. Of hands in chronic alcoholic insanity, opium or morphia insanity, acute delirium, general paralysis, insanity of paralysis agitans, saturnine insanity, senile insanity, epileptic insanity, insanity with exophthalmic goitre, syphilitic pseudo-general paralysis. Of lips in general paralysis, saturnine pseudo-general paralysis, sometimes in syphilitic pseudo-general paralysis, acute delirium, acute mania. Of tongue in general paralysis (from very early stage), saturnine and syphilitic pseudo-general paralysis, though often absent in latter; organic dementia.

Trophic Disturbances (Special).—In acute delirium, general paralysis, syphilitic dementia, organic dementia, epileptic dementia, melancholia, terminal dementia (Spitzka), anergic stupor.

Unconsciousness, Apparent.—See "Apparent Unconsciousness."

Unsociability.—In melancholia; phthisical insanity; climacteric insanity; masturbational insanity, especially as regards the opposite sex.

Unworthiness, Ideas of.—In melancholia; lactational insanity.

Upper Lip, Swelling of.—Frequently in chronic hysterical insanity at menstrual periods.

Urine.—Increased in general paralysis; urates increased and chlorides and phosphates diminished in general paralysis; urine scanty in severe delirium tremens; rich in phosphates in anergic stupor; often fœtid in third stage of general paralysis; loaded with lithates in gouty insanity; in diabetic insanity the sugar sometimes alternates with the attacks of insanity; albumen and sugar should be looked for in traumatic insanity (Clouston).

Vacant Expression.—See "Facial Expression, Vacant."

Vacillation.—In second stage of masturbational insanity.

Vague Fears.—Worse at night in insanity of cyanosis from bronchitis, etc.

Vanity.—In partial exaltation.

Vascular Changes.—In organic dementia and organic melancholia, atheroma, etc.; fatty and fibrous changes in terminal dementia; dilatations, atheroma, etc., in alcoholic insanity; dilatations, tortuosities, atheroma, etc., in general paralysis; temporary local spasm in neurasthenia.

Vaso-Motor Paresis.—In anergic stupor (acute dementia).

Verbigeration.—Word-making. Talking in sounds belonging to no known language. Spitzka gives as an example, "Risti pili chinko ti ki ti.........chichotitonifor tikohoforchink." In *katatonia*, epileptic insanity, hysterical insanity, chronic confusional insanity (chronic mania), pubescent insanity (Spitzka).

Vertigo.—See "Giddiness."

Violence.—In epileptic insanity; acute mania; alcoholic insanity; impulsive insanity; sometimes in imbecility; some cases of chronic mania; some cases of senile insanity; agitated melancholia; persecutory monomania; monomania of jealousy; chronic hysterical insanity; outbursts of, in expansive syphilitic insanity; acute rheumatic insanity; traumatic insanity; some cases of insanity of myxœdema; insanity with ex-ophthalmic goitre.

Visual Hallucinations.—See "Hallucinations."

Vocal Cord or Cords Paralysed.—Sometimes in confirmed general paralysis.

Vociferation, Abusive.—In maniacal form of saturnine insanity; expansive syphilitic insanity.

Voices.—See "Hallucinations" (auditory) and "Hears Voices."

Voluntary Movement, Little or no.—In stuporous melancholia.

Voracity.—In fully developed mania; confirmed general paralysis; some cases of terminal dementia, cretinism.

Wander Abroad, Disposition to.—Mania errabunda. In moral insanity; early general paralysis; pubescent insanity; early monomania.

Weakness.—In diabetic insanity; neurasthenia and neurasthenic insanity.

Wealth, Ideas of.—In most cases of general paralysis; ambitious monomania; sometimes in acute and chronic mania.

Weeping.—In melancholia; senile melancholia; chronic alcoholic insanity; motiveless in insanity from coarse brain disease and in general paralysis; motiveless and alternating with laughter in choreic insanity; in climacteric insanity, especially at menstrual periods; tearless in some cases of delirium of young children; frequent fits of, in agitated melancholia.

Weighing Own Judgments.—In folie du doute.

Wet in Habits.—(*a*,) *From inattention, indifference, perverseness, or loss of sense of decency.*—In terminal dementia; idiocy; puerperal insanity; acute and chronic mania; third stage of general paralysis. (*b.*) *From unconsciousness.*—In anergic stupor; epileptic insanity (in fits, etc., and during paroxysmal excitement). (*c*,) *From paralysis or atony of neck of bladder.*—In organic dementia; third stage of general paralysis; terminal dementia. (*d*,) *From loss of normal feeling of discomfort caused by full bladder.*—In terminal dementia; melancholia; acute delirium. (*e*,) *From two latter combined.*—In terminal dementia; melancholia; acute delirium; senile dementia.

Wildness and Intractability.—In epileptic idiocy.

Will, Affections of the.—See "Abulia" and "Hyperbulia." Abulia in neurasthenia.

Words and Acts (Own), Paying Extreme Attention to.—In folie du doute.

Work, Distaste for.—In organic melancholia; prodromal symptom of acute delirium; often prodromal of melancholia; prodromal of erotic and religious forms of delusional insanity; in neurasthenic insanity.

Wringing Hands.—In agitated melancholia.

Wrists Flexed.—In advanced apathetic terminal dementia; sometimes in saturnine insanity, paralytic idiocy, and organic dementia.

Write, Inability to.—See "Loss of Ability to Write."

Writing Altered.—In general paralysis; acute mania; mania with delusions; alcoholic insanity; in third stage of general paralysis it becomes shapeless and hieroglyphical.

CHAPTER III.

INDEX OF MENTAL DISEASES, WITH THEIR SYNONYMS AND SYMPTOMS.

ABDOMINAL DISORDERS *(Insanity from)*.

See "Gastro-Enteric Insanity" (Sibbald) in section on "Hypochondriacal Melancholia." Schroeder van der Kolk describes a case of catarrh of the bladder with great dysuria and at times complete anuria, in which mental symptoms arose, viz., "Violent nervous symptoms, hallucinations of hearing, and subsequently of seeing also" (Bucknill and Tuke, "Psychological Medicine," p. 369). "Melancholia is the form of mental disorder which we most frequently witness in combination with hepatic derangement of a chronic character" (*op. cit.* p. 370). (See "Bright's Disease, Insanity of.")

ADOLESCENT INSANITY.

Clouston ("Mental Diseases," p. 534) gives the period of its occurrence as being that between the ages of 18 and 25, notably between 20 and 25. This form of insanity is marked by exaltation, loquacity and restlessness, sleeplessness, anorexia, viciousness, the maniacal period being sometimes preceded by a stage of depression. The disease is marked by a periodicity and a tendency to remission recalling a similar character of the nisus generativus in both sexes. The attacks of mania seem to have a special relationship to the function of reproduction (Clouston, *op. cit.* p. 540). Masturbation or illicit sexual intercourse seems to be a frequent forerunner or concomitant, or both, of the derangement.

Clouston gives the characteristics of the mania of this form of insanity as: (1,) Very acute though seldom delirious; (2,) Of short duration, patients soon apparently recovering; (3,) Constant tendency to relapses; (4,) Frequent neurotic heredity.

ANÆMIC INSANITY.

This mostly takes the form of mild melancholia, sometimes with an element of stupor. Occasionally there is an alternating acutely maniacal and melancholic condition (Clouston, "Mental Diseases," p. 592).

BRIGHT'S DISEASE *(Insanity of)*.

This is met with in chronic Bright's disease with contracted kidneys, enlarged heart, and dropsical tendency.

"The symptoms present are mania of a delirious kind, with extreme restlessness, delusions as to persons round the patient, an absolute want of fear of jumping through windows, or other actions that would kill or injure. The symptoms are characterised by remissions, during which the patient is quiet, rather composed in mind and rational, but very prostrate in body" (Clouston, *op. cit.*, p. 596).

CATALEPTIC INSANITY.

The intellectual faculties of cataleptic patients are often found far from sound. The intellect is narrow, presents lacunæ, and is incapable of much development.

Frequently the psychical troubles amount to actual insanity. In the mild form this consists of a more or less considerable depression of the faculties. The perceptions still exist, but they are less precise than in the normal condition.

In a more severe form a condition of stuporous melancholia exists. This is in some sort a semi-cataleptic state left by the cataleptic attack, and which persists during the intervals. The maniacal or expansive form is very rarely observed.

Finally, in the most severe form intelligence, consciousness, and memory are completely abolished, and the termination is permanent dementia (Bra, "Manuel des Maladies Mentales," p. 93).

CHOREIC INSANITY.

Not to be confounded with *choreomania*, that is, the *Epidemic Dancing Mania* of the middle ages (Spitzka, Bucknill and Tuke, etc.). Symptoms of choreic insanity (Bra, "Man. des Mal. Ment.," p. 90, *et seq.*). According to Marcé there are in the choreic mental state four elements, sometimes occurring singly, generally combined : (1,) Troubles of moral sensibility. There is a change of character betraying itself by excessive irritability, angry outbursts, inexplicable oddity of behaviour, rapidly alternating, and motiveless laughing and weeping. Sometimes there is a melancholy condition with a feeling of anguish and anxiety; (2,) Intellectual disorders. These soon appear. The memory first fails. The patients become incapable of fixing their attention on any subject. They become dull and indifferent. At the same time the ideas are very mobile and extremely futile; (3,) Hallucinations are very frequent. These occur on awaking, during dreams, and especially in the state intermediate between sleeping and waking. They have always a painful and terrifying character. Often solely visual (animals, lights, etc.), sometimes they are auditory (voices, slamming doors, etc.), olfactory (odour of sulphur, phosphorus, etc.), gustatory (taste of poisoned food, etc.), or even tactile (formication, pricking, etc.). Hallucinations are most frequently met with in hysterical chorea, but may sometimes accompany pure chorea; (4,) Sometimes maniacal excitement with incoherence and hoarse and inarticulate cries. Under the influence of the visual and auditory hallucinations the intellectual faculties may present a condition amounting to actual mania which in half the cases is rapidly fatal, death being preceded by formidable ataxic disturbances. Even in the more fortunate cases intellectual troubles of varying duration often remain.

CIRCULAR INSANITY (FOLIE CIRCULAIRE).

It most often begins with depression, sometimes with a maniacal access (Bra, "Maladies Mentales," p. 44). Bra describes three varieties :—

(1,) *Folie circulaire*, properly so called, in which the attacks of excitement and depression follow each other without interruption (Marcé).

(2,) *Folie alternante*, in which there is a very short lucid interval between each phase (Falret).

(3,) *Folie à double forme*, in which the attacks of mania and melancholia are separated by an interval often considerable (Baillarger). Both the maniacal and melancholic phases of circular insanity are characterised by the absence of delusions and by the complete retention of consciousness by the patient.

In the maniacal period the ideas preserve in general a certain sequence. There is not the incoherence of ordinary mania. There is a tendency to mockery, to complaint making, to mischievousness, to be epigrammatical. The patients are cunning and fertile in expedients of every sort. A patient, a typical case of folie circulaire at Haydock Asylum, was overheard plotting with another patient to make false accusations against the young and new nurses, as she said they would not be believed if they made them against the older ones. This patient was highly educated and accomplished, and in her maniacal periods expressed herself well both orally and in her letters which were numerous and voluminous. She belonged to a good family. Speaking of this form of insanity Clouston ("Mental Diseases," p. 237) says, "Another remarkable fact is that by far the greater number of persons who suffered from it were persons of education, and far more than a due proportion of them were persons of old families. I never met with a fine case in a person whose own brain and whose ancestors' brains had been uneducated." A physical peculiarity about the patient above mentioned was that she possessed a well developed moustache and beard.

Bra (*op. cit.*, p. 45) says some patients evince the possession of grand ideas, set numerous projects on foot, become prodigal, and make purchases in all directions. In the melancholy period, on the contrary, the same patients are remarkable for their sordid avarice.

Sufferers from folie circulaire are remarkable also for a very great tendency to alcoholic excesses, for a moral perversion, in fact, leading to all sorts of excesses. Sometimes there is sexual perversion impelling the patients to make advances to persons of their own sex. With regard to the trophic disturbances, the patients become thin during the melancholy phase and increase in weight during the maniacal. Some patients suffer from a periodical anguish, an intense dyspnœa presenting some analogy to angina pectoris (Bra, *op. cit.*, pp. 45–46). Each phase may last one, two, or six months or more. They are sometimes regular and equal in length, but more frequently irregular (Clouston).

Towards the end of the attack, maniacal or melancholic, the patient wakes up as if from a dream and gradually becomes possessed of his or her mental faculties (Bra).

"The order of each cycle varies in different patients; the mania may precede the melancholia or *vice versa*. Both may be of a mild type, and both may be very severe; or one may be slight and the other intense." "As a rule the mania and melancholia correspond to each other in intensity. Where the cycle is of brief duration, lasting a few days or weeks, both are apt to be very well marked; where it is of a duration of months, both are apt to be of a mild type" (Spitzka, "Insanity," p. 272).

Clouston (*op. cit.*, p. 230) only found one or two cases out of forty that were absolutely regular. He says (p. 236) about one half of his cases followed a more or less regular monthly periodicity. About one-third obeyed the law of seasonal periodicity, all in an irregular way. The remaining sixth he could bring under no known law on account of their irregularity.

The cases described by Clouston (pp. 217–230) correspond to the French folie alternante and folie à double forme in most instances, and he gives the order of the phases as mania, melancholia, sanity, followed again by the same cycle, but with differing duration of the component parts and of the whole cycle. He states (p. 236) that in his experience at least 90 per cent. of the cases begin with maniacal exaltation.

CLIMACTERIC INSANITY.
Symptoms.

Incipient.—Loss of energy, bodily and mental, failure of courage, annoyance at trifles, groundless fears, sleep dreamy and broken, appetite diminished, bowels costive, complexion less fresh, skin often muddy and more pigmented than usual, patient dislikes going into company, symptoms aggravated at menstrual periods; then patient apt to suffer from real mental depression, irritability of temper and sleeplessness; there is then also weeping (Clouston, " Mental Diseases," pp. 555–556).

Developed Stage.

The headaches, giddiness, flushings, flashes of light, etc., which usually precede or accompany the climacteric, now disappear. The thinness, flabbiness of muscle, and pigmentation of incipient stage get worse. Frequently there are skin irritations, causing picking and scratching; constipation, anorexia, insomnia and diminution of capacity for work; depression more real and continuous; morbid apprehensiveness; loss of self-control or fear of the loss of it; loss of natural affection, often repugnance to husband; vague suicidal impulses. In the worst cases these suicidal feelings are strong, and attempts are frequent though generally feeble. Sometimes there are impulses to injure husband or children. Hallucinations of hearing are frequent.

May terminate in (1,) Acute excited melancholia, exhaustion, and death; in (2,) A sort of shy uselessness or "paralysis of energy"; or (3,) In complete recovery.

General type the same in men as in women. Impairment of spontaneity, courage and energy. Insomnia, depression, restlessness, hypochondriasis. Loss of flesh. Avoidance of society. There may be suicidal longings and desires usually not very intense (Clouston, "Mental Diseases," pp. 556–557 and 560).

Dr. Merson's groups of symptoms of climacteric insanity ("West Riding Medical Reports," vol. vi., 1876); (Bucknill and Tuke's "Psychological Medicine," 4th ed., pp. 366–367):—

(1,) Cases characterised by simple depression without hallucinations of the senses, or intellectual derangement. In some, extreme nervous irritability and hyperæsthesia of sensation, almost amounting to hallucinations.

(2,) Depression also the prevailing condition, but along with this great emotional and intellectual disturbance, hallucinations of the senses not uncommon, and some vague delusions of a depressing kind nearly always present.

(3,) Delusions of suspicion and persecution the most prominent symptoms. In most cases hallucinations of the senses and outbursts of excitement not unusual.

"An intense craving for drink may be the prominent, and perhaps the only symptom which characterises the morbid condition of the system" (Bucknill & Tuke, p. 366).

Skae considers as pathognomonic of this form of insanity both in men and women, "a monomania of fear, despondency, remorse, hopelessness, passing occasionally into dementia" (Bucknill & Tuke, *loc. cit.*).

COARSE BRAIN DISEASE *(Insanity from)*.

I.—ORGANIC DEMENTIA (PARALYTIC DEMENTIA), (PARALYTIC INSANITY, CLOUSTON).

There is usually paralysis or muscular weakness, and often aphasia at some period of the illness. Clouston (p. 380) says, "Paralytic Insanity, or Organic Dementia, is that form of mental disturbance that accompanies and results from gross brain lesions, as apoplexies, ramollissements, tumours, atrophies, and chronic degenerations of the brain." Its symptoms vary according to the position, kind and intensity of the pathological process. But it is typically a dementia, an enfeeblement, a lessening of the mental power super-added to some sort of motor paralysis. Along with this enfeeblement there may be, and there usually is, a certain amount of depression at first, followed afterwards by a mild exaltation and emotionalism of a childish kind, this gradually passing off and leaving the patient, if he lives long enough, forgetful, helpless, and torpid.

(1,) *Form associated with Apoplexy or Ramollissement.*—This is the most typical form, and follows apoplexy from rupture of a blood vessel in one of the great basal ganglia, or supervenes on embolism or thrombosis, causing local starvation of brain tissue and ramollissement. "Motor restlessness is a special characteristic of the worst class of cases." In the ordinary hemiplegic cases the alteration of speech is the most characteristic motor symptom, "it is a thick articulation, not a tremulous speech." Every word is imperfectly pronounced, and not merely those at the end of a sentence. Before, or during speech the labial and facial muscles do not quiver, "though the tongue usually trembles when put out." Where there are apoplexies or similar lesions of the convolutions themselves, the speech symptoms are more like those of general paralysis, and there are often epileptiform, epileptic and congestive attacks. Clouston does not believe that complete aphasia can co-exist with perfect integrity of the intellectual faculties, and says he has never seen such co-existence. An organic dement with no nervous heredity will be "calm, reasonable, and quite manageable, though forgetful, torpid, and emotional," while one with a bad nervous heredity will become, under the same conditions, "restless, depressed, noisy, and sleepless." Most organic dements are treated at home, as single cases, or in workhouses; only "the noisy, the restless at night, the very dirty, the troublesome find their way into asylums, motor restlessness necessitating special nursing and special rooms."

In one case, E. M., under my care at Haydock Asylum, there was no motor paralysis, except some indistinctness of articulation. Before admission there was a history of a fit, described in the statement as epileptic. On admission, sense of smell feeble, vision of left eye defective. The patient was listless, stupid, and very irascible. After death there was found a large patch of softening in the prefrontal region encroaching on the third frontal convolution and the island of Reil (left side). The left anterior cerebral artery was much diminished in calibre as compared with the right.

In another case in which there was softening of the inferior surface of the right temporo-sphenoidal lobe, there was extreme dementia. The gait was somewhat feeble, and both legs were emaciated to an equal extent, and were equally weak. The patient complained of an itching sensation in the skin of the chest and abdomen (a symptom mentioned by Charcot as characteristic of softening of this region).

These cases are given because they are exceptional, and somewhat difficult to diagnose; the most common motor symptom in organic dementia being well marked hemiplegia.

(2,) *Form caused by Brain Tumours.*—"Intense cephalalgia is undoubtedly the most common sensory symptom." Next to this, "optic neuritis and blindness are the most common symptoms." "The motor signs are paresis and paralysis local and general, convulsions local and general, and congestive attacks."

The most common mental symptoms are, "first, irritability and loss of self-control, and change of disposition; *then* depression, with or without excitement; *then* confusion, loss of memory, muttering to self, loss of interest in all things, perhaps delirious attacks; *then* drowsy half-consciousness, *ending* in coma and death."

The bodily and mental symptoms are affected by the kind position and mode of growth of the tumour, and there may be no symptoms or only vague ones. Some tumours cause direct and reflex brain irritation; others give rise to destructive lesions, especially ramollissement (Clouston, p. 388).

II.—ORGANIC MELANCHOLIA.

Clouston (*op. cit.*, p. 107) says, "In some of these cases I have seen the mental symptoms the very first to appear, long before the paralysis or even before great bodily weakness made its appearance. A paralysis of the sense of well-being and the enjoyment of life, a difficulty in coming to decisions, a loss of mental energy, an intolerance of the usual work if not an actual incapacity to do it well, a tendency to make slight mistakes in small things, a loss of memory and a sub-acute mental pain, I have seen to exist for two years before a man showed any diagnostic signs of brain ramollissement or tumour. The melancholia is usually of the simple type, seldom assuming the excited, delusional, or distinctly suicidal form. I have seen it of the hypochondriacal kind in a few cases."

In a case reported by me in "Brain," 1882, depression of spirits, suicidal tendency, defective memory for recent events, complete left hemiplegia, were some of the outward and visible signs of softening affecting the cortex of portions of the right frontal, parietal, and temporo-sphenoidal lobes, and adhesion of the arachnoid over the occipital lobe.

CONFUSIONAL INSANITY (PRIMARY).

PRIMARY CONFUSIONAL INSANITY.   ACUTE CONFUSIONAL INSANITY.

Synonyms.—Hallucinatory Delirium, Hallucinatory Psychoneurosis, Wahnsinn (Krafft-Ebing), Acute Primäre Verrücktheit (Westphal), Hallucinatory Confusion, Mania Hallucinatoria (Mendel), Delusional Stupor (Newington), Frenosi Sensoria Acuta (Morselli), Paranoia Psiconeurotica.

Spitzka (p. 161) says, "This disorder is rare, and develops rapidly on a basis of cerebral exhaustion." He further says, "The patients suffering from this psychosis, after a rapid rise of their symptoms during a period of incubation rarely exceeding a few days, present hallucinations and delusions of a varied and contradictory character. The delusions resemble those of mania, and more often those of melancholia, but no emotional state is associated with them. The patients assert in the same breath that their property is being stolen, and that they are going to take part in some great state affair."

"The speech in confusional insanity is characteristic, although there is no richness of ideation, as with mania. The sentences are left uncompleted, and are entirely irrelevant as well as incoherent."

"Delusions of identity are very common. The patients believe they are not in the same place, or they recognise as old acquaintances persons to whom these bear no resemblance. It is noteworthy that a large number of the patients are aware that a change has taken place, that they are no longer their former selves, and they may be able to give—by snatches, it is true—a tolerably fair account of the circumstances preceding the outbreak of the disease. But as the latter develops, the patients cease to recognise their position, or to complain of the "head trouble," whose existence they previously admitted, and at most they speak of their former selves in the third person, or manifest a confused variety of double consciousness. When the hallucinations, as is frequently the case, preponderate from the beginning, the disorder we have here considered is termed by some *Acute Hallucinatory Confusion*. Recovery is gradual, the patient becoming progressively clearer; his somatic complaints, such as headache, then occupy his attention more than his incoherently recounted delusive troubles, and finally reason is entirely restored. In only a small proportion of cases does the insanity remain and the patient become permanently deteriorated, his disorder appearing as a form of *chronic* confusional insanity."

Krafft-Ebing ("Lehrbuch der Psychiatrie," p. 396) states that insomnia or unrefreshing sleep with frightful and terrifying dreams, nervous irritability, feeling of oppression, headache, vertigo, irascibility, and confusion of ideas are constant phenomena of the developing psychosis. When the disease has reached its acme there are illusions and hallucinations of all the senses, especially in acute cases, those of sight predominating. So that consciousness soon becomes seriously blurred; the patients are confused and have no idea of their position.

CONFUSIONAL INSANITY (CHRONIC).

Chronic Confusional Insanity (Spitzka).

Synonym.—Chronic Mania.

Ordinarily called Chronic Mania, see Spitzka, "Insanity", p. 102. Some maniacs and melancholics lose their characteristic emotional state, but retain their delusions, which become fixed, though not truly systematised as in monomania (delusional insanity). In a small proportion of acute confusional cases the insanity remains, constituting a form of chronic confusional insanity. The delusions in chronic confusional insanity are not elaborate, and not defended with skill and a show of judgment. "The delusions resemble ruins left over from the destruction of the more elaborate and multitudinous if less fixed delusions of mania and melancholia, around which the gathering tide of a slowly progressing dementia rises till the assertion of the delusions becomes a mere parrot-like repetition, and is finally buried under the same levelling sea of dementia which closes the history of all these primary psychoses entering the domain of the secondary deteriorations.

"The weakening of the logical power and the memory accounts for the frequent observation in these patients of a change in their sense of identity."

"The appetite and assimilation as well as the sleep become normal, or nearly so, and not unfrequently the patients become very stout."

"While the general nutrition does not always suffer in the terminal deteriorations, certain trophic disturbances are quite common. Hæmatoma auris, cutaneous eruptions, premature grayness, and fatty and fibrous changes of the blood vessels are frequent accompaniments" (Spitzka, pp. 170-171).

CONSECUTIVE INSANITY.

I.—Post Febrile.

Nasse's classification of post-febrile insanity (Bucknill & Tuke, 4th edit., p. 371): (1,) The immediate result of the fever itself; (2,) Constituting a prolongation of the delirium when the fever has subsided; (3,) Arising during convalescence.

The last class being more especially intended by the term *Post-Febrile Insanity*. Scarlatina, small-pox, typhus, intermittent fever, measles, erysipelas, the acute anginas, cholera, acute rheumatism, are the febrile conditions in and after which it occurs most frequently (Griesinger, "Die Pathologie und Therapie der psychischen Krankheiten," p. 186, *et seq.*; Clouston, p. 600).

Symptoms of (1,) and (2,) (Savage, "Insanity and Allied Neuroses," p. 435):—"The patient passes through the first few days of febrile disturbance naturally, then becomes sleepless, chattering and often amorous; refusal of food is very common, and rapid exhaustion follows." "After the mania has lasted for a variable number of weeks, depression or partial dementia is well marked, there being some vague dread or other in the former case, and in the latter listlessness with loss of memory, disregard of friends and relations, and neglect of the decencies of life" (Savage, *op. cit.*, p. 436).

Savage states ("Insanity," etc., p. 216) that he has often met with cases following fevers in which loss of memory was the most marked evidence of weakness of intellect. He has frequently met with patients whose memory was seriously damaged for a longer or shorter period after typhoid fever. Now and then such cases do not recover but pass on from one stage to another till they become absolutely and hopelessly weak-minded.

The same result may follow rheumatic fever. With regard to the relation of certain cases of insanity to intermittent fever, Griesinger states that most of the cases occur during convalescence from, or even several months after, the fever; others intermit with the paroxysms of the fever; others replace them. The form is mania with delirium. It is very beneficially influenced by quinine.

Symptons of (3):—*Prodromal.*—irritability, change of manner, perhaps some childishness (Bucknill and Tuke, 4th ed., p. 371). *Fully developed.*—Many cases are examples of melancholia and mania. Some present the ordinary features of delusion of persecution with hallucinations of hearing and (not so frequently) of sight. Some cases pass into dementia, not complicated with paralysis (Bucknill and Tuke, *loc. cit.*). See above-mentioned forms of insanity. Primary Confusional Insanity is sometimes developed during the exhaustion following fevers.

II.—CONSECUTIVE TO LOCAL INFLAMMATORY DSEASES.

Pneumonia, pleurisy. Savage thinks it does not essentially affect the form of insanity whether the mental disturbance is due to the above diseases or to rheumatic, scarlet, or typhoid fever (p. 436).

Griesinger ("Die Pathologie und Therapie der psychischen Krankheiten," 4th ed., p. 191, *et seq.*) describes two cases of pneumonic insanity. In one there was transient mental disorder of a delirious nature with ringing in the ears and vertigo. In the second case the delirium passed into acute mania with slight left facial monoplegia. Afterwards there were visual and auditory hallucinations. Then profound dementia with weakness of left side, and mydriasis. The dementia diminished, and four months after admission the patient's condition was simply one of slight mental weakness.

CYANOSIS FROM BRONCHITIS, CARDIAC DISEASE AND ASTHMA *(Insanity of)*.

"This is a form of delirium, with confusion, hallucinations of sight, sleeplessness, sometimes suicidal impulses and vague fears.

These symptoms are usually worst at night, and often end in mental torpor passing into coma. It is more commonly seen in persons of advanced age than in young people (Clouston, p. 589).

DELIRIUM, ACUTE (ACUTE DELIRIUM).

Synonyms.—Delirium Grave (Spitzka); Grave Delirium; Acute Delirious Mania; Typhomania (Luther Bell).

PSYCHICAL SYMPTOMS.

Precursory (not always present); change of character, lassitude, distaste for work, an indifference more and more pronounced to the patient's previous occupation, then loss of sleep, cephalalgia, apprehensions, presentiments of evil, inquietude, distortion of surrounding objects and persons. Then supervene ideas of persecution, poisoning, etc., with a train of hallucinations of a terrifying nature, blood, flames, corpses, precipices, etc.

PSYCHICAL SYMPTOMS OF FULLY DEVELOPED STAGE.

Maniacal excitement which is violent and continuous; all the members are in motion, great loquacity with the most absolute incoherence. Incapable of attention, the patient flies in terror from all who approach him, will hear nothing, and defends himself with energy. All his acts bear the stamp of fear, and are determined by terrifying hallucinations.

At the end of the disease, when the temperature has attained its maximum, the maniacal excitement disappears, giving place to depression and fatal coma.

PHYSICAL SYMPTOMS.

Face pale, of an earthy hue, the malar prominences red, the countenance expressionless and resembling that of a drunken man.

The pulse is frequent, may reach 150; it is small and irregular. The temperature oscillates between 102°F. and 104°F. The tongue is black and dry, the teeth and lips covered with sordes, the breath malodorous. Great thirst. Obstinate constipation. There is excessive motor agitation, grinding of the teeth, cramps; often embarrassed speech and exaggerated reflexes.

In fatal cases all these symptoms are accentuated, the respiration becomes panting, the temperature rises to 106°F., deglutition and articulation become impossible, diarrhœa succeeds the constipation; finally subsultus tendinum supervenes. Among the terminal conditions are found pulmonary and renal congestions, ecchymoses and bedsores. A gradually increasing drowsiness ends in the extinction of life. The course of the disease is not always regular, but is sometimes marked by remissions (Bra, "Manuel des Maladies Mentales," pp. 162–164).

DELUSIONAL INSANITY.

Synonyms.—Monomania or Monopsychosis of Clouston; Delusional Monomania of Spitzka; Paranoia of Krafft-Ebing; Partial Delirium of Bra. Primäre Verrücktheit (Griesinger); Délire Partiel, Délire Systématisé, Folie Systématisé (Morel); Monomanie Intellectuelle (Esquirol); Paranoia Universalis (Arndt); Chronischer Wahnsinn (Schüle); Paranoia Originaria Degenerativa (Morselli).

Five Forms : (1,) Persecutory and suspicious ; (2,) Ambitious; (3,) Religious ; (4,) Erotic ; (5,) Jealous.

PERIOD OF INCUBATION. PERSECUTORY (much the most common) and AMBITIOUS FORMS. (See "Delusions," Chapter II.)

Mental inquietude and excitement. The ideas are mobile and painful. The patient takes no interest in anything which is not connected with his puerile and apparently insignificant preoccupations. The most trivial events are remarked by him and assume in his eyes an extraordinary importance. A person who turns in the street, who sings, coughs, or uses his pocket-handkerchief is immediately suspected of having made a sign of contempt, or having so acted intentionally to give offence. Or in the ambitious form the patient fancies the passers-by speak of him as some great personage (Bra, "Manuel des Maladies Mentales," pp. 49 and 52).

Spitzka ("Insanity," pp. 301–302) believes that although monomania may develop after any deep and sudden injury to the nervous system, yet in the vast majority of cases it is based on an inherited taint of insanity or on a transmitted neurotic vice, and that in this larger number of cases there are usually noted before the actual outbreak of the disorder, anomalies of character, of the general nervous functions, and of the somatic constitution. In a form of monomania depending on gross structural defect and named by Sander "Originäre Verrücktheit," and by Krafft-Ebing "Originäre Paranoia" (primitive or congenital paranoia) the subjects are noted to be peculiar from infancy; they entertain vague aspirations, are excessively egotistical, and the non-recognition of their supposed importance or abilities leads them to consider themselves the subject of persecution. In others the egotism is so great that the most ridiculous failures are not capable of dispiriting them, but, on the contrary, are accepted as proofs of a divine mission which is to be carried on in a state of perpetual martyrdom. Hallucinations frequently develop as the disorder progresses. With these there are symptoms of neural disorder, similar to those found in imbecility and idiocy; some have epileptiform, others choreiform, and still others quite peculiar and indefinable movements of an "imperative" kind. Peculiarities in pronunciation and inability to pronounce certain consonants in childhood have been noticed in others. There are in addition defects in the bodily conformation of a similar kind, although usually of lesser degree than those characterising idiocy and imbecility; the head is often asymmetrical or deformed, the teeth are sometimes badly developed, and there may be club-foot, strabismus, or atrophy of one side of the body.

PRODROMAL STAGE, RELIGIOUS AND EROTIC FORMS.

These forms rarely commence suddenly. There is a gradual transformation of the individual mostly at the period of puberty. The manifestation of mental aberration is preceded by a feeling of inquietude, a dislike for work, a profound aversion to life and its enjoyments, and a constant pre-occupation with ideas about sin. The patients neglect their usual duties in order to read pious books and devote themselves to religious exercises. To these prodromata are added sexual excitement and erotic ideas; these latter by their frequency constitute one of the great distinctive characters of religious insanity (Bra, *op. cit.*, p. 56).

I.—PERSECUTORY AND SUSPICIOUS FORM (Monomania Persecutoria, Délire des Persécutions).

Fully developed stage. This is by far the most common form of delusional insanity or monomania. The delusion or delusions may simply arise out of some false idea accepted without discussion or out of false sensations (hallucinations).

In fully developed persecutory delusional insanity the delusions are systematised. Magnan in his lectures at St. Anne's, speaks of a primary period of uncertainty and restlessness, and a secondary one of systematised delusion.

Spitzka (p. 313) writes, "Delusions of persecution are the most common ones in delusional monomania. There is a marked difference between these delusions and the delusions of persecution found in melancholia. While the melancholiac believes that he is pursued or punished because he is a weak, cowardly, bad, or criminal person, the monomaniac believes that he is persecuted from motives of envy, and as a rule, he develops exalted delusions of his personal importance or worth, side by side with those of persecution."

The most common delusions are that certain persons are plotting against the sufferer, have evil designs on him, annoy him, poison his food, or persecute him in some way, or in various ways.

Auditory Hallucinations.—These are generally the first hallucinations to show themselves. The patient hears nicknames and insulting phrases, at first in a low voice, afterwards in a loud one. The voices are in the workshop and in the street, especially at night. The voices come from all directions, from people passing in the street; and indoors, from the chimney, the keyhole, the ceiling, the walls, the floor.

Olfactory Hallucinations are less common than those of hearing. The patient perceives disagreeable or disgusting odours.

Gustatory Hallucinations are common. The food tastes badly, hence arises the delusion that it is poisoned.

Cutaneous Hallucinations generally accompany the auditory; they seldom appear alone (Bra, p. 51). The patients feel itchings, formications, sensations of heat which they attribute to magnetism, electricity, etc.

Visceral Hallucinations or Illusions.—The patients fancy that people tear out their liver, heart, or testicles.

Sexual Hallucinations or Illusions.—People cause the sufferers to feel voluptuous sensations, or practise masturbation on them. These symptoms may be unilateral.

Visual Hallucinations are rare and even denied by some authors.

The patient is at first astonished at the persecutions and wonders why he is subjected to them. By degrees the vague feeling disappears, and having discovered his supposed enemies, the patient either (1,) flies or hides from them; (2,) commits suicide; or (3,) acts strongly on the defensive and offensive (Bra, *op. cit.* p. 52).

II.—Ambitious Form (Ambitious Monomania (Esquirol); Megalomania).

This may succeed the persecutory form or develop during it, the patient thinking he must be of consequence to attract so much attention.

It may precede the persecutory form, the latter becoming developed owing to the patient's extravagant claims not being recognised. It may develop primarily, but most frequently it follows auditory hallucinations. Thus a patient had heard voices which said, "There is the king of France," and he became Henri de Bourbon (Magnan, Bra, *loc. cit.*).

These patients are the emperors, kings, queens, princes, dukes, lords, generals, and colonels of asylums.

The delusions in ambitious delusional insanity are systematised; they are distinct and fixed, and the patient can reason logically with regard to them though his arguments are based on false premises. The patients also speak, act, and dress, if permitted, conformably as they believe, to the character they represent. They will wear straw crowns, tin medals, and hoop-iron swords. One patient, a farmer, called himself "Captain A—— of the sappers and miners," and talked about his "sweetheart, Victoria," to see whom he walked all the way from the North of England to Windsor.

III.—Religious Form (Theomania).

Developed Stage.—Religious monomania.—(1,) Expansive, characterised by ambitious ideas and egotism. The patients believe themselves to be prophets, apostles, the Messiah, etc. They have a mission to save society, and to wrest it from its iniquity. They dress, speak, and act accordingly; visual hallucinations, and more rarely auditory and gustatory soon appear. They see angels, hear the voice of God, and perceive intoxicating odours. They may commit self-mutilation, suicide, arson, assassination.

(2,) *Depressive.*—The patients accuse themselves of having committed all sorts of crimes, theft, adultery, murder, abominable acts : of being the cause of all the evil in the world. Others believe themselves possessed of the devil. They evince impulses to mutilation, suicide, and homicide; hallucinations are rare. (Bra, *op. cit.*, pp. 57–60). Sexual ideas and delusions founded on them are very common in religious delusional insanity.

IV.—Erotic Form (Erotomania).

Developed Stage.—The patient, most frequently a female, evinces a platonic affection for some person of the opposite sex (a clergyman, a doctor, a person in authority), and connected with this becomes possessed of delusions.

Acting on these, the patient writes to the object of her affections (who may be married), follows him about, tries to reside in the house with him, and accuses others of preventing this object being accomplished. A patient at Haydock Lodge had thus acted with regard to a curate previously to admission, and afterwards transferred her affections to one of the doctors of the asylum. When in private care after leaving the asylum, she escaped and travelled many miles to see the doctor. In all other respects the patient was quite rational, though quarrelsome and somewhat egotistical, and ambitious. Physically, she was dwarfed and deformed. This form would include the so-called Ovarian, or Old Maids' insanity of Skae, described by Clouston ("Mental Diseases," p. 478).

Spitzka ("Insanity," p. 27) says, "Systematised delusions of an erotic character are found as the leading symptoms of the so-called 'Erotomania.' This perversion is not necessarily accompanied by animal sexual desire, and the adjective erotic is here used in its classical sense. The patient, noted in his adolescence for his romantic tendencies, construes an ideal of the other sex in his day dreams, and subsequently discovers the incorporation of this ideal in some actual or imaginary personage, usually in a more exalted social circle than his own. He then spins out a perfect romance . . . . . . expansive or depressive delusions are incorporated with the erotic ones. As a rule, the affection for the adored object remains as chaste and pure as it begins . . . . . . hallucinations are woven into the delusive conception, which consequently assumes such a predominating position in the patient's mental horizon, as to entirely overshadow it."

V.—JEALOUS FORM.

Savage ("Insanity," etc., p. 263) speaks of this as a "dangerous and troublesome form of delusional insanity." There is a fixed delusion that the husband or wife, as the case may be, of the patient, is unfaithful, and the most trivial occurrences are taken as proof of his or her guilt. This delusion leads to constant annoyance, threats, and violence. Of several cases of this nature at Haydock Asylum, one manifested some weakness of the limbs of one side.

DEMENTIA, TERMINAL (TERMINAL DEMENTIA).

Synonyms.—Secondary Dementia (many authors); Terminaler Blödsinn (Krafft-Ebing).

The mental defect in terminal dementia may vary from a mere loss of memory, usually of recent events, to the nearly complete extinction of mind (Spitzka, p. 169). When the patient becomes calmer and stouter, without real mental improvement, dementia is to be feared (Bucknill and Tuke, p. 183).

In the severe forms of terminal dementia, the patient cannot answer questions however simple; in the mild forms he may be able to answer questions, but cannot sustain a conversation. The power of calculation is diminished or lost, and the memory, especially for recent events, is defective to a pathological extent; *e.g.*, the patient cannot remember the year, the month, or the day of the week; cannot remember his own age, and if told his age, cannot calculate the year of his birth. Forgets what he says immediately after he has said it, and cannot tell what he had for his last meal. Has no idea of place or time; does not know where he is or how long he has been there.

May be (1,) agitated, or (2,) apathetic (Krafft-Ebing, Spitzka).

(1,) *Of the Agitated or Active Form.*—This form of terminal dementia, succeeds chronic mania, and agitated melancholia; the patients are restless and talkative, and even obtrusive or destructive. Speech and acts are disconnected and senseless. Fragments of delusions are retained. There is frequently a change in the patient's sense of identity, in consequence of weakening of the logical power and memory. This form of terminal dementia is progressive and of longer duration than the apathetic (Spitzka, p. 170).

(2,) *Of the Apathetic Form.*—This form of dementia is a sequel of anergic stupor, stuporous melancholia, and violent outbreaks of maniacal furor (Spitzka, p. 168). Countenance expressionless. Extensors not innervated, and flexors, therefore, predominate. Some patients have to be dressed, undressed and fed. Others are docile and assist the attendants in various ways. Others lounge listlessly about, sit or stand in one place all day, performing rhythmical movements, or vociferating some set phrases (Spitzka, pp. 168-169).

Some dements are "wet" and "dirty" from loss of the sense of decency; others simply from loss of the normal feeling of discomfort caused by a full bladder and loaded rectum, or from muscular weakness, or both the latter combined. Some have to be fed forcibly, others eat automatically, and others have ravenous appetites. In all there is more or less loss of memory of recent events. Old recollections may be retained, but new impressions cannot be registered (Spitzka, p. 169).

DEPRIVATION OF SENSES *(Insanity from).*

Clouston (p. 603) mentions a case in which a patient became melancholic and suicidal coincidently with loss of sight from cataract, and improved greatly after a partially successful operation.

"It is very common indeed for those who are deaf to become quiet, depressed, and irritable. It is also common for such persons to become subject to hallucinations of hearing, and so insane as to need to be sent to asylums."

DIABETIC INSANITY.

Clouston describes two cases, and summarises as follows (*op. cit.*, p. 595): "They were both melancholic. They both imagined they had no money and were ruined, and could not pay their debts. They both had a disinclination to take food. They both wanted in affection for their children. They both were thin and weak. They both had a tendency to sores on extremities, with small healing power; but the one was more resistive and dogged, the other more passive, inattentive, and utterly uninterested in anything in the world. Death in both cases occurred rather suddenly."

Savage, in a paper read before the Medical Society of London ("Lancet," Nov. 29, 1890), stated that diabetes and insanity alternate in families (one generation being diabetic, and the next insane or *vice versa*); and also in individuals, acute diabetes being replaced by acute melancholia, this latter giving place again to diabetes, which was again replaced by temporary mental depression. In two cases, elderly men (diabetic), became melancholic shortly before the fatal termination, and the sugar disappeared. Besides patients with diabetic relatives, Savage had ten patients in Bethlem, who were both insane and diabetic. "Nearly all the cases with diabetes and insanity were melancholic."

EPILEPTIC INSANITY.

Synonyms.—Folie épileptique, Epileptisches irresein, Pazzia epilettica.

Mental State of Epileptics not Actually Insane.—They are irritable, irascible and lazy. They are often dipsomaniacal, and in general erotic. They are mostly gloomy, morose, taciturn, dreamy, and cowardly. Some are extremely docile and suave, servile and obsequious, speak with facility, and have a habit, peculiar to themselves, of talking quite close to the face of the person with whom they are conversing. All epileptics pass rapidly from anger to suavity; mobility is the dominant note of their character (Bra, "Manuel de Maladies Mentales," pp. 73-74).

Epileptic insanity (fully developed) may be divided into acute and chronic.

ACUTE EPILEPTIC INSANITY.

The three principal characteristics of acute epileptic insanity are: (1,) Impulsive acts. These are sudden, brusque, automatic, and may express themselves in the form of crimes and misdemeanours, *e.g.*, murders, suicides, assaults, indecent exposures, etc., etc. They may, or may not be preceded by an aura. Huchard records a case in which an epileptic called out, "Mother, run away, I must kill you!" (Bra, p. 79); (2,) Loss of memory. This is generally absolute, but occasionally the patient may have a vague recollection of what has happened; (3,) The resemblance to each other of all the attacks in the same patient (Huchard, Legrand du Saulle; Bra, p. 79). This form may precede or follow the attack, may occur in the intervals between the convulsive crises, or may even replace the latter:—

(1,) *Before the Convulsive Attack.*—Prodomal symptoms: Various auræ, *e.g.*, the sensation of cold air, of tickling, or of swelling, palpitations, anorexia, painful trembling of certain muscles, formication, gyratory impulses, cursatory impulses; the patients throw themselves forwards or backwards, or turn on their own axes. At other times these symptoms are accompanied or are replaced by an intellectual aura. Some patients suffer from insomnia, become irascible, have suicidal ideas or a presentiment of the attack; sometimes on the contrary they possess great serenity of mind.

Developed attack: Hallucinations of all the senses frequently occur, those of sight being most common. Some epileptics perceive sweet tastes and strong odours. These patients under the influence of their hallucinations, suddenly commit insensate acts of which they retain not the slightest recollection. One will upset everything in his neighbourhood, and then fall; another will take up a glass and fling it to a distance; others pronounce inconsequent words, or are seized with an access of sudden rage.

(2,) *Post-Convulsive, Acute Epileptic Insanity.*—More common than the ante-convulsive form. Bra describes two forms (p. 77): (*a*,) The "petit mal intellectuel"; (*b*,) The "grand mal intellectuel" (Epileptic mania or epileptic furor). The intellectual petit mal occurs most frequently between the ages of eighteen and twenty. There are three distinct varities (Bra, *loc. cit.*), viz.: (a,) The *Depressive*, characterised by hebetude, confusion of ideas, profound malaise, extreme prostration, inquietude, and a sensible loss of memory; (β,) The *Expansive*, with cerebral excitement, loquacity, ideas of satisfaction, bursts of laughter, fits of gaiety, a passion for travel and long excursions, and an incessant desire for change of place; (γ,) The *Mixed*, in which the epileptic fits, as Billod has observed, are preceded by a melancholy state, and followed by a maniacal one. Finally, a sort of persecutory maniacal variety has been observed, in which the patients believe themselves surrounded by enemies, and commit crimes under the influence of these ideas. The intellectual grand mal (epileptic mania) (epileptic furor), most frequently commences suddenly. It is a furious and violent outburst of maniacal form, with blind impulses to destruction, murder, incendiarism, etc. Threats, complaints, and blasphemies are uttered in quick succession; the muscular force is increased, and the gestures are violent. Epileptic maniacs destroy automatically, and stop at nothing. Afterwards they preserve in general no recollection of what they have done.

(3,) *Attacks of Insanity* may occur between the epileptic attacks (intervallary, Spitzka), or even in the total absence of convulsive seizures; they may precede the latter by a long period, and themselves constitute the epilepsy. This is the "Epilepsie Larvée," of French authors. In it the mental symptoms constitute the sole epileptic manifestation. In these cases, periodical excitement is followed by prostration and stupor; there is excessive irascibility, acts are commited suddenly, and apparently as the result of irresistible impulses; there is a tendency to homicide and to suicide, or the patients may suffer from terror-inspiring hallucinations (Grasset; Bra, "Manuel des Maladies Mentales," p. 78). A form of epilepsy called Epileptic Vertigo, is characterised by a temporary absence or short eclipse of reason. The patients commit unconscious acts, *e.g.*, indecent exposure, micturating in public, making incoherent speeches, etc., etc. These are according to Herpin, veritable fits of automatism, and of somnambulism. These attacks are characterised by their suddenness, and by the loss of memory experienced by the patient of what has occurred during them (Legrand du Saulle, Trousseau, Lasègue, Magnan; Bra, "Maladies Mentales," p. 76).

CHRONIC EPILEPTIC INSANITY, OR EPILEPTIC DEMENTIA.

This is a consequence of repeated attacks of acute epileptic insanity, or of epileptic fits, continued through a succession of years. It is manifested by a diminution of memory and of power of attention. The patients forget everything, and become childish in their actions. They become furiously addicted to onanism (Bra, "Manuel des Maladies Mentales," p. 80). The special sense reactions are abnormally slow in epileptic insanity (Lewis).

EXOPHTHALMIC GOITRE (*Insanity with*).

The chief characteristics of a typical case described by Savage (p. 413) were excitement, incoherent talking, violence, destructiveness, and sleeplessness. Afterwards she was filthily dirty in her habits. Became dull and sleepy before death.

The principal symptoms of Graves' disease (Basedow's disease) may be divided into: (1,) The glandular, enlargement of the thyroid; (2,) The circulatory, palpitation of the heart and rapid pulse; (3,) The ophthalmic, exophthalmos and proptosis; (4,) The nervous, tremor and insomnia; (5,) The electric, diminished electrical resistance. The latter has been demonstrated in many cases at the Golden Square Throat Hospital, by Wolfenden. All the symptoms may not be present at first in any given case.

Savage says (p. 415), "Besides ordinary insanity associated with this condition, I have met with several cases of general paralysis." He further observes that in some cases of recurrent mental disorder, prominence of the eyes, rapid pulse, and a somewhat enlarged neck were among the earliest symptoms of the recurrence.

FOLIE À DEUX.
Synonym.—Folie Communiqué.

(1,) Folie imposée, in which weak-minded patients acquire *delusions* from stronger-minded ones, generally monomaniacs (see " Delusions, Spurious "), but do not suffer from hallucinations.

(2,) When two or more people living together drift simultaneously into the same form of insanity, such cases are called by the French alienists cases of *Folie simultanée;* but (3,) when, as is sometimes the case in *folie à deux, insanity* is acquired by one person and then from him by another, it is called *folie communiqué* (Ireland, " The Blot upon the Brain," p. 201, *et seq.*); (Marandon de Montyel, " Rev. in Allg. Zeit. f. Psych.," 1881.)

FOLIE DU DOUTE (DOUBTING INSANITY).
Synonyms.—Geistestörung durch Zwangsvorstellungen; Abortive Verrücktheit; Pseudomonomanie; Paranoia Rudimentaria Ideativa (Morselli).

The patients attacked by this form of mental disorder generally belong to the better classes of society, and are rarely found in public asylums.

At first the patient is gloomy, timid, dreamy, and punctilious. He loses self-confidence, reads and re-reads what he writes, pays extreme attention to his words and acts, weighs his judgments, repeats constantly to himself the same words, and is always interrogating himself. He becomes possessed by a certain idea or set of ideas, and around this his intellectual operations concentrate themselves.

According to Legrand du Saulle the subjects which most frequently form the pabulum for this *psychological rumination* are: God, the birth of Christ, the creation, nature, the virgin, life, the human intellect, the sun, the moon, the stars, thunder, the difference of the sexes, the conformation of the genital organs, copulation, sleep, sudden death, precipices, the forgiveness of sins, omissions at confession, the size of animals, the dimensions of objects; glass, gold, silver and copper money, rabid dogs, pins, door latches, paper or pencils.

According to Ball, in a general manner these patients may be divided into five categories, viz.—The metaphysical, the realistic, the scrupulous, the timid, the counters or reckoners, or enumerators, to which may be added those affected with the touch craze *(délire du toucher)*: (1,) The *metaphysical* patients are always demanding the why and the how concerning the most inexplicable things, the creation, God, man, the external world and their origin and destination; (2,) The *realists* limit themselves to trivial questions, the dimensions of an object, the height of a person, etc.; (3,) The *scrupulous* ones pass their lives in the incessant fear of not having told the truth, of having had bad thoughts, of having wronged others, etc.; (4,) The *timid*, when they write a letter, read it several times to be sure there are no mistakes; if they shut a door they return four or five times to be certain that they have not left it open; if they take a carriage they search the cushions and pockets minutely; if they go out they are always afraid of seeing somebody fall out of a window, etc., etc.; (5,) The *counters* enumerate all the objects they see, buttons on coats, books in a library, windows of houses, trees in the boulevards, etc.; (6,) Those who suffer from the *touch craze* (*délire du toucher*) have a constant fear of having touched some object which was dirty (mysophobia) or contained poison, or even of contaminating others by touching them; hence result incessant ablutions which never in the patient's opinion produce perfect cleanliness. Then there are other forms in which the sufferers doubt the existence of themselves or of the external world; others, again (somewhat resembling agoraphobia) in which the patient cannot pass a tree or even cross the doorstep.

At first the patients are conscious of their condition, say their fears are absurd, but that they cannot divest themselves of them. They have an irresistible desire to be reassured, and select some friend or physician of whom they incessantly ask the same questions (Bra, "Manuel des Maladies Mentales," pp. 62–67). Krafft-Ebing (p. 529) considers *folie du doute* to be a form of mental degeneration developing on a neurasthenic basis.

GENERAL PARALYSIS OF THE INSANE.

Synonyms.—Paretic Dementia (Spitzka), General Paresis, Diffuse Interstitial Periencephalitis; Dementia Paralytica (Krafft-Ebing); Periencephalo-Meningitis Diffusa Chronica (Calmeil); Paralysie générale des Aliénés, Paralysie générale progressive, Folie paralytique; Paralyse der Irren, Allgemeine Paralyse der Geisteskranken, Paralytischer Blödsinn; Paralytic Dementia, Progressive General Paresis; Psicopatia paralitica (Morselli).

SYMPTOMS—PRODROMAL.
I.—Physical or Somatic.

Neuralgia, either (*a*,) general, characterised by its mobility, flying from the trunk to the lower extremities, or (*b*,) localised, affecting a single nervous branch. The most frequent forms are sincipital; temporo-frontal and occipital; then follow cardialgia, epigastralgia, and rachialgia. There are besides various painful sensations, of heat, of cold, of pressure, etc. There are often complaints of formication and pricking sensations in the skin; sensations of electric currents in the head. Some individuals feel themselves as light as birds, and have a desire to walk a great deal; others are dull, heavy, fatigued, and are not relieved by rest in bed. The eyes are often injected and the glance unusually vivacious. There may be hummings, whistlings, sounds of bells in the ears, vertigo and temporary losses of consciousness ("absences" or "stunnings").

The principal circulatory troubles are palpitations, flashes of heat to the head, sudden redness of the face.

There are often disturbances of the functions of the digestive organs, anorexia, eructations, constipation.

Dysmenorrhœa, and amenorrhœa are often noticed, the latter more frequently than the former (Voisin, "Traite de la Paralysie Générale des Aliénés," pp. 6-8).

Julius Mickle ("General Paralysis of the Insane," 1st ed., p. 4) says, "Some embarrassment of speech is occasionally the very first sign noticed, and Austin, on the authority of Phillips, speaks of extreme contraction of the pupils, and, on his own authority, of a fixed or unsymmetrical condition of the pupils as being frequently prodromal; whereas, Griesinger observes that the pupils are sometimes irregular for years before the onset of the disease; while trembling of the limbs has been placed in the same heralding category." In his 2nd ed. Mickle lays stress on epileptiform and apoplectiform seizures as forerunners of general paralysis.

II.—Psychical or Mental.

(1,) Absence or diminution of sleep at night is very frequent; what sleep is obtained is disturbed by dreams and nightmares, and is only very slightly refreshing. At the same time there is a tendency to sleep after food; (2,) A change of character and of habits is always observed; this may be simply an exaggeration of the patient's previous tendencies, or it may be a completely new and opposite character. There are three varieties of this change: (a,) Depressive, (b,) Expansive, (c,) Demented. The first variety is the most frequent, and Doutrebente says that it appears to be constant. Lasègue also says that sadness opens the march of the disease (Voisin, "Paralysie Générale," pp. 8-10). In the depressive variety the patient becomes sombre, sad, melancholy; in the expansive, gay, enterprising, loquacious, perhaps witty, foolishly confidential, putting the world *au courant* with his projects and ideas, satisfied with everything, starting ambitious projects which fail; (3,) Change in the affections, persons previously dear to the patient become hateful to him; (4,) Unusual and useless activity; (5,) Change of occupation; (6,) Frequently excess in eating and drinking; (7,) Sexual excesses in some cases; (8,) Extraordinary egotism, often revolting; (9,) Inconsistencies in speech and acts, generosity, just debts being unpaid, selling necessary articles to buy unnecessary ones; sordid avarice accompanied by a habit of purchasing articles of secondary utility; (10,) Thefts are frequently committed in the demented or debilitated form; (11,) Loss of memory for recent events is the first failing; proper names are early forgotten; the speech is slow, wrong words are used unconsciously; instead of names, substitutes as "thingumy," "what do you call it," "what's his name," etc., are employed; sentences are broken off abruptly; (12,) The attention in the debilitated or demented form cannot be sufficiently fixed; when the patients write they leave out letters and words; they cannot continue steadily at any occupation; they cannot manage household affairs or business so well as formerly; some calculate badly and give wrong change; reasoning power and judgment are also weakened (Voisin, "Traité de la Paralysie Générale des Aliénés," 1879, pp. 10--18).

Savage ("Brit. Med. Jour.," Apr. 5, 1890) gives as "warnings" of general paralysis, muscular fatigue, ataxic symptoms, *temporary aphasia*, change in handwriting with difficulty or fatigue in writing; alteration of facial expression, the lines of the face being wiped out and giving rise to an appearance of fatness; ptosis, and external strabismus; slight attacks of giddiness, or loss of power, temporary loss of sight or of feeling (so called slight "fainting fits"), epileptiform fits (if epileptic or epileptiform fits occur at irregular intervals in a middle-aged man, mental changes such as loss of memory occurring with each fit and expression being permanently affected, general paralysis is almost certainly present). Such fits may precede the ordinary symptoms of general paralysis by several years.

Headache, facial neuralgia, other forms of neuralgia, headache, accompanied by tenderness or feeling of lightness in the head, sciatica (double, with change in habits, etc., not to be forgotten), rheumatic pains, temporary loss of sight, optic neuritis, affections of hearing, alterations in the senses of taste and smell, defect of colour perception in artists, loss of the finer relations of time in musicians.

Loss of power of social accommodation, memory for recent events and for engagements defective (especially if any fainting or other fits have occurred), loss of power of attention, want of persistence, restlessness, motor or mental. In cases where the mind suffers before the body, the above intellectual defects nearly all occur in greater or less degree.

Moral weaknesses or faults; stupid stealing, and thoughtless indecency (exposure of the person being more frequent than assaults).

Instability; tremor of the finer muscles, uncertainty of gait, and tendency to fall; restlessness, endless things being started and set aside on the first fresh suggestion; almost always abnormal reaction to alcohol or drugs, the patient being very easily intoxicated or poisoned.

Change of *temper* and *character* are probably the most constant symptoms of early general paralysis. Change of character with instability of purpose and some motor weakness occurring in a middle-aged man almost always point to general paralysis.

Hypochondriasis may be one of the warnings; alternations of buoyancy and depression; marked hysterical or hystero-epileptic fits in middle-aged men are even more alarming than epileptic fits.

Sudden outbreak of mania is a frequent precursor of general paralysis. General paralysis may quickly succeed an attack of acute delirium (acute delirious mania, grave delirium, "brain fever").

In addition there are vaso-motor troubles, with a history of syphilis or brain injury. Take note of early fatigue, fainting or other fits, loss of smell, vague optic disc changes, unaccountable knee phenomena, unusual headaches, neuralgia and sciatica, change of character, progressive loss of the highest control, moral lapses and instability in various forms.

INTERMEDIATE STAGE.

When there are mental symptoms without certain diagnostic somatic characteristics.

This period is often absent. It sometimes commences brusquely without a previous prodromal period. Or it may commence insidiously. It may be said to have commenced when the mental trouble of the prodromal period has become sufficiently marked to merit the name of insanity. This mental condition may resemble: (1,) Melancholia with stupor; stuporous melancholia, melancholia attonita; (2,) Hypochondriacal melancholia; (3,) Lypemania (melancholia); (4,) Mania; (5,) *Folie circulaire.*

It may consist of modifications of the sentiments and instincts.

Exaggeration of the altruistic sentiments and intellectual debility may be observed at this period.

Voisin (p. 21) fixes the maximum limit of this period at two years, and says if somatic symptoms appear after a longer period of mental aberration than this, we have had to do in the first instance with some form of simple insanity. He says, however, that a remission lasting two years may occur between the subsidence of the mental symptoms and the commencement of the somatic troubles.

CONFIRMED GENERAL PARALYSIS.
First Stage — Somatic Troubles.

The five most valuable signs, because the most constant, ordinarily the earliest, and the most persistent, are: (1,) Loss, or diminution of the sense of smell on one or both sides (Voisin, p. 39). Mickle ("General Paralysis," 1st ed., p. 16) states that his experience does not agree with that of Voisin as to the loss of smell, but he admits that hallucinations of smell are very rare in general paralysis (" comparatively infrequent ") (2nd ed., p. 21); Griesinger says (p. 404) that after the disease has lasted a long time, and only then, the senses of smell and taste become blunted so that, for example, wine can not be distinguished from water; (2,) Febrillary tremors of the muscles of the face, lips, and tongue. Griesinger (" Path. u. Therap. Psych. Krank.," 4th ed., p. 402) says, "The *Tongue* is always the organ whose movements first present irregularity. The patient commences to speak with difficulty, to articulate somewhat inaccurately, and to stammer. The tongue, when protruded, does not incline to one side, but tremulous and occasionally convulsive movements of it are observed "; (3,) Tremulousness of speech; (4,) Inequality, excessive dilatation, excessive contraction, want of mobility of the pupils; (5,) Fever with special characters, *i.e.*, it occurs every eight, ten, or fifteen days, suddenly, lasts one or several days, is of low range, higher in the evening, and disappears as suddenly as it comes.

The Accessory signs are:—

I.—Sensory. Anæsthesia (cutaneous) occurs often at the commencement; hyperæsthesia, temporary, quite at commencement of disease.

Griesinger (*op. cit.*, p. 404) states that cutaneous sensibility is blunted in all cases at the commencement, and in some cases afterwards abolished, and that occasionally there occur transitory states of extreme hyperæsthesia in which the slightest touch excites the most extended reflex movements, convulsions of all the voluntary muscles.

Pains, often like neuralgic pains of the trunk and limbs; sometimes vague. The pains generally disappear when the disease becomes pronounced.

The muscular sense, the heat sense, and the electric sensibility may be abolished, or over-excited, altered or hallucinated.

Visual acuteness may diminish to the extent of blindness. There may be visual hallucinations and illusions.

Hyperacousia sometimes occurs and auditory hallucinations are frequent and tumultuous.

All these sensory troubles included in the accessory signs are common to the intermediate and the true first period.

II.—Motor. Temporary paralytic attacks lasting from several hours to several days. Persistent attacks which may be (1,) General; (2,) Incomplete; or (3,) Progressive. The gait is peculiar in consequence, the legs are widely separated, the toes turned out, and the tread heavy, the patient stumbling and tripping.

Griesinger (*op. cit.* p. 402) says, "At the same time that the speech becomes difficult, more frequently not until soon after, a change in the *gait* of the patient is observed; he does not lift his legs properly, walks stiffly, involuntarily deviates to one side when attempting to walk straight forward, and easily stumbles if the ground be at all uneven—for example, when going over a step."

The physiognomy is expressionless; the eyebrows are raised up on the forehead towards the middle of their arch, or fall over the eyes like a moustache. The generalised paralysis is rare at the commencement of the disease, so also is the unilateral. The latter is most conspicuous when the patient walks, by the lateral inclination of the body, and by the elevation of the opposite shoulder. On the side of the low shoulder will be generally noticed a slight deviation of the facial traits; the furrows are less marked, the eye is less open, the labial commissure is lowered, the tongue is protruded to the weak side, and the uvula deviates. This unilateral paralysis is generally temporary.

Ptosis sometimes occurs; exophthalmia and external strabismus are not frequent, but are sometimes met combined. Internal strabismus has been observed at the beginning of the disease.

Paralysis of one or more, rarely of both, vocal cords sometimes occurs; constipation arising from the muscular weakness is frequent.

Tremors of the lower and upper extremities, ataxic in nature. Manual inability, leading to awkwardness in delicate manipulations and to alterations in writing, the letters being smaller and less regular. With regard to the lower limbs the patient can stand steadily but walks too rapidly; attempted movements are over done or insufficiently done.

Champing of jaws, with movements of mastication; rigidity of the trunk muscles; grinding of the teeth. The first of these three symptoms belongs rather to the second stage, the other two to the first (Voisin, *op. cit.*, p. 62).

III.—Modifications in the character of the pulse. Compressible under the finger. The tracings show a certain elevation of the ascending line; the frequent presence of a plateau, oscillations in the line of descent, and a variable degree of dicrotism. As the disease progresses this condition becomes accentuated.

IV.—Modifications of the Urine. Density increased, more or less, either during depression or during excitement. Quantity of urine augmented. The quantity of urea is generally augmented, and that of the chlorides and phosphates diminished (Voisin, *op. cit.*, p. 64).

CONFIRMED GENERAL PARALYSIS.
First Stage—Psychical Troubles.

These are only the exaggeration of those which are met with during the prodromal and intermediate periods. Most frequently there is mental aberration, although occasionally there is only enfeeblement of the intellectual faculties accompanied by a perversion of the sentiments.

When there is mental aberration it has always special characters; the ideas are multiple, mobile, absurd, and contradictory; the mental state may be expansive or depressive, the former being much more frequent.

In the expansive form there are: (1,) Ideas of satisfaction; (2,) Ideas of grandeur, differing sensibly from those of other megalomaniacs; (3,) Ideas of wealth; (4,) Ideas of exaggeration; (5,) Ambitious ideas (second and third) combined with hypochondriacal ideas, giving rise to very odd conceptions.

The depressive form may take the guise of lypemania or of hypochondria.

The lypemania may assume: (1,) The form of melancholia agitata; (2,) Of stuporous melancholia; (3,) Of religious melancholia; (4,) Of persecutory delusional insanity; (5,) Of delusions of poverty.

The hypochondriacal mental aberration may be sub-divided into: (1,) Delusions concerning obstruction of organs; (2,) Delusions of negation of existence; (3,) Micromaniacal, the patients believing themselves to be little infants or dwarfs.

The hypochondriacal aberration has characters which are only exceptionally met with outside general paralysis; *e. g.*, it appears suddenly, it is exceedingly absurd, it is mobile. When the intellectual trouble takes the form of dementia, this may vary from slight loss of memory for recent events (hardly perceptible to strangers) to dementia closely approaching stupor.

Whatever the intellectual troubles may be, there are simultaneously modifications in the sentiments and the instincts, and the actions are stamped with a character of improvidence and absurdity (Voisin, *op. cit.*, pp. 68-108).

CONFIRMED GENERAL PARALYSIS.
Second Stage.

Any one of the accessory signs may now acquire major importance. Various affections of *sensibility* and of the organs of sense are observable, anæsthesia, hyperæsthesia, weakness of sight with few hallucinations, abolition of smell and taste. Tortuosity of retinal arteries. The patients *eat* in a disgusting manner, swallow without discrimination, eat with voracity, and do not masticate sufficiently, so that they are often choked. Hemiplegia is rare, and in general the muscular force is preserved.

There are phenomena of *vaso-motor* paralysis, flushings of the face, and the pupillary phenomena. (See "Somatic symptoms," 1st Period.)

The patient requires assistance in walking, dressing and eating. He cannot turn round without staggering or even falling.

The *ataxy* differs from that of tabes dorsalis; it is not increased by closing the eyes. There are *tremors*; silent agitation. The *speech* is slow, drawling, hesitant from the cerebral lesions; stammering, stuttering, and tremulous from the bulbar. There may be also mutism. The *writing* is irregular and shaky. The *lingual* tremors become more and more accentuated; the other first period symptoms continue. There may be *fever* slight and continuous, or sharper and temporary.

The mental aberration may be expansive or depressive, or often both combined; may be hypochondriacal. The ideas become more and more absurd, and less and less numerous and mobile. There may be dementia without delusions, and then the course is slow.

The affective sentiments become enfeebled, and finally abolished. The instincts lose their vivacity, and the sexual appetite is lost never to return (Voisin, *op. cit.*, pp. 109-143).

CONFIRMED GENERAL PARALYSIS.
Third Stage.

(1,) *Cutaneous Anæsthesia.*—Sometimes so complete that the patient can be burnt without feeling it.

(2,) The *organs of sense.*—Very variable. In some cases optic nerve atrophy has been observed, but the most frequent lesion seems to be tortuosity of the retinal arteries (Voisin and Galezowski).

(3,) *Ataxia.*—In some cases the patient can neither walk nor stand. The tongue is protruded and withdrawn with difficulty.

The *speech* becomes more and more hesitant and tremulous until it becomes unintelligible, and the patient speaks very little. Sometimes complete mutism supervenes, either on account of extreme ataxy of the muscles, or complete abolition of the intellectual faculties, or degeneration of the lingual muscles. The *writing* is absolutely shapeless and hieroglyphical.

(4,) *Paralysis.*—Formerly apparent, now becomes real. May be general or unilateral; more frequently the latter. The patient becomes first *wet* and then *dirty* in consequence of dementia rather than paralysis. But a temporary congestive flushing of the face is often accompanied by incontinence of urine owing to paralysis of the vesical sphincter.

(5,) *Muscular Contractures.*—In general paralysis without complications these are not persistent; they may last a day or a week, and then disappear from day to day.

(6,) *Psychical Troubles.*—The delusional condition has generally subsided into dementia, and when it does persist it takes the character of the depressive as often as that of the ambitious form; these may replace each other from day to day, or even be simultaneous. The patients mutter such words as millions, jewels, castles, etc. Coquetry persists to the end with the females, whereas the males make use of such words as emperor, king, etc. Anything bright attracts the patients, and they collect pieces of rubbish under the delusion that they are jewels, gold, etc. The memory is entirely lost, both for recent and long past events, so that the patients forget even the meal times, though they remain very greedy and very voracious.

(7,) *Affective Sentiments.*—These have entirely or almost entirely disappeared. Or occasionally there may be unaccountable antipathies.

(8,) *Cachexia.*—This is manifested by a dull tint of the skin which becomes dry, desquamates furfuraceously, or becomes greasy and emits a repulsive, acrid, ammoniacal characteristic odour. The hair falls. If the patient is not well nursed he becomes affected with impetigo, and with pediculi capitis and corporis. The temperature is subnormal. The arterial diastole is hardly perceptible. The urine is often fœtid.

(9,) *Trophic disturbances.*—Bedsores; these will sometimes heal in one situation whilst they are forming in another.

(10,) *Alterations in the blood.*—It is fluid, viscous, does not coagulate, or coagulates with difficulty; the clot which floats in a brownish serosity is soft, diffluent, and tears readily. Voisin has discovered in some cases bacteria and vibriones (*op. cit.*, p. 161). There will sometimes be found angular globules and crystals of urate of soda. The alterations of the blood may also be observed in other lunatics who have fallen into a cachectic state.

(11,) *Othæmatomata.*—More frequent in the second and third stages of general paralysis than in mania or epileptic insanity. In the sane they disappear quickly. Voisin (p. 162) believes they result from violence of some kind, and acquire their peculiar character and persistency from the cachectic condition and altered blood of the patient. (See "Hæmatoma Auris," chap. II.)

(12,) *Mucous Hæmorrhages.*—There are no ulcerations of the membranes, but there is considerable vascularity. Hæmorrhages may occur from the vaginal, intestinal, nasal, and buccal mucous membranes, towards the termination of the disease (Voisin, *op. cit.*, pp. 143–167).

Voisin describes five forms of general paralysis:—

(1,) Acute General Paralysis in which the course is rapid, the stages are confounded, and death occurs early as a rule. It may suddenly attack an apparently healthy person without any warning.

(2,) The Common form of general paralysis in which the mental state is generally expansive and ambitious. Often accompanied by epileptiform and apoplectiform attacks.

(3,) The form in which symptoms of dementia predominate (Paralytic Dementia). It is the chronic form *par excellence*, and is accompanied by few somatic troubles.

(4,) The Senile form connected with atheroma of the arteries. In its course it is next in rapidity to Form 1. It is very rare.

(5,) The Spinal form in which the medullary troubles dominate the scene, and the intellectual are of secondary importance. It is very irregular in its manifestations.

Another division of General Paralysis is into four forms, three of which depend on the character of the mental symptoms, and the fourth on their absence or insignificance: (1,) The expansive form; (2,) The depressive or melancholy form; (3,) The demented form; (4,) The somatic form.

GESTATIONAL INSANITY.

Synonym.—Insanity of Pregnancy.

Generally melancholic in character but may be maniacal. *Mild form.*—Mental depression or mental apathy not amounting to stupor; loss of interest in matters formerly interesting; loss of conscious affection for husband and sometimes for children; a slight weariness of life; a fear of something going to happen; timidity; disinclination for social intercourse (Clouston, p. 518).

Severe form.—In this there may be delusions of suspicion as to poison; dislike of the husband; dipsomania; kleptomania; mendacity; obscene language; suicidal tendency; occasionally homicidal impulse.

HYPOCHONDRIASIS.

(See Melancholia, Sub-Division Hypochondriacal Melancholia.)

HYSTERICAL INSANITY.

Synonyms.—Folie Hystérique, Hysterisches Irresein, Pazzia Isterica.

Mental state of hysterical patients not insane.—In childhood these patients are remarkably intelligent and imaginative, learn readily and imitate cleverly; they are exceedingly impressionable, very coquettish, unblushingly mendacious, and extremely eager to attract attention. They suffer from *migraine*, headache, *insomnia*, *nightmare*, hallucinations, *night terrors*, and sometimes from gastralgia, *ovarian hyperæsthesia*, and palpitation.

In *adults* in whom hysteria has actually developed, the character is infantile and extremely mobile, the patients passing rapidly from laughter to tears, from anger to amiability, from loquacity to mutism, displaying what Huchard calls ataxie morale.

They are very prone to denounce others, and are much given to opposing, contradicting, and arguing.

They have a great desire to be talked about and consequently simulate various strange and unusual symptoms, injuries, and diseases.

They manifest sometimes a sort of cerebral laziness with quietude and moral anæsthesia.

A few are actually sensual, more are erotic but stop short of actual sensual gratification, others seem to be averse to sexual intercourse but indulge in solitary vice, whilst many evince no sexual excitement whatever.

Hysterical delirium may break out before, during, or after the hysterical convulsive attack or may replace it (hystéric larvée). This delirium is characterised by exaltation, mobility of ideas, perversion of sentiments, illusions, hallucinations (the colour red generally predominating), automatism, abulia, irresistible impulses, simulation, maniacal agitation or stupor, mutism, sometimes a strong erotic tendency, and nearly always retention of consciousness, the patient remembering everything she has said or done. The exaggeration and prolongation of this condition constitutes acute hysterical insanity.

(1,) *Acute Hysterical Insanity.*—In a hysterical person, in consequence of some emotion, of some menstrual trouble, of weakness after an illness, often after some incomplete hysterical attack, there is extreme agitation which develops into a veritable attack of acute mania. This often appears to replace the convulsive attack which is completely absent. In most cases the intellectual faculties are little affected, and the patients remember what they have done during the attack. A passive acute form of a depressive character has also been observed.

(2,) *Chronic Hysterical Insanity.*—May be either *maniacal* or *melancholic*. It may be a sequela of the acute form, or it may arise through progressive aggravation of the peculiar hysterical character. The patients become more sombre or very violent and passionate; then they emaciate, become anæmic, and even fall into marasmus. They are constipated, dyspeptic, and suffer from dysmenorrhœa and irregular menstruation. Finally chronic vesania is not slow to appear.

There is sometimes classical chronic mania with all its symptoms, or profound melancholia may develop with refusal of food, and even suicidal tendency, or, again, but more rarely, there is a form of simple mania with egotistical ideas. With many patients there is an erotic or slightly religious tinge in the mental derangement, but by no means with all.

In chronic hysterical insanity there are often exacerbations at the menstrual periods; these are frequently accompanied by migraine, swelling of the upper lip, and intestinal troubles (Bra, "Maladies Mentales," pp. 88–89).

Savage ("Insanity," p. 87) says, "Hysteria usually occurs in women, but I have seen grave hysteria in young men; and although I have never met with true hemianæsthesia and paraplegia in hysterical young men, yet I have seen some cases of globus hystericus, so that the man passed from the condition of the hysterical girl into that of the hypochondriacal man." He also observes, "Hysteria may colour other mental affections, that is, an exaggeration of any one of the perversions seen in hysteria may become a delusion."

Spitzka ("Insanity," p. 257) writes, "Hallucinations are frequent in chronic hysterical insanity, and usually of the kind described by Wundt as fantastic hallucinations of hypochondriacs, being the outcome of the patient's fancy and fears."

Griesinger ("Die Pathologie und Therapie der Psychischen Krankheiten," 4te Anflage, s. 185) states that the melancholic or maniacal forms of the chronic insanity of hysterical patients are developed from easily noticed though at first moderate changes of character; sadness of disposition, a degree of egotism not formerly evinced, valetudinarianism, great indecision and abulia, impatience, violence, and irascibility.

IDIOCY.

(Including Congenital Imbecility, and Cretinism.)

Esquirol defines idiocy as "A condition in which the intellectual faculties are never manifested; or have never been developed sufficiently to enable the idiot to acquire such an amount of knowledge as persons of his own age, and placed in similar circumstances with himself, are capable of receiving. Idiocy commences with life, or at that age which precedes the development of the intellectual and affective faculties, which are, from the first, what they are doomed to be during the whole period of existence" (Bucknill and Tuke, "Psychological Medicine," p. 150).

"The degraded condition of the idiot is very clearly displayed in his vacant stare, in the thick everted lips, the slavering mouth, the irregular, crowded, and decayed teeth; the gums often swollen, the frequent strabismus, the ill-formed, generally large ears, and the absence or defect of one or more of the senses—sight, hearing, speech, taste, or smell. His staggering walk is also very striking; yet he seems as if he must be in motion if he is on his feet; and even if seated has a difficulty in balancing himself. There is a general want of symmetry; the limbs are frequently contracted or paralysed; the fingers are long and slender; the grasp of the hand feeble or powerless, while the extremities are often cold and bluish from imperfect circulation. Psychologically, we may regard the idiot with M. Séguin, as badly served by imperfect organs, the instincts limited, but imperious, sensation and reflex action taking the place of attention, comparison, judgment, memory, foresight, and will" (Bucknill and Tuke, *op. cit.*, p. 151).

Esquirol speaks of three degrees of idiocy, viz.: (1,) That in which isolated words and short phrases are used; (2,) That in which only monosyllables or certain cries are used; (3,) That in which speech is absent.

Bucknill and Tuke (*op. cit.*, p. 152) speak of three classes of idiots, viz.: (1,) Those who exhibit nothing beyond the reflex movements known as the excito-motor; (2,) Those whose reflex acts are consensual or sensori-motor, including those of an ideo-motor and emotional character; (3,) Those who manifest volition, whose ideas produce some intellectual operation, and consequent will.

Dr. Down classifies idiots ethnologically; Caucasians, Ethiopians, Malays, Mongolians, being all found typified among them.

Dr. Ireland's classification of idiocy is: (1,) Genetous or congenital idiocy; (2,) Microcephalic idiocy; (3,) Eclampsic idiocy; (4,)

Epileptic idiocy; (5,) Hydrocephalic idiocy; (6,) Paralytic idiocy; (7,) Cretinism—see post; (8,) Traumatic idiocy; (9,) Inflammatory idiocy; (10,) Idiocy by deprivation, *i.e.*, the loss of two or more senses (Bucknill and Tuke, *op. cit.*, p. 157).

I.—GENETOUS OR CONGENITAL IDIOCY.

Characters: Often dwarfish; deformities common, especially a highly vaulted palate; teeth irregularly placed and subject to decay; deficient growth of finger nails, clubbed fingers; squinting and rolling of the eyes; cyanosis, deficient cardiac valves, lobulated form of kidneys.

II.—MICROCEPHALIC IDIOCY.

The circumference of the head may be as small as $14\frac{1}{8}$ inches. These patients are quick and quarrelsome; they "improve under training, and have more physical and moral energy than is common with idiots of other classes."

III.—ECLAMPSIC IDIOCY.

Suceeds infantile convulsions chiefly during dentition. The child, although motion and sensation (general and special) are uninjured, is capable of very little education.

IV.—EPILEPTIC IDIOCY.

"If during the intervals between their epileptic seizures, they learn anything, a new attack is apt to erase it from their memory; they are generally wild and intractable, and, indeed seem to be on the boundary between imbecility and insanity" (Quotation from Dr. Ireland in Bucknill and Tuke's "Manual," p. 159).

V.—HYDROCEPHALIC IDIOCY.

In twelve cases the head did not exceed 24 inches in circumference. Dulness of touch and deafness not uncommon. Albers, of Bonn, says that mental obtuseness and paralysis indicate effusion in the lateral ventricles, and restlessness and mental derangement in the sac of the arachnoid.

VI.—PARALYTIC IDIOCY.

Paralysis may take the form either of paraplegia or of hemiplegia. Mentally these patients are generally tractable and docile.

VII.—CRETINISM (See below).

VIII.—TRAUMATIC IDIOCY.

Due to injuries by attempts at abortion, and by the use of the forceps. The mental impairment may be trifling or severe, permanent or temporary.

IX.—INFLAMMATORY IDIOCY.

Cases following so-called attacks of brain fever. In one case described by Dr. Ireland, the patient possessed all his senses and normal sensibility, and was learning to read and write.

X.—IDIOCY BY DEPRIVATION.

Idiocy owing to the absence of two or more of the principal senses.

IMBECILITY.

Synonym.—Congenital Imbecility.

A minor degree of mental deficiency than idiocy. Imbeciles think, feel, and speak, and are capable of acquiring a certain amount of education. Some imbeciles are affectionate; many are passionate; many have a strong tendency to theft; some are shrewd and jocular; not a few are dangerous, prone to homicide, and incendiarism; occasional ones devote their lives to works of benevolence.

Hoffbauer's Three Classes:—

I.—Imbeciles incapable of forming a judgment on a new subject, but capable of judging regarding subjects familiar to them. The memory is very weak, although a certain routine of occupation is observed with scrupulous exactness; they do not talk much to themselves; they are liable to sudden paroxysms of anger.

II.—Those who are even less able to judge and act, in regard to their accustomed occupations; they are exceedingly confused in regard to the place in which they are and the person with whom they converse, and are very generally at fault in regard to their ideas of time.

III.—Those who have delusions of the evil intentions of others, and are not only passionate, but suspicious and misanthropic; they frequently talk to themselves (Bucknill and Tuke, *op. cit.*, pp. 162--163).

CRETINISM.

"An arrested development of the nervous system and bodily organization generally, either before or after birth, due to a local cause, as the condition of the soil, water, air, etc., and marked by characters which in some respects distinguish it from ordinary idiocy "(Bucknill and Tuke, *op. cit.*, p. 164).

"A certain combination of symptoms may allow us to prognosticate, in childhood, the future development of cretinism. In well marked cases it is stated that after the fifth or sixth month the child presents the following symptoms: The development of the body proceeds very slowly; the child, though weak, is remarkably stout, and appears swollen; the colour of the skin is sometimes dusky, sometimes yellow, sometimes natural; the head is large; the fontanelles widely separated and sometimes all the sutures disjointed; the expression is stupid, the appetite is voracious, and much time is passed in sleep; the belly is swollen: the extremities are generally attenuated; the neck is thick, without, however, being always goitrous; teething is not completed for many years and is accompanied by an offensive salivation, and frequently by convulsions. Usually the child cannot stand before its sixth or seventh year, and it is then that it begins to articulate certain sounds, if not deaf from birth; the voice is hoarse and shrill, and words are spoken with difficulty. The development of cretinism, strictly speaking, occurs about seven, but it is clear that all its main features were present long before" (Bucknill and Tuke, *op. cit.*, p. 167).

Some cases are really congenital. Dr. Guggenbühl's classification is: (1,) Congenital cases; (2,) Rachitic cases: (3,) General atrophic cases; (4,) Hydrocephalic cases.

The classes generally spoken of by authors are: (1,) Cretins, manifesting only vegetative functions and deprived entirely of reproductive and intellectual faculties, including the power of speech; (2,) Semi-cretins, possessing the power of reproduction, and some faculty of speech; intellectual faculties limited to corporeal wants; (3,) Cretinous, having intellectual faculties superior to the former, and able in some degree to apply to trade and other employments (Bucknill and Tuke, *op. cit.* p. 168).

The crania of the second and third class (megalocéphales) are more capacious than those of the first. Some cretins are brachyocephalic, and in some the head is conical. Face almost unchanged from puberty to old age, eyes expressionless, generally strabismus, very large zygomatic arch, mouth remarkably large, lips thick, the lower one hanging down; superior maxilla prominent, the inferior small, retreating, and obtuse-angled. Deafness is very frequent.

Feet disproportionately large, abdomen prominent, resting upon lank, attenuated legs, head sometimes cumbrously large, drooping over an ill-developed thorax. The thyroid gland may be enlarged or there may be supra-clavicular fatty swellings without any bronchocele.

Many cretins are only 3 feet high; they are mostly under 4 feet, and rarely exceed 4 feet 11 inches, though they may attain even 6 feet (Bucknill and Tuke, *op. cit.* pp. 168–170).

IMPULSIVE INSANITY.

Synonyms.—Inhibitory Insanity, Psychokinesia or Hyperkinesis of Clouston; Emotional or Affective Insanity (Maudsley); Paranoia Rudimentaria Impulsiva (Morselli).

The loss of the power of inhibition is the chief and by far the most marked symptom. Action in cases of morbid impulse may take place from a loss of controlling power in the higher regions of the brain, or from an over-development of energy in certain portions of the brain, which the normal power of inhibition cannot control. In the former division consciousness may be absent, the ego, the will, being non-existent for the time. Murders committed during somnambulism, hypnotism, or epileptic unconsciousness are the most perfect examples. In other cases consciousness is present but power of self-restraint absent; as when imbeciles and dements appropriate articles (Clouston, "Mental Diseases," 1st ed., p. 317).

These states of morbid impulse may be momentary or constant, slight or most intense.

Morbid impulses may present themselves in several forms of mental disease, *e.g.*, the destructive impulse in mania; pyromaniacal impulse in pubescent insanity, epileptic insanity and imbecility; the homicidal impulse in epileptic insanity, climacteric insanity, melancholia; the suicidal impulse in alcoholic insanity, puerperal and gestational insanity; the wandering impulse (mania errabunda) in early general paralysis, and in pubescent and periodical insanity; planomania in simple mania and moral insanity.

In *Impulsive Insanity*, properly so called, a morbid impulse constitutes the most prominent or even the only perceptible symptom, and there is neither delusion, exaltation, depression nor enfeeblement; dipsomania and kleptomania are examples.

The principal varieties of morbid impulse are: (1,) Destructive mania, the impulse to destroy; (2,) Dipsomania, the impulse to drink intoxicating liquors. The dipsomaniacal fits (which are periodical) are ushered in by sadness, moroseness, headache, præcordial anxiety and dyspepsia, then the desire to drink becomes irresistible, and the patient never stops drinking until the attack ceases or the patient is isolated (Bra, "Maladies Mentales," p. 69, quoting Marcé). After several days of excess alcoholic delirium appears with painful hallucinations, tremor, insomnia, gastric troubles, amnesia, etc., (Bra, *op. cit.* p. 70). The dipsomaniac only indulges when the impulse to drink seizes him; the drunkard drinks whenever he has the opportunity. When the dipsomaniacal impulse is frequently yielded to, chronic alcoholic insanity is induced; (3,) Homicidal mania, the impulse to kill; (4,) Kleptomania, the impulse to steal; (5,) Lycanthropia, the impulse to act like a wild beast; (6,) Necrophilism, the impulse to exhume and eat dead bodies (Clouston, *op. cit.* p. 317). Spitzka ("Insanity," p. 43) defines necrophilism as a name given to a desire to violate dead bodies and would classify it as a propensity, not as an impulse ; (7,) Nymphomania, uncontrollable sexual impulse in the female. Should be distinguished from erotomania in which there are delusions but not necessarily animal sexual desire; (8,) Planomania, the impulse to wander from home and throw off the restraints of society (Clouston, *op. cit.* p. 317); (9,) Pyromania, the impulse to set things on fire. Is usually exhibited at or shortly after puberty (Spitzka); (10,) Satyriasis, uncontrollable sexual impulse in the male; (11,) Suicidal mania, the impulse without any depression to commit suicide. Several of the above morbid impulses may be combined in the same patient; (12,) General psychokinesia, impulsiveness in all directions (Clouston, p. 319).

Spitzka (p. 271) looks upon dipsomania as a form of periodical insanity. He also says (p. 37) that "pyromania, like kleptomania, may be a leading manifestation of periodical insanity."

Maudsley considers impulsive insanity and *moral insanity* to be two sub-divisions of emotional or affective insanity. (See "Moral Insanity.")

KATATONIC INSANITY.

Synonyms.—Katatonia (Kahlbaum), Pazzia Catatonica (Morselli).

Spitzka (p. 149) defines katatonia as "a form of insanity characterised by a pathetical emotional state and verbigeration, combined with a condition of motor tension." He says the initial stage resembles that of ordinary melancholia. Then there is a period of almost cyclical alternation of atony, a peculiar excitement, confusion, and depression merging into a state of mental weakness approaching, if not reaching, the degree of a terminal dementia. Any one of these phases may be absent. In many cases the initial stage is accompanied by cramps, chorea-like movements of the facial muscles and epileptiform and hysterical convulsions.

In some cases the initial depression is accompanied by self-reproaches relating to masturbatory excesses, and very frequently disappointment in love determines the morbid ideas of the patient. On this basis fear of poisoning, delusions of persecution and dread of committing unpardonable crimes crop up.

The excited stage presents, as it were, a connecting link between agitated melancholia and delusional insanity. Some of the patients present exaggerated, others diminished, self-esteem. The delirium not rarely assumes an expansive tinge. All katatoniacs exhibit a peculiar pathos, either in the direction of declamatory gestures and theatrical behaviour, or of an ecstatic religious exaltation. "Frequently the patients wander about imitating great actors or preachers, and often express a desire and take steps to become such preachers and actors." In addition to the manufacturing of words and sounds resembling words or verbigeration (see "Verbigeration," Chap. II. of this work), there is a tendency, noted by Kahlbaum, to use diminutive expressions. The hallucinations are commonly of a depressive character, the Devil, hell fire, blood, "droves of dogs," etc., etc. Yet there is rarely the profound painful emotional state of true melancholia. The facial expression often indicates rather a silly hilarious tendency even in the atonic states, and this reaches its acme in the excited phases. The acts and ideas in the latter states are exceedingly monotonous, and the tendencies are oppositional and destructive; hence refusal of food and refusal to leave his bed, distinguishing the katatoniac from the maniac. "Occipital headache of an occasionally severe character is said to be characteristic of katatonia by Kahlbaum." The most striking phenomena of the disorder are its cataleptic periods. The catalepsy is typical and extreme. It may last for days, weeks, or months. Some writers deny the existence of katatonia as a pathological entity.

LACTATIONAL INSANITY.

Synonym.—Insanity of Lactation. Included by some authors in Puerperal Insanity.

Generally melancholic in character but occasionally maniacal, and still less frequently assuming the form of dementia.

Early Symptoms.—Pallor, emaciation, headaches, uneasy sensations at the top of the head, shortness of breath, palpitation, sense of weakness and sinking, giddiness, flashes of light before the eyes, lassitude, nervous irritability.

Developed Stage.—Depression, sleeplessness, loss of self-control, lethargy and stupidity, or suicidal tendency with delusions of suspicion, apprehensiveness, ideas of unworthiness, delusions as to personal identity, hatred of relatives, hullucinations of sight, smell and hearing (Clouston, p. 510; Savage, p. 380; Bucknill and Tuke, p. 364).

MANIA.

Synonyms.—Psychlampsia (Clouston); Tobsucht, Manie.

Defined by Clouston ("Mental Diseases," 1st ed., p. 143) as mental exaltation or delirium, usually accompanied by insane delusions, always by a complete change in the habits and modes of life mental and bodily, by a loss of the power of self-control, sometimes by unconsciousness, and loss of memory of past events, and almost always by outward muscular excitement, all those symptoms showing a diseased activity of the brain convolutions. In this definition Clouston includes acute delirium (acute delirious mania, delirium grave).

Mania may be divided into: (1,) Simple mania; (2,) Acute mania; (3,) Delusional mania; (4,) Chronic mania; (5,) Transitory or ephemeral mania.

SIMPLE MANIA.

Synonyms—Mania sine delirio (Griesinger); Maniacal Exaltation (Krafft-Ebing); Hypomania (Mendel).

Loquacity, especially about the patient's own private affairs, the patient not having been previously markedly loquacious; impaired judgment; increased egotism; fickleness, restlessness, unsettled conduct, foolish manner; motiveless action; diminution of power of self-control. These symptoms being not merely transitory, but lasting for days or weeks.

Griesinger (*op. cit.* p. 302) says, "The frequent states of incompletely developed mania are of great practical importance."

Clouston (*op. cit.*, p. 146) states that the greater number of cases of so called "*Moral Insanity*" are cases of simple mania. In the severe forms of simple mania there may be insomnia or muscular or sexual excitement. There may be extreme mendacity, absurd boastfulness, disgusting obscenity. Clouston (*op. cit.*, p. 160) is of opinion that the *primäre verrücktheit* of the Germans may at first usually be classed as simple mania; also that the *folie raisonnante* of the French corresponds in a general way to the milder cases of simple mania. But, although the symptoms resemble each other, *primäre verrücktheit* and *folie raisonnante* are degenerative states, and should not, strictly speaking, be confounded with simple mania, which is a psychoneurosis, (Krafft-Ebing, Spitzka).

According to Clouston (*op. cit.*, p. 160), simple mania is very often the first stage of acute mania.

ACUTE MANIA.

(1,) It may commence as simple mania; (2,) It may begin quite suddenly; (3,) It is often preceded by a melancholic stage (4,) It sometimes begins by a delusion out of which the extravagancies arise; (5,) Sometimes it begins by emotional exaltations and perversions; (6,) Sometimes by intellectual exaltations and perversions; (7,) Sometimes by both; (8,) It may begin by alterations of habit, appetite, and propensity (Clouston, *op. cit.*, p. 163).

Premonitory Symptoms.

It commonly has premonitory symptoms bodily and mental, such as headaches, a confused feeling in the head, muscular fidgetiness, unrest of body and mind, a feeling that something is going wrong, or that something dreadful is to happen, a sensation as if the head were about to burst; an impulsive desire to do something, to break glass, to be violent to those within reach; usually disturbed sleep, with unpleasant dreams; the temperature may rise to 100° before the patient becomes maniacal (Clouston, loc. cit.).

Developed Stage.

Great restlessness and muscular agitation; complete change of emotional state, this often becoming very joyous; rapid and uncontrolled passing of the ideas through the mind; vivid kaleidoscopic mental pictures of the past; scraps of former life and experience suggested by chance associations; tendency to talk constantly whether any one is present or not, passing from one thing to another and soon becoming incoherent; manner quite changed, jolly or fierce; sometimes ceaseless laughing, or scolding, or swearing; sometimes auditory and visual hallucinations, and consequent conversations in loud tones; sometimes hallucinations or illusions of smell and touch; the senses may be hyperæsthetic at first, but afterwards become dulled; there may be rhythmical movements; frequently a tendency to shut the eyes to exclude external impressions; there is perversion or paralysis of the affective sensibility: those previously most liked are now most disliked; those most trusted previously are now the objects of suspicion; those formerly most intimate are now shunned; the patient may shout, sing, run about wildly, or attack those near him; if he writes at all he writes very incoherently; the face is flushed, the eyes glisten, the eyelids are widely dilated, showing the sclerotic above and below the cornea (Clouston, op. cit., pp. 163-170).

In acute mania there may be delusions. These may be fixed or fleeting, and may assume various forms, suspicious, ambitious, etc.; the patient may fancy himself galvanized, poisoned, etc.; or he may believe himself to be a king, an emperor, etc. Illusions as to personal identity are common.

When the acute symptoms pass off there may be prostration, depression, stupor, or apparent mental enfeeblement resembling dementia. Certain mental peculiarities remain permanently in many cases (Clouston, op. cit., p. 179).

In a case of acute mania under the care of the writer, there was first a short febrile attack without mental symptoms; then insomnia and severe headache, vertigo, relieved by the sitting posture and the application of ice; then hyperacousia with delusions of suspicion; then exaltation with improvement of memory, great loquacity, and a tendency to write very long letters; then incoherence, restlessness, motor excitement, and a tendency to use blasphemous language and threaten violence, though these threats were never put into execution. The case ended in recovery after a duration of about six months. After the fever of invasion subsided, the temperature remained normal.

The less severe forms of acute mania are called *sub-acute mania* by some alienists.

DELUSIONAL MANIA.

Maniacal general symptoms centring round a fixed delusion or set of delusions. The delusions constitute the most prominent symptom. They are not systematised as in delusional insanity (monomania).

CHRONIC MANIA

Synonym.—Chronic Confusional Insanity.

Acute and sub-acute mania may pass into a chronic condition of which the most prominent characteristics are: restlessness; want of affection; want of self-control; always weakness of judgment; often weakness of memory; sometimes incoherence, destructiveness or violence, or all three; sometimes delusions which are fixed but not systematised, and hallucinations of various kinds. The patients may be wet and dirty. They often sleep very little. Many are able to work and are industrious in various ways. Chronic mania is often only a stage on the way to terminal dementia.

TRANSITORY OR EPHEMERAL MANIA.

Synonyms.—Transitory Insanity, Transitory Frenzy.

A rare form of maniacal exaltation. It comes on suddenly, is usually sharp in its character, and accompanied by incoherence, partial or complete unconsciousness of familiar surroundings, and sleeplessness. It may last from an hour up to a few days (Clouston, *op. cit.*, p. 202).

MASTURBATIONAL INSANITY.

Synonyms.—Masturbatory Insanity, Insanity of Masturbation.

"It comes on in youth; it generally begins by an exaggerated and morbid self-feeling, or by a shallow, conceited introspection, or by a frothy and emotional religious condition, or by a restless and unsettled state, with foolish hatchings of philanthropic schemes. There is no continuity or force in any train of thought or course of action. Then comes a melancholic stage of solitary habits, disinclination for company, especially that of the other sex; irritability, variableness of mood, hypochondriacal brooding, vacillation, and perversion of feeling towards near relations; suicide is often thought of, and oftener talked of, but masturbation makes most of its victims too cowardly to kill themselves. Then an acute attack follows, usually of a maniacal kind. This may end in recovery or may run quickly into a dementia that is masturbational in character, being solitary, unsocial and subject to impulses, sometimes homicidal—a sort of masturbational hyperkinesia—all these being incurable. With these mental symptoms there are usually well marked bodily signs of the disease. The patient is thin, pale, and pasty, with a cold, clammy skin, a haggard face, and an eye that never looks straight at you. He has weakness in the back, pains in the head, palpitation of the heart, impaired sight, muscular relaxation, and sometimes spermatorrhœa" (Clouston, "Ment. Dis.," 1st ed., pp. 484-485).

Spitzka ("Insanity," p. 54) says that false taste and smell perceptions are almost characteristic of masturbatory insanity, and that they are of bad import and indicate rapid deterioration. The patients "smell dead bodies, putrefying and filthy substances, noisome gases, seminal discharges, etc."

MELANCHOLIA.

Synonyms.—Psychlampsia (Clouston); Lypemania, Lypothymia.

Prodromal Period.

The invasion is rarely sudden. There may be a vague sadness, an indefinable *malaise*, a dislike of work, puerile fears (Bra, *op. cit.*, p. 23).

Clouston (*op. cit.*, p. 127) says that melancholia "begins in nearly all patients as simple lowness of spirits, and lack of enjoyment in occupation and amusement, and loss of interest in life; this may be premonitory of the disease by months or even years." There is at first a painful mental state which may continue in the form of a vague feeling of anxiety, oppression, dejection, and gloom; generally this feeling passes into a single painful perception, false ideas arise, the intellect becomes slow and sluggish, and the thoughts monotonous and vacant (Griesinger, *op. cit.*, p. 213).

Sub-divisions: (1,) Simple melancholia; (2,) Hypochondriacal melancholia; (3,) Delusional melancholia; (4,) Agitated melancholia; (5,) Suicidal melancholia; (6,) Religious melancholia; (7,) Stuporous melancholia; (8,) Chronic melancholia.

SIMPLE MELANCHOLIA (FULLY DEVELOPED).
Physical Symptoms.

The countenance expresses inquietude, distrust, indifference, self-effacement, or inertia. The attitude is almost always immobile; the eyes are fixed on the ground; there is an insurmountable aversion to movement of any kind. The voice evinces want of energy; it is weak and indistinct, and the words are badly articulated. There is anorexia; there are also sitophobia and constipation. The pulse is slow and small, and the temperature lowered; the respirations are slow and irregular. The general sensibility and the special senses are enfeebled. The secretions are diminished; sexual appetite diminished; continuous insomnia and obstinate cephalalgia (Bra, *op. cit.*, pp. 23–25). The movements are slow, languid, and feeble.

Psychical Symptoms.

A state of mental pain; a profound feeling of ill-being; of inability to do anything; of depression and sadness. Impressions formerly agreeable, now excite pain; irritability and irascibility. There is either perpetual expression of discontent or a resort to complete solitude; dislike, often absolute hatred, of family, relatives and friends. There may be slowness, monotony, hesitancy or even a total absence of reaction of the will to impressions. Some melancholics are discontented, others indifferent, others completely self-absorbed, others again who call themselves miserable creatures and say everything is too good for them. The mind is occupied by a few ideas and only a few monotonous complaints are uttered; desire for intellectual intercourse diminished. Patient speaks timidly, hesitatingly, in a low tone with frequent self-interruptions, or he may sit perfectly mute.

The melancholic insane ideas have one essential character, that of passive suffering, of being controlled and overpowered (Griesinger, p. 227, *et. seq.*).

HYPOCHONDRIACAL MELANCHOLIA.
Synonym.—Hypochondriacal Insanity.

Change of disposition without any assignable cause, dejection, peevishness, suspiciousness, extreme sensibility, and a disposition on the part of the patient to connect everything with himself. Everything wearies the patient, and he is very easily fatigued.

At first there are many remissions, and the paroxysms assume the form of an irritable, restless, and distrustful disposition, or a mental apathy which may produce weariness of life, or anxiety which may proceed to despair and loss of self-control.

There may be morbid sensations, formications, sensations of heat and cold, of bursting of head, of emptiness, etc. These sensations are anxiously watched and their importance exaggerated. The hypochondriac often feels his pulse, examines his tongue and his excretions. He reads medical books and changes his medical adviser frequently. In the higher grades volition is altogether absent. In the more advanced stage everything that does not fall within the circle of pre-occupying ideas is without the slightest interest, of perfect indifference, and soon forgotten.

These patients are therefore often extremely absent-minded and forgetful. They are very loquacious upon the one subject of their affection, but are little inclined to converse about anything else. In the most severe grades the patients are dull, morose, and almost incapable of any intellectual exertion (Griesinger, p. 217, *et seq.*).

The hypochondriacal melancholic or insane patient, as distinguished from the mere hypochondriac or sane patient, has real and intense mental depression that he *cannot throw off*; he loses his self-control, outrages decency openly, practises things that will soon end his days, or threatens to take his own life, and cannot at will withdraw his mind and speech from his delusion. There is, however, no line of demarcation. Hypochondriasis is often the first stage of hypochondriacal melancholia (Clouston, *op. cit.* pp. 55–56).

Savage (*op. cit.* p. 131) states that there are three classes of cases of ordinary hypochondriasis seen in an asylum besides the sufferers from general hypochondriasis who complain of syphilis or hydrophobia or believe themselves about to die. Savage here evidently uses the term hypochondriasis as synonymous with hypochondriacal insanity. These three classes are:—

I.—Brain Hypochondriasis affecting both men and women, and in which the patients believe the brain to be dried up or changed in some way. This form is liable to occur about the climacteric.

II.—Gastric Hypochondriasis which he subdivides into (1,) Patients who complain of obstruction or disease about the throat; (2,) Those who complain of similar feelings and uneasiness at the pit of the stomach; and (3,) Those whose complaints are referred to the lower bowel. These cases are furnished chiefly though not solely by persons of mature years, and more frequently by men than women.

The Gastro-enteric Insanity of Sibbald ("Quain's Dictionary," p. 724) would seem to be a sort of symptomatic link between Savage's Gastric Hypochondriasis and Simple Melancholia. Sibbald writes: "In addition to the mere depression caused by anæmia, there is associated with such affections a peculiar anguish of mind and tendency to self-accusation, which is often of the most distressing nature. Refusal of food is often a prominent symptom. The intellectual perversion is slight, and seldom so prominent as in other acute insanities. Relief of the bodily symptoms is generally accompanied by a return to sanity." The affections most frequently causing it are irritation and catarrh of the mucous membrane; constipation; stricture or other causes of distension of the viscera; and epigastric tumours (See "Abdominal Disorders, Insanity of"). Hypochondriasis is not infrequently induced by real bodily disorders.

III.—Sexual Hypochondriasis. Middle-aged men who fancy themselves impotent, and youths who, having given way to masturbation, fancy they are suffering from spermatorrhœa. Sexual hypochondriasis is rarely met with in women.

Krafft-Ebing looks upon hypochondriacal melancholia as a primary affection not developing out of hypochondriasis, and gives as two of its forms melancholia syphilophobica and melancholia hydrophobica. According to him (p. 592, *et seq.*) mental weakness or hypochondriacal paranoia may develop out of hypochondriasis.

Clouston (*op. cit.*, p. 55) remarks: "In hypochondriacal melancholia a sense of ill-being is substituted for the healthy pleasure of living, but the ill-being is localised in some organ and function of the body. The patient's depressed feelings all centre round himself, his health, or the performance of his bodily or mental functions. He is all out of sorts, he cannot digest his food, his bowels will never act, his kidneys or liver are wrong, he has no stomach, his heart is weak, and he asks you to feel his pulse which is just going to stop beating. He is paralysed and will not move a limb till he forgets his fancy for a moment; he cannot think because his brain is made of lead; he is made of glass and will break if roughly handled. There are no limits to the fancies of the hypochondriac or the hypochondriacal melancholic."

In a case under my care the patient laboured under the delusion that he suffered from obstruction of the bowels, and would converse about nothing else. He might answer a few questions probably quite correctly, but he immediately reverted to his favourite topic. He was a man rather past middle age, stout and healthy looking, with somewhat florid complexion and without any gloominess of facial expression. He took his food well and went daily for long walks with a number of other patients; but he did not occupy himself in any other way, and would have lain in bed all day had he been allowed. His bowels acted fairly regularly, and when the attendants affirmed that he had recently had an evacuation, he either denied the assertion or said it was only something that had lodged below the obstruction.

Another patient fancied his mouth and throat were full of long hairs, and he would give anyone the end of a hair as he imagined, and walk backwards many yards declaring he could see the hair coming out of his mouth as he retreated. He both felt and saw the hairs, and at times his agitation was extreme with weeping and motor excitement. He refused food and had to be fed forcibly. His countenance expressed great anxiety and misery, and at first sight his case closely resembled one of agitated or anxious melancholia.

DELUSIONAL MELANCHOLIA.

This includes the group of cases of melancholia in which delusions (or a delusion) are the most prominent mental symptom (Clouston, *op. cit.*, p. 63). The delusions are very various, such as "having committed the unpardonable sin," "being extremely poor," "being very wicked," "being poisoned;" "that wife, or husband and children are being burnt alive or otherwise killed," "that he or she (the patient) has committed murders or unnatural crimes," "that death is impending," "that it is sinful to take food," "that the patient is unworthy to eat," "that every morsel taken brings the patient and all about him or her so much nearer ultimate starvation," "that it is in fact wasteful and extravagant to take food."

AGITATED MELANCHOLIA.

Synonyms.—Active Melancholia (Savage); Mélancolie Anxieuse (Bra); Melancholia Agitans (Griesinger); Excited (Motor) Melancholia (Clouston).

There is a confused tumult of thought expressed by physical restlessness. The thoughts and movements are monotonous and of little variety. The state differs from mania through the paucity of ideas and want of fertility. Sometimes the patient keeps up a perpetual motion, breaks out into frequent fits of weeping, and constantly wrings his hands (Griesinger, *op. cit.*, pp. 234–235).

There is extreme agitation, and an imperious desire for movement (Bra, *op. cit.* p. 26).

"It is characterised by restless misery as seen in the constant picking of fingers, pulling out of hair, and a tendency to strike or damage anything that appears to be an obstacle to its free exhibition. Generally in these cases there is some marked delusion, and most commonly this delusion is connected with the idea that someone else is going to be injured on her account" (Savage, *op. cit.*, pp. 175-176).

SUICIDAL MELANCHOLIA.

In which the suicidal tendency is marked and prominent. Clouston (*op. cit.*, p. 112) believes that tendency to suicide exists in some form or other, in wish, intention, or act, in four out of five melancholics, and we can never tell when it is to develop in any patient.

Savage (*op. cit.*, p. 190) observes that the most suicidal patients are those who believe they are to be injured, and that suicidal tendencies are most marked in the early morning. He also states that heredity plays an important part in the causation of the suicidal tendency.

The suicidal melancholic (as distinguished from the subject of impulsive insanity) arrives at the determination to commit suicide: (1,) From a calmly and carefully reasoned out plan because he is not going to recover, or because his death will benefit his children, or from other similar motives; (2,) From the desire to escape imaginary torture or persecution; (3,) In consequence of some depressive delusion or delusions, or of auditory hallucinations ("Voices").

Clouston (*op. cit.*, p. 118) states that in his experience the greatest danger of suicide is near the commencement of the attack of melancholia. In some cases of suicidal melancholia there exists also a *homicidal tendency.*

RELIGIOUS MELANCHOLIA.

Melancholia Religiosa (Griesinger).

Intense despondency as to religious condition. Delusions as to eternal damnation, as to having led wicked lives and neglected the services of religion, and caused others to do so. Self-accusations of hypocrisy and impurity. Tendency to suicide. In a case of "Religious Delusional Melancholia," described by Clouston (*op. cit.*, p. 82), the first symptoms were "mental confusion and depression, and falling off in bodily looks, appetite and strength, and her head feeling queer." On admission, there were mental depression, impairment of memory, dulling of sensory functions, impairment of reflex functions, and various delusions of a religious kind (that she was the greatest sinner alive, and had committed many and unpardonable sins).

STUPOROUS MELANCHOLIA.

Synonyms.—Melancholia with Stupor (Savage, Griesinger); Melancholic Stupor (Clouston); Mélancolie avec stupeur (forme active) of the French authors; Melancholia attonita.

The patient remains in one position and attitude, with little or no voluntary movement. He resists being moved, and compresses his lips firmly when food is placed against them. The features are contracted, and the countenance expressive of wretchedness and misery or terror. The complexion is sallow or yellow. The cutaneous sensibility is diminished. The pupils are generally dilated and immobile. Food is refused and nutrition fails. The excretions are deficient. Insomnia, suicidal tendency, mutism, which is generally absolute for the time being. Consciousness is wholly or partially retained, and the patient as a rule remembers and can describe the frightful delusions and terrifying hallucinations frequently present in this condition. These hallucinations often cause sudden fits of agitation and give rise to dangerous impulses.

Stuporous melancholia is often developed rapidly. It may be primary, or succeed other forms of melancholia, or constitute an interlude in their course.

Any of the foregoing forms of melancholia may be acute with regard to the severity of the symptoms, or *chronic* in duration.

Savage (*op. cit.*, p. 203) speaks of *Recurring Melancholia*, and says cases of it differ only in degree from those in which there are several distinct attacks of melancholia, each followed by complete recovery.

MENTAL DETERIORATION (PRIMARY).

Synonym.—Simple Primary Dementia.

SYMPTOMS.

Prodromal.—Lack of energy, both physical and mental. Insomnia, or unrefreshing dreamy sleep, the dreams relating to the patient's daily occupations and cares. Dyspepsia. Signs of functional or organic heart disorder, or of the prodromal period of Bright's Disease may be noted. Often premature grayness (Spitzka, "Insanity," p. 164).

Fully developed.—Absent-mindedness, failure of memory, lack of attention, general inertia, impaired ability of acquiring new impressions. Syllables and words are omitted in writing; important engagements are broken, articles of value mislaid, expenditures unrecorded, and with the intensification of all these symptoms complete fatuity may be developed.

With the above symptoms, the muscular power is enfeebled, the articulation is affected, the pupils are unequal, and the temperature is subnormal, while the patient generally complains of a sensation of pressure and fulness in the head (Spitzka, "Insanity," pp. 164–165).

Savage's *Partial Dementia* would seem to be included in this form of mental disorder.

METASTATIC INSANITY.

Clouston (p. 599) says the typical rheumatic insanity is essentially a metastatic insanity. He speaks of cases where the healing of an old ulcer was followed by an attack of insanity; of erysipelas of the face "striking inwards" and causing an attack of acute mania; of the disappearance of syphilitic psoriasis being followed by melancholia, and its reappearance by mental recovery.

MORAL INSANITY.

Synonyms.—Manie sans Délire; Folie raisonnante; Monomanie affective; Gemüthswahnsinn; Moralisches Irresein (Krafft-Ebing).

It "is manifested by insane actions and conduct, rather than by insane ideas, delusions, or hallucinations." Dr. Maudsley gives moral insanity and impulsive insanity as two sub-divisions of "emotional or affective insanity" (Blandford, "Quain's Dict.," p. 727).

Blandford describes two forms, acquired and congenital:—
(1,) Acquired Moral Insanity.—"A disorder of mind shown by an entire change of character and habits, by extraordinary acts and conduct, extravagance or parsimony, false assertions, and false views concerning those nearest and dearest, but without absolute delusion. Such a change may be noticed after any of the ordinary causes of insanity. It may follow epileptic or apoplectic seizures, or may be seen after a period of drinking. Its approach is gradual, as a rule, rather than sudden, and the extraordinary character of the acts may not at first be so marked as subsequently. . . . . . . Such insanity, of course, varies in degree. When it is well marked, and the conduct is outrageous, there will be no difficulty in the diagnosis. But it may be less marked; it may consist of false and malevolent assertions concerning people, even the nearest, of little plots and traps to annoy others, in which great ingenuity and cunning may be displayed. And there will be the greatest plausibility in the story by which all such acts and all other acts will be explained away and excused." Blandford says further, "It may be necessary to prevent a man from squandering all his property—a common

symptom in this variety—or from wandering from home and absenting himself no one knows where, or keeping low company."

(2,) Congenital Moral Insanity (Moral Imbecility); Moralische Idiotie; Originärer Moralischer Schwach—und Blödsinn (Krafft-Ebing).—"The congenital moral defect occasionally met with in persons who have been from birth odd and peculiar, and incapable of acting and behaving like other people. They can hardly be called idiots or imbeciles, for they may exhibit a considerable amount of intellect and even genius in certain special directions. We shall generally find that they are the offspring of parents strongly tainted with insanity, epilepsy or alcoholism, and many in childhood are the subjects of fits, chorea, or other neurosis. They are incapable of being instructed like other boys and girls; are often frightfully cruel towards animals, or their brothers and sisters, and seem utterly incapable of telling the truth or understanding why they should do so . . . . . These are the persons who commit crimes, and become the chronic inmates of prisons; and it is most difficult for medical and other prison officials to say how far they are responsible and how far not" (Blandford. *loc. cit.*).

MYXŒDEMA *(Insanity of)*.

Savage ("Insanity," p. 418) remarks that a sense of suspicion and of injury is common in these cases. In the cases he quotes the principal symptoms are delusions of suspicion and persecution; hallucinations, sometimes of sight, sometimes of hearing; violence and threatening in one case, restlessness, sleeplessness, and much depression in another; feebleness of memory, dulness of perception. Then there are the physical signs—Cretinoid physiognomy, slow and muffled speech, impaired sight and hearing, hard and dry skin, subnormal temperature. Atrophy of thyroid. The tendency is to pass slowly but steadily into dementia.

NEURASTHENIA AND NEURASTHENIC INSANITY.

NEURASTHENIA.
(Beard, Arndt, Krafft-Ebing.)
SEVERE OR CONSTITUTIONAL FORM.

As a rule it is developed gradually, the early symptoms being relieved by rest and sleep. There are at first phenomena of exhaustion; lassitude; weakness; dislike of mental work and difficulty in performing it; craving for sleep, nourishment, drink, and even the means of excitement and enjoyment. The patient is overwhelmed by the fear of impending illness of a serious nature. Symptoms of irritation are soon added; there is irritability, and what sleep the patient obtains is light, dreamy, interrupted, unrefreshing. Vaso-motor symptoms also occur early; local hyperæmia (rush of blood to the head, palpitation, feeling of oppression, etc.), and local anæmia (vascular spasm amounting to local asphyxia, feeling of cold, etc.). The most important symptom, however, is the feeling of broken strength, bodily and mental, with the resulting depression and fear of impending illness, a fear amounting finally to inconsolable nosophobia. Neurasthenia may be cerebral, spinal, visceral, or sexual. For Causes, see Chap. IV.

Cerebral Neurasthenia.—There may be mental incapacity, psychical anæsthesia, mental blindness and deafness, and even amnesic aphasia and agraphia. The never-failing mental depression is, unlike melancholia, reactive; nevertheless there are fleeting transitions to that disease. Imperative conceptions (fixed ideas) are frequent, and are sometimes accompanied by suicidal or misanthropic tendencies. A feeling of pressure on the head is hardly ever absent, and it is accompanied by nosophobic ideas of cerebral ramollissement and threatened insanity. Asthenopia is frequent and cystospasm is not rare.

Spinal Neurasthenia.—In this form the patients rapidly becoming feeble feel depressed, complain of paralgiæ in skin, muscles, and joints, are exhausted after very slight exertion, suffer from palpitation, sudden sweating, and feelings of oppression and anxiety; sleep is disturbed by startings. There may be paræsthesiæ and even local anæsthesia. Spinal irritation is especially frequent, and gives rise to an exceedingly obstinate idea that the spinal cord is diseased.

Amongst the *Visceral Neurastheniæ*, Neurasthenia Cordis is the most prominent. In this form there are attacks of cardiac disturbance and intervallary symptoms. During the attacks there is a feeling of arrest of the heart's action; there are pain, pressure, and vibrations in the cardiac region. Fear of apoplectic attacks aggravates the patient's condition. During the intervals the patient is languid, exhausted, emotional, nosophobic. In Neurasthenia Gastrica there are dyspeptic troubles with pressure on, and rushes of blood to, the head; there are somnolence, palpitation, etc., etc.

Sexual Neurasthenia.—There are nocturnal emissions, and on attempting coitus ejaculation is premature. There is fear of disease of the spinal cord. These depressing influences induce psychical impotence. There are paralgiæ and neuralgiæ in the region of the lumbo-sacral plexus. Spinal neurasthenia becomes developed, and even general neurasthenia may supervene, accidental circumstances determining which form (cerebral, gastric, etc.) is at any given time predominant. The sexual neurasthenic is shy, oppressed-looking, self-depreciative with hypochondriacal depression, ataxiaphobia, indolent attitude, lowered muscular tone, pale complexion with well-nourished body, tremor to helplessness and motor ataxy when he knows he is being observed. There are also dyspepsia, flatulence, constipation, great alteration in the frequency of the pulse under the influence of emotion or bodily exertion (it will sometimes run up to 120 under these influences), fits of general vascular spasm with violent palpitation and paroxysmal anguish and oppression in the cardiac region. To these symptoms must be added the peripheral genital and the lumbar cord neuroses with aggravation of the psychical troubles by repeated ejaculations. In women the general phenomena are the same as in men.

THE NEURASTHENIC PSYCHOSES.

There are two groups, psychoneurotic and degenerative. The psychoneurotic group is composed of cases developing on a basis of neurasthenia which is acquired, episodical, or at all events not constitutional. The psychoses of this group may be transitory or protracted. The protracted cases take the form of melancholia, stupor, and acute confusional insanity.

NEURASTHENIC PSYCHONEUROSES.

I.—*Transitory Neurasthenic Insanity* appears sometimes as the culminating point of neurasthenia cerebralis, and is accompanied by external signs of inanition and exhaustion, such as tremor and subnormal temperature. There are sensory disturbances extending to loss of perception, blurring of consciousness amounting in some cases to unconsciousness with corresponding defects of memory, loss of speech and movement, anxiety. Isolated delirious ideas appear in this almost or quite stuporous condition, and give rise to dreamy, odd actions. The pupils are wide and sluggish. The pulse is small and wiry.

II.—*Protracted Neurasthenic Psychoneuroses.* Most cases of anergic stupor (acute dementia) and acute confusional insanity (Wahnsinn) belong, according to Krafft-Ebing, to this category. Maniacal symptoms rarely develop on a neurasthenic basis, melancholic ones very frequently. Masturbatory melancholia is one of these forms of melancholia characterised more by restrained mental action than by psychical pain. Melancholia masturbatoria develops on a basis of sexual neurasthenia. Nosophobic symptoms (fear of insanity, tabes, or incurable impotence) are never absent during the stage of incubation. In the developed disease there is great self-depreciation. The seldom failing olfactory hallucinations lead the patient to believe he emits an evil odour. Attempts at suicide are common, so also is mutilation of the genitals. There may be attacks of angina pectoris vaso-motoria, at night. In severe cases there are often observed uncleanliness, disgusting habits (eating excrement, rain worms, the contents of spittoons, etc.), fixed ideas, and a religious delusional state (the Messiah, etc).

NEURASTHENIC DEGENERATIVE PSYCHOSES.

I.—Mental disorders, of which the chief symptoms are *Imperative Conceptions* (fixed ideas, idées fixes, Zwangsvorstellungen). (See "Folie du doute," "Partial Emotional Aberration.") Krafft-Ebing holds the view that these cases are developed on a basis of neurasthenia nearly always constitutional and hereditary. In the rare cases in which the disease is acquired through certain causes (see "Etiology") recovery may take place in time. This psychosis may be accompanied by hysteria or hypochondriasis, and melancholia may occur episodically.

II.—*Neurasthenic Paranoia* is a form of paranoia (monomania, delusional insanity) in which the delusions are founded on the sensations occurring in the neurasthenic neurosis. The patient's thoughts are disturbed through the cunning machinations of enemies; his dyspeptic troubles are the result of attempts to poison him, etc. Hallucinations are developed later, and these are especially numerous when sexual neurasthenia is the basis, as in *paranoia masturbatoria*. In this form there are hallucinations of smell and hearing, illusions of sight and hearing, physical persecutory delusions, fits of apprehension, etc.

III.—Most cases of *Melancholic Folie Raisonnante* (q.v.) (Krafft-Ebing, "Lehrbuch der Psychiatrie," p. 517, *et seq.*).

OVARIAN OR OLD MAID'S INSANITY.

"The disease usually occurs in unprepossessing old maids, often of a religious life, who have been severely virtuous, in thought, word and deed, and on whom nature just before the climacteric takes revenge for too severe a repression of all the manifestations of sex, by arousing a grotesque and baseless passion for some casual acquaintance of the other sex, whom the victim believes to be deeply in love with her, dying to marry her, or aflame with sexual passion towards her, or who has actually ravished her after having given her chloroform. Usually her clergyman is the subject of this false belief." "Such patients are all of them between thirty-five and forty-three, and the reverse of sensuous in appearance" (Clouston, "Mental Diseases," p. 478). Out of ten of Clouston's cases, seven have had clergymen as their supposed wooers or seducers.

OXALURIA AND PHOSPHATURIA (*Insanity of*).

Clouston (p. 597) observes, "All writers on the urine have noticed the hypochondriasis, depression of mind, want of energy and originating power, and the irritability that so often go along with the presence of much oxalate of lime, or phosphates in the urine." He says further (p. 598), "I think there is scarcely enough evidence to show whether this condition of the urine is a cause, or an effect of the brain state."

PARALYSIS AGITANS (*Insanity of*).

When paralysis agitans is of long continuance, the facial expression becomes heavy and sad, the intellect blunted, the memory unreliable, sleep difficult to obtain and disturbed. The patient is irritable, fickle, and difficult to get on with.

Ordinarily, the symptoms go no further, but in some cases real insanity, with illusions and hallucinations, develops, and usually takes the form of simple melancholia, or anxious melancholia. Occasionally, delusions of persecution are observed. These mental troubles are intimately associated with the disorders of sensation, and increase, diminish, or disappear with them. They are amenable to the same treatment (Bra, "Manuel des Maladies, Mentales," p. 94).

PARTIAL EMOTIONAL ABERRATION.

In this form there is insane feeling, usually fear or apprehension, limited to certain positions or surroundings. The principal varieties are agoraphobia, claustrophobia, and mysophobia. Less important forms are anthropophobia, monophobia, and astraphobia.

Agoraphobia.—In this variety, there is a dread of being alone in a wide space. If the patient finds himself in this position, he suffers from intense dread, palpitation, giddiness, cold perspiration. He tries to seize any person or object near him. If taken into a small apartment, he becomes quite well.

Claustrophobia.—The same sensations are felt in a confined or shut place.

Mysophobia.—Morbid fear of defilement, causing the patient to refuse to touch other persons. When touched by others, he will rub or wash the part touched (see *folie du doute*). In some cases the patients are afraid that they themselves will defile others, but this approaches melancholia of a hypochondriacal nature.

Anthropophobia.—Morbid dread of society.

Monophobia.—Morbid dread of being alone.

Astraphobia.—Morbidly intense fear during thunderstorms. (Robertson, "Finlayson's Clinical Manual," pp. 351–352).

Krafft-Ebing (*op. cit.* p. 70) considers that in these cases the patient is seized by the idea (imperative conception, Zwangsvorstellungen, fixed idea), that he cannot cross the street, square, etc., and thereupon gets into a nervous condition with extreme fear. These symptoms are the expression of an irritable weakness of the nervous centres, and are founded on a neurasthenic basis constitutional or acquired. (See "Neurasthenia," etc.). The well-known impulses, also, to throw self or others from towers, etc., are generally felt when the brain is exhausted. Partial emotional aberration and folie du doute are generally described together as "Mental Disorder through Fixed Ideas."

PARTIAL EXALTATION OR AMENOMANIA (Rush).

Bucknill and Tuke under the above name, describe a form of insanity closely resembling the ambitious form of delusional insanity (Ambitious Monomania), but differing in that the exalted emotional condition is primary, and the delusions of grandeur are secondary, arising from the emotional state. It is the gay, partial insanity or monomania proper of Esquirol; the chæromania of other French writers.

Esquirol says, "Amongst monomaniacs, the passions are gay and expansive, enjoying a sense of perfect health, of augmented muscular power, and of a general sense of well-being. This class of patients seize upon the cheerful side of everything; satisfied with themselves, they are content with others. They are happy and joyous, and communicative. They sing, laugh and dance, controlled by vanity and self-love. They delight in their own vainglorious convictions, in their thoughts of grandeur, power and wealth. They are active, petulant, inexhaustible in their loquacity, and speaking constantly of their felicity. They are susceptible and irritable; their impressions are vivid, their affections energetic, their determinations violent; disliking opposition and restraint, they easily become angry, and even furious" (Bucknill and Tuke, p. 236).

This form of insanity includes religious exaltation or excitement, without delusions, illusions, or lesion of intelligence.

There are ecstatic expressions of happiness, pride and haughtiness, with a violent deportment; or, selfishness, want of natural affection, variableness of spirit, irregular mental habits. Religious revivals afford many examples of this form of insanity.

Sankey holds pleasurable exaltation to be a mere symptom, and few authors treat it as a separate form of insanity.

PELLAGROUS INSANITY.

This is a form of insanity associated with pellagra, and not met with in Britain. It is characterised by mental symptoms usually indicative of anæmia—great depression, frequently with tendency to suicide, passing on to chronic dementia. It is most frequently met with in Italy" ("Quain's Dictionary," p. 728).

Morselli ("Malattie Mentali," vol. i., p. 437) gives four forms of pellagrous insanity, viz., supra-acute pellagra (pellagrous typhus), pellagrous melancholia, pellagrous dementia, and pellagrous pseudo-general paralysis.

PERIODICAL INSANITY.

The attacks of insanity recur more or less regularly with lucid or sublucid intervals. There is degenerative taint shown by : (1,) Bad family history of the majority of the patients. ; (2,) The presence of somatic stigmata ; (3,) The coincidence of the beginning of the disorder with certain physiological periods, such as puberty, and climacteric ; its exacerbations often following other physiological periods, such as menstruation.

A form occurring at, or near, the menstrual periods has been called *menstrual insanity*.

The patients are much influenced by barometric, and seasonal conditions.

The periodical outbreaks are more abrupt both in their commencement and termination than in simple mania and melancholia. They also reach their acme sooner and are shorter. There are moral or affective perversions, and certain propensities and impulses not usually found in the simple insanities.

When the disease is fully established, the attacks in the same patient are almost exactly similar to each other, displaying the same propensities, impulses, delusions, hallucinations, and language.

Divided into Periodical Mania, and Periodical Melancholia :—

(1,) Periodical Mania may begin abruptly (usual manner), or may be heralded by such symptoms as palpitation, vertigo, neuralgia, or by a short period of depression.

The maniacal explosion is marked by angry excitement, moral perversion, and *délire des actes* (sexual excesses, indecent assaults and exposures ; thefts, incendiarism, wandering abroad, etc.). Outbreaks of angry excitement of a violent and dangerous character are frequent. There are illusions, occasionally hallucinations, rarely delusions.

Sometimes a single morbid propensity, such as sexual perversion, is the most prominent feature. Morselli (" Manuale di Semejotica delle Malattie Mentali," vol i., p. 214) gives the following forms of sexual aberration and perversion, viz., Masturbation (solitary sexual vice in the male); Clitorism (solitary sexual vice in the female); Pederasty (unnatural intercourse of men with boys); Sodomy (unnatural intercourse between adult males); Tribadism (unnatural sexual relations between females); Bestiality (coitus with an animal); Sapphism (cunnilinguism); mouth pollution, etc. To these he would add the various malthusian methods of so-called preventive intercourse, including conjugal or biblical onanism. Alcoholic stimulants are not well-borne, and there is a craving for them. Krafft-Ebing includes maniacal folie raisonnante in periodical mania.

(2,) Periodical Melancholia.—"Periodical melancholiacs are the most persistent, cunning, and successful of all suicidal lunatics" (Spitzka, "Insanity," pp. 267–271). Many mild cases never reach asylums (Krafft-Ebing).

In the lucid or sub-lucid intervals (which may last weeks, months or years) of periodical insanity, most of the patients are "nervous," hysterical, or morbidly irritable. As time goes on, the character becomes permanently changed, the patients become more irascible, their energies are diminished, their emotions blunted (Spitzka).

Krafft-Ebing and Spitzka include circular insanity under the head of periodical insanity.

PHTHISICAL INSANITY.

Clouston, who has specially studied this form of insanity, and who, in fact, gave it its name, states ("Clinical Lectures on Mental Diseases," pp. 461–463) that there may sometimes be an acute stage at first, but that this is not common, and is always short.

"Most frequently the disease begins by a gradual alteration of disposition, conduct, and feeling in the direction of morbid suspicion of those about the patient, a morbid fickleness of purpose, an unsociability, an irritability and an entire want of buoyancy and proper enjoyment of life. Along with this there is loss of weight, indigestion, intolerance of fat, want of enjoyment of food, perversion of taste in regard to food, and a bad colour of the skin. There may or may not be any chest symptoms present; most frequently there are not. Then comes the acutest part of the attack, if there is such a stage in the case. The patient gets sleepless and mildly melancholic, or maniacal, the bodily state running down all the time. The organic enfeeblement that characterises the disease is often shown by refusal of food. The patient thinks he is being poisoned, this no doubt being the con-volitional misinterpretation of the pain and uneasiness of indigestion. In a way he is often poisoned, for his food is badly digested and assimilated, and the subjective symptoms accompanying this are not unlike some kinds of poisoning. After a little, the patient becomes irritable, sullen, unsociable, and suspicious, his state varying from time to time. The intellectual processes are not so much enfeebled, as there is a disinclination to exercise them. There are occasional unaccountable little attacks of excitement. The patient is disinclined to amuse or employ himself; he looks on any attempt to do so as persecution, and as being prompted by hostile motives. There is some depression, but no intense mental pain. The patient associates with no one, and the kindnesses of relatives merely call forth reproaches. If the patient lives long, he becomes more silent and apparently demented, but he can always be roused out of this for a short time. Complete typical dementia does not usually occur. If there is any tendency to periodicity the remissions and aggravations are not regular or complete. Bodily, he cannot be fattened; he looks sallow and haggard, his circulation is poor, his pulse weak, and anything like tone is entirely absent. There is no muscular energy, and a strong disinclination to exertion. The appetite is poor and capricious. Colds are taken very easily. The patients loose weight, and are all round worse in cold weather. The temperature tends to be low, until the lungs become affected, and then there is an insidious evening rise, which is perhaps the only sign of the presence of a bodily disease. In very many of the cases—one half of the number according to my experience—the chest symptoms are at first latent, even after the lungs have become markedly affected. There is no cough, or spit, or pain. I have often happened to notice that a patient labouring under phthisical insanity (and this applies to cases of dementia, and many cases of acute insanity too) was breathing a little more quickly than normal, or was looking more pinched, or was falling off his food, or his pulse was quicker and weaker than usual, or he had a hectic-looking spot on one cheek, or his skin felt hot; and on examining the chest, in consequence of some such indication, I have found extensive broncho-pneumonia, or consolidation, or breaking up of the lung tissue. The progress of the lung disease varies much in different cases, in some being rapid and causing death in a few months, and in others going on for years if the conditions, food, and hygiene are favourable. I have seen such cases in the very feverish stage before death, when the temperature rose over $102°$, rouse up wonderfully, and even cease to manifest the morbid suspicions; but such cases are exceptional. It would seem as if in these cases, the high temperature and quickened circulation stimulated the anæmic, and ill-nourished convolutions to increased and almost normal mental activity."

Sankey ("Lect. on Ment. Dis.") asserts that phthisical insanity is only insanity occurring in a phthisical patient.

PODAGROUS OR GOUTY INSANITY.

The section headed "Insanity and Gout," in Bucknill and Tuke's "Psychological Medicine," contains the following paragraphs (p. 381):—

"In the 'Annales Médico-Psychologiques,' 1869, Dr. Bertheir records twenty-two cases in which the two diseases (insanity and gout) were associated.

"One was a case of Stupor; one, Delusional Insanity of a melancholy character; two, Suicidal Melancholia; three, Simple Dementia; four, in which the features of the malady were not well defined; five, Dementia Paralytica; six, General Mania.

"Of these, eight have been observed by the author himself, and in six of them hereditary predisposition was ascertained.

"In twelve cases, the insanity was consecutive to disappearance of gout; in eight cases it alternated with it; in two cases it accompanied the gouty condition. The great majority occurred among males. He draws the following conclusions :—

"(1,) If the gout has a marked action on the mind of its victims, and a special predilection for the nerves, it may, under the influence of the predisposition, become the source of every kind of neurosis, and chiefly those affecting the sight.

"(2,) The psycho-neuroses dependent on the gouty diathesis, are sometimes, and more commonly metastatic and alternating, and sometimes connected with a specific condition which disposes the system to the development of a latent or larval vesania.

"(3,) Gouty insanity, though generally associated with fixed gout, will, when its study has been completed, be frequently recognised in union with wandering or anomalous gout.

"(4,) Sometimes the gouty symptoms disappear and become lost in the insanity, which then passes into the chronic and incurable state of dementia.

"(5,) Gouty insanity must henceforth be regarded as having an established place in science, and is to be classed along with dartrous, syphilitic, rheumatismal, etc.

"(6,) It shows a preference for the form of general mania.

"(7,) The diagnosis of gouty insanity is to be drawn from the heredity, the antecedents of the patient, the connection of the insanity with gout, and the presence in the urine of the characteristic chemical ingredients.

"Outside the walls of asylums, cases are frequently met with which are marked by symptoms of unfounded dread, especially on awakening from sleep early in the morning, in which there is a gouty diathesis, and suspicion is aroused that there is a causal connection between the bodily condition, and the mental anguish. This suspicion is confirmed by the marked success of treatment, founded upon the supposition."

POST CONNUBIAL INSANITY.

Depression, suicidal tendency, and stupidity (Clouston, p. 607).

PUBESCENT INSANITY.

Synonyms.—Hebephrenia; Insanity of Puberty; Insanity of Pubescence; Hebephrenic Katatonia (Kahlbaum).

"The insanity of puberty in both sexes is characterised by motor restlessness. Such patients never sit down by night or day, and never cease moving. There is noisy and violent action, sometimes irregular movements, or, in the few melancholic forms and melancholic stages of the maniacal cases, cataleptic rigidity. The mental symptoms consist most frequently of a kind of incoherent delirium, rather than any fixed delusional state. In boys, the beginning of an attack is frequently ushered in by a disturbance in the emotional condition—dislikes to parents, brothers or sisters, expressed in a violent open way; there is irrational dislike to, and avoidance of, the opposite sex. The manner of a grown-up man is assumed, and an offensive "forwardness" of air and demeanour. This soon passes into maniacal delirium, which, however, is not apt to last long. It alternates with periods of sanity, and even with stages of depression" (Clouston, *op. cit.*, pp. 531–532).

Spitzka (*op. cit.*, p. 176) says this psychosis begins with a period of sadness without depth, and in which the patient may suddenly burst out in causeless laughter, or even make silly jokes.

After this, there are vague or blind propensities, fickleness of purpose, malice to surroundings. Then the intellect gradually weakens, and the patient, who is generally a confirmed masturbator, will pass into a terminal dementia, marked by occasional furious outbreaks. He (Spitzka) states that everything about these patients is shallow, and even unreal. He gives the age of its occurrence as the period between the fifteenth and twenty-second years.

Clouston means by "puberty," the initial development of the function of reproduction, or its first appearance as an energy of the organism; and by "adolescence," the whole period of twelve years from the first evolution up to the full perfection of the reproductive energy (*op. cit.*, p. 535).

Spitzka's pubescent insanity, therefore, covers part of the ground included in Clouston's adolescent insanity, as the symptoms given would indicate.

PUERPERAL INSANITY.

This may assume the form of mania, melancholia, or dementia (Savage, *op. cit.*, p. 371).

In one half the number of cases, the disease begins with the first week after confinement, and in three-fourths of them, within the first fortnight (Clouston, *op. cit.*, p. 494).

Dr. Burrows found the third and fourth day the most obnoxious to the disease (Bucknill and Tuke, p. 357).

I.—PUERPERAL MANIA.
Early Symptoms.

Alteration of facial expression, dull, self-absorbed look. Sleeplessness, irritability, some depression. Alteration, diminution, or suppression of the lochia, often, but not always; less frequently the milk is diminished, or its secretion suppressed (Bucknill and Tuke, *loc. cit.*). Patient takes a dislike to her husband, or child; refuses food, or takes it reluctantly; complains of unpleasant smells. There is headache, or an uneasy sensation in the head.

Fully Developed Stage.

"Excitement, chattering, incoherent, blasphemous, or amorous talk." "Sleeplessness, anxiety, aversion to relations, erotic tendencies, mistakes of identity, with hallucinations of smell and taste, and refusal of food" (Savage, *op. cit.*, p. 373).

"She is restless; her eyes are brilliant. She expresses foolish fancies, such as that she is poisoned, that there is some one under the bed. She takes a violent dislike to the doctor, or the nurse, or the child . . . . . . She gets violent, and needs to be held in bed. Impulsively, and without set intent, she attempts to commit suicide, or tries to kill her baby, or to throw herself out of the window. She seems as if she had a supernatural strength; yet when you feel her pulse it is weak and thready, her face looks haggard, her temperature has risen to 100° or more, her womb is tender on pressure over the abdomen, and she will not look at food" (Clouston, *op. cit.*, p. 495).

Bucknill and Tuke (p. 359) say the abdomen is in most cases tolerant of pressure, and the pulse is accelerated and usually irritable in character. Patients are frequently wet and dirty, and destructive.

Puerperal mania only differs from ordinary mania in the predominance of certain symptoms, such as constant babbling, homicidal impulse, and the obscene and erotic character of the delirium (Bra). The presence of suicidal impulse, the refusal of food, the absolute sleeplessness, the perversion of natural affection are also characteristics of puerperal as compared with ordinary mania.

II.—PUERPERAL MELANCHOLIA.

Sleeplessness, anxiety and dread, are followed by delusions in reference to husband or children, and associated with hypochondriacal or other similar symptoms. Melancholic symptoms generally, but not always, come on later after delivery than attacks of mania (Savage, *op. cit.*, p. 377).

III.—PUERPERAL DEMENTIA.

Savage (p. 379) observes, "In some cases after delivery, the patient slowly becomes apathetic; she takes little or no notice of her child, and may be slightly emotional; her indifference becomes more and more marked, till it is recognised as a mental disorder. She neglects her personal cleanliness, and has to be tended like a child. This condition may slowly pass off, or it may be but the early symptoms of an incurable state of *dementia.*"

Clouston quotes several cases of puerperal insanity, in which the pulse attained 120 or considerably more, and the respirations were 56 or 60. Of sixty cases, the temperature was under 99° in thirty-four; between 99° and 100° in twelve; over 100° in fourteen, and in five of these over 103° (*op. cit.*, p. 506).

Sankey denies the existence of puerperal insanity as a morbid entity.

REASONING INSANITY (FOLIE RAISONNANTE).

Griesinger, writing of melancholia with persistent excitement of the will, mentions (*op. cit.* p. 275) that some of the cases of this nature are adduced by authors as examples of emotional insanity, mania sine delirio, folie raisonnante, or Prichard's moral insanity. He further speaks of maniacal folie raisonnante as a state of incompletely developed mania which may either end in recovery or pass into gay weak-mindedness or foolishness (*moria*), if the exaltation, as yet moderate, does not increase and explode outwardly as mania, or increase less demonstratively with the development of fixed delusions until it becomes delusional insanity.

Clouston, as already stated, considers that folie raisonnante corresponds in a general way to the milder cases of simple mania.

Krafft-Ebing, whose most recent classification (see end of Chap. I.) is one of the best lately proposed, uses folie raisonnante as a term synonymous with "Constitutional Affective Insanity," a degenerative functional psychosis. He also, in treating of disturbances of the will, says that the frequent occurrence of cases in which the patients talk sensibly, and are able to excuse their extremely foolish actions with wit and shrewdness, has given rise to the erection of a special form of insanity, the so-called folie raisonnante. He subdivides it into *maniacal* and *melancholic* folie raisonnante. Cases of the former display well-marked periodicity, and are in reality cases of periodical mania (*délire des actes*, perversions, etc.). (See "Periodical Insanity," maniacal form.)

Melancholy Folie Raisonnante.—It occurs most frequently in females. (For causes, see Chap. IV.) There is a constant ill-humour, a permanent depressive condition which expresses itself in irritability, discontentedness, quarrelsomeness, abusiveness, proneness to illtreat surrounding persons. Such patients are abulic, spiritless, joyless, incapable of sustained bodily or mental work, unhappy, desponding to the extent of tædium vitæ. There are remissions and exacerbations, and the symptoms are more strongly pronounced at the menstrual periods. This disease is thus distinguished from a mere evil disposition. There may be, though there rarely are, fits of apprehension and delusions of persecution. Neuropathic symptoms (neurasthenia, spinal irritation, hysteria) may complicate the paroxysms of apparent ill-humour and irritability.

RHEUMATIC INSANITY.

This may be divided into cerebral rheumatism, and rheumatic insanity properly so-called.

CEREBRAL RHEUMATISM in the great majority of cases appears during the course of the rheumatic affection, from the fifth to the twentieth day. Its frequency is not in constant relation with the intensity of the rheumatism; ordinarily, however, it accompanies the grave forms of acute articular rheumatism. It sometimes commences suddenly, the articular pains diminishing a short time before the attack. At other times it is announced by matutinal elevation of temperature, profuse sweats, slight mental aberration, most frequently of the melancholic type, frontal cephalalgia, change of character, alteration of facial expression. Hallucinations, vertigo, embarrassment of speech resembling that of general paralysis, dysphagia, and *choreiform* movements often open the scene.

The *Severe or Supra-acute* form is characterised by the rapidity of its course, and suddenly fatal termination. Death is sometimes preceded by asphyxia, coma (comatose, depressive, or apoplectic form), sometimes by violent delirium or convulsions (meningitic form).

The *Acute* or less *Severe* form is preceded by a *prodromal* period of variable duration from a few hours to a few days, and manifests itself by great *loquacity*, extraordinary exaltation, alternating with moments of depression and melancholy. There are visual hallucinations, and, more rarely auditory, olfactory, or gustatory, *violence*, suicidal attempts, *tremor*, *convulsions*, *dyspnœa*, and sometimes analgesia.

The *temperature* in the rectum reaches 109° F., 111° F., or even 112° F.

The *Pulse* is *small* and *frequent*. This state lasts from a few hours to a few days, and frequently terminates in coma and death.

Sometimes the delirium appears in irregular attacks, coming on most frequently at night; the prognosis is then more favourable (Bra, *op. cit.*, pp. 100–102).

RHEUMATIC INSANITY properly so-called (sub-acute, or chronic, or vesanic form of cerebral rheumatism, Bra); Prolonged rheumatic encephalopathy (Griesinger; Rheumatic form of consecutive insanity of several authors). It occurs most frequently at the termination of the rheumatic affection, or during convalescence. Most frequently it is announced by a *change* of *character*, often coincident with the diminution of the articular pains. The patient is restless, suspicious, irascible, and suffers from insomnia.

Visual Hallucinations appear.—Those of the other senses are less frequent. The patient believes himself surrounded by spies and enemies; he accuses his friends and relations, believes they are killing him, sees flames, fancies himself in a furnace, etc. There are then periods of remission and exacerbation which alternate in a regular manner.

Sometimes the mental aberration assumes the *maniacal form*; then there are excessive loquacity, extreme animation, restlessness, destructiveness, noisiness, howlings, hollow eyes, drawn features; the intellect is greatly enfeebled; the speech is jerky, and there may be absolute incoherence. The *pulse* is frequent, the *temperature* normal, or nearly so. Sometimes there are choreic movements, spasms, or convulsions.

In the great majority of cases, however, the mental state assumes the *melancholic form*. The patient is completely prostrated, and *refuses food*; with fixed eyes, he is self-absorbed, and appears to be overwhelmed, does not speak, and seems completely unconscious of what is taking place around him.

It is a veritable condition of stuporous melancholia (mélancolie avec stupeur).

It is well to remark that in chronic rheumatic insanity, the general symptoms which accompany the intellectual troubles are only slightly accentuated. They never attain the violence observed in cerebral rheumatism, properly so-called (Bra, *op. cit.*, pp. 102–103).

SENILE INSANITY.

Including: (1,) Senile mania; (2,) Senile melancholia; (3,) Senile dementia.

I.—SENILE MANIA may come on suddenly or slowly. In the former case, an initial stage of stupidity and peculiarity is succeeded "by constant talking, shouting, incoherence, loss of memory, loss of attention, sleeplessness, and, above all, by a constant motor restlessness by night and day, but especially *by night*" (Clouston, *op. cit.*, p. 573).

In the latter case (the attack coming on slowly) the memory fails, the patient becomes stupid and confused, then suspicious, then restless, then unmanageable, then violent. The speech is senile; there is great motor restlessness, especially at night (Clouston, *op. cit.*, pp. 574–575).

II.—SENILE MELANCHOLIA begins with failure of memory, irritability, exaggerated opinions of self, morbid suspicions, sleeplessness, restlessness, and lack of self control. Afterwards, slight transient attacks of hemiplegia.

As in the other forms of senile insanity, there is atheroma of the vessels, the radials being hard and cord-like, and the temporals tortuous. There is the *senile speech* described by Clouston (p. 568) as "a slight indistinctness of speech, a want of motor activity, and perfect co-ordination in the articulatory muscles, a change in the tone of the voice in the direction of feebleness, a difficulty in finding words, a tendency to stop in the middle of sentences, an omission of words, especially nouns"; a "mixture of aphasic, amnesic, and paretic symptoms." Afterwards, there is complete loss of memory for recent events. There are fits of moaning, groaning, and tearless weeping, without any apparent cause. There are at times sudden suicidal attempts or homicidal attacks. All the worst symptoms come on at night. Clouston would classify as melancholia cases where there are the outward signs of mental pain (p. 570). Savage (*op. cit.*, p. 202) considers that there is a condition of painful action and sensation, such as may be described by the term senile melancholia. "It appears sometimes rather suddenly, as the result of some family distress or domestic loss." He further says (p. 203), "I saw one doctor who suffered from constant subjective annoyances through his ears, his nose, and his skin, for months before his fatal attack of apoplexy. In some other cases of senile melancholia, mental or bodily hypochondriasis, with great emotional disturbance, is met with; and it is not unknown for patients of advanced years to destroy themselves, being convinced they have outlived their time."

III.—SENILE DEMENTIA.—Spitzka (*op. cit.*, p. 172) considers this to be the only characteristic form of senile insanity.

According to him the symptoms are, increased egotism, penuriousness, enfeeblement of memory, especially for recent events, unreasonable prejudices, frequently profound moral deterioration, coarse and filthy language being used, and filthy or intemperate habits being indulged in. In addition there may be, especially in males, a pathological sexual desire, a senile satyriasis, manifesting itself in indecent assaults on girls or infants, or in absurd and ridiculous marriage plans. Some have unsystematised ambitious delusions, but the majority have depressive delusions, and rare instances are on record where senile dements have committed suicide.

The most common delusions relate to their property. They suspect they are being defrauded or robbed. They are consequently disposed to be lachrymose, and to evince a restless and purposeless activity. Some of them roam about at night continuously, either purposelessly, or watching for thieves. There may be hallucinations and illusions of a painful nature. The patient finally becomes fatuous, and may be voracious and dirty, and die with apoplectiform or paralytic symptoms. In addition to arterial sclerosis, there are often observable a marked arcus senilis, opacities of the vitreous body, and sometimes cataract.

Tremor is an invariable symptom. Other symptoms which occur in some cases are marked hyperæsthesia, vertigo, anorexia, paraparesis, hemiparesis, disturbances of speech, and epileptiform attacks.

SOMNAMBULISM (*Pseudo-Insanity of*).

"Most bad and confirmed sleep-walkers have a neurotic heredity, or a nervous temperament, or both, though it is fortunately quite certain that few of them ever become insane. Acts of violence, homicide and suicide may be done in a state of somnambulism" (Clouston, p. 608).

STUPOR, ANERGIC.

Synonyms.—Acute Dementia; Acute Primary Dementia; Stuporous Insanity (Spitzka).

SYMPTOMS.—*Prodromal:* Either in consequence of masturbation, starvation, or exhausting discharges (1,) gradually increasing apathy and want of energy, or (2,) sudden invasion after hæmorrhage, shock, etc. (Spitzka, p. 159).

Fully developed: A loss of facial expression; a marked vaso-motor paresis, so that the extremities are blue and cold; a lowering of the trophic energy, so that sores are apt to form and even gangrene may occur; the reflex functions of the cord are markedly diminished, and the higher reflex functions of the brain almost in abeyance; no muscular resistance; no delusions; complete unconsciousness, and of course no after memory of events that occurred during its persistence (Clouston, p. 301).

The patient is in a state of immobility and does nothing of his own initiative. Sensibility is impaired as much as mobility, so that even the cautery may not be perceived by the patient. Food to be swallowed must be pushed well back into the pharynx. The pupils are dilated and react poorly, the heart's action is greatly enfeebled, the pulse tardy, small, and frequent, the temperature is slightly lowered and the extremities are cold, while œdema of the feet is constantly, and of the hands and face sometimes observed (Spitzka, p. 159). Wet and dirty in habits. The saliva dribbles from the mouth. Urine rich in phosphates, and physiological discharges of skin and uterus suppressed (Spitzka, *loc. cit.*).

SYPHILITIC INSANITY.

Fournier (Bra, "Man. des Mal. Ment.," p. 106) refers the clinical types of cerebral syphilis to six forms, viz.: the cephalalgic, the congestive, the convulsive or epileptic, the aphasic, the paralytic, the *mental.*

MENTAL FORM.

The mental form (Bra, *op. cit.*, p. 110, *et seq.*) is most frequently associated with various phenomena, congestive, epileptic, paralytic, etc., but it may exist alone. There are two varieties, the depressive and the expansive.

I.—DEPRESSIVE VARIETY.

This may be again subdivided into simple intellectual depression and intellectual depression with incoherence.

(A,) *Simple Depression.*—This is a form of intellectual asthenia. The power of attention is diminished and intellectual operations are less rapidly performed. There is a dulness, a slowness, a laziness of ideation. There are in consequence constant mistakes, awkwardness, oversights, and forgetfulness. The patient is vaguely conscious of his condition.

At the same time there is a progressive change in his character and habits, and there is an unusual inequality of temper.

Amnesia is an important symptom. It may come on gradually, interruptedly, or suddenly. In most cases the defects of memory are at first slight and insignificant, but they gradually increase, and the oversights, errors, and omissions become more frequent, until a veritable amnesia is established, the memory for long past events alone persisting.

(B,) *Depression with Incoherence.*—The patient commences a thing and does not finish it ; forms a project and abandons it ; rises and lies down without knowing why. There is a delusional state which is general and non-systematised, referring to all the acts of life and not to one in particular, and without any predominant fixed ideas.

With the exception of momentary attacks of excitement the condition is calm and tranquil. Occasionally there is a melancholic condition with somewhat systematised delusions ; a persecutory delusional condition. Still more rarely there are cases marked by suicidal ideas.

II.—EXPANSIVE VARIETY.

When slight the patient may not actually be insane, but merely the subject of insanity in the germ (Fournier). The patient is then exalted, loquacious, prone to exaggerate, irritable, unreflecting.

In the severe form there is a true maniacal condition. The facial expression is wild and the eyes wander.

There is motiveless laughter ; there are sudden outbursts with abusive epithets, menaces, vociferations. The words and acts are extravagant ; there is constant agitation ; unceasing loquacity. The patient becomes indifferent to all and to everything ; neglects his affairs ; wanders at random ; dresses and undresses without motive. There are outbursts of dangerous violence, sometimes going the length of homicidal attempts. There is complete *insomnia.*

There are observed, though rarely, hallucinatory phenomena of various kinds. Where the delirium is a concomitant of the epileptic form of cerebral syphilis there may be impulsive phenomena.

SYPHILITIC PSEUDO-GENERAL PARALYSIS.

Fournier gives as the principal differences between this and real general paralysis :—

(1,) In syphilis the intellectual troubles frequently follow one or several apoplectiform or epileptiform attacks, while in general paralysis the alteration of the faculties ordinarily first attracts attention.

(2,) Syphilitics have rarely an ambitious expansive delirium. They are embruted; stupid; sometimes extravagant, but always relatively humble and modest in their delirious conceptions.

(3,) Tremor is more rarely seen in syphilitics; it is neither so intense nor so permanent; it has not the appearance so characteristic of the tremor of the lips and tongue in general paralytics. In general paralysis the tremor is jerky and shifts rapidly from one group of muscular fibres to another. There is also very fine fibrillary tremor often requiring close scrutiny for its perception.

(4,) In syphilis the paralyses are much more accentuated than in general paralysis. Partial paralyses are more frequent.

(5,) The evolution of syphilitic general paralysis is less regular than that of ordinary general paralysis.

(6,) In syphilis there is, from the earliest periods, a more considerable alteration of the general condition; there is often a very pronounced cachexia.

(7,) The lesions differ. In syphilis those of the meninges predominate; in general paralysis, those of the cortex.

(8,) Cure, possible in syphilis, is almost impossible in ordinary general paralysis.

TOXIC INSANITY.

I.—ALCOHOLIC INSANITY.

A.—ACUTE ALCOHOLIC INSANITY.

"The most frequent form of the affection is *(a,)* violent maniacal delirium, known as mania a potu, with a tendency to homicidal acts. In some cases the mental disorder takes *(b,)* the melancholic form, and it becomes necessary to guard specially against the strong suicidal tendency which generally characterises it" (Sibbald, " Quain's Dict.," vol. i., p. 723).

B.—DELIRIUM TREMENS.

(a,) Apyretic.—The onset may be sudden but is most frequently preceded by a prodromal period characterised by præcordial pain, gastric irritability, cephalalgia, sensory disturbances, hyperæsthesia, hyperacousia. Insomnia is almost always present, and there is generally a condition of considerable depression.

At the end of several days the real attack commences and is marked by the following symptoms: Terrified expression; injected sunken eyes; excessive agitation. Incessant incoherent babbling difficult to understand, yet the patient may at times answer pressing questions quite pertinently. Hallucinations soon appear, the visual being most frequent, the patient seeing cats, rats, wild beasts, etc. Sometimes there are hallucinations of hearing or of the general sensibility; there are buzzings in the ears, sensations of burning, pricking, formication, etc.

The hallucinations have certain special characters: (1,) They appear at first at night. Once established they may either disappear in the daytime to re-appear the following night, or they may be continuous. When disappearing (during recovery) they continue at night after they have ceased by day; (2,) They are painful, often aggressive, the moral impressions produced by them varying from astonishment to profound terror. All the patient's surroundings seem to threaten him. The fear of being poisoned is frequent and causes the refusal of food which is so often troublesome; (3,) Their mobility. The things which form the objects of the hallucinations are constantly moving and displacing each other; hence the rapidity of ideas and actions in alcoholics who are quickly by turns frightened, anxious, unquiet, suppliant or aggressive (Magnan, " L' Alcoolisme et son action sur l'intelligence," p. 121).

The hallucinations are infinite in their variety but they often reflect either the daily occupations or the predominant preoccupation of the moment, with a special preference for that which is painful or disagreeable. The patient may fancy he hears abusive epithets, or the voices of his relations and friends; or he may see his wife surrounded by people who are outraging her before his eyes, or he may fancy himself in prison, before a tribunal, etc., etc. The condition may be maniacal, melancholic, or stuporous, the first form being the most frequent (Magnan, *op. cit.*, pp. 122-124).

Perversions, illusions, and hallucinations of the senses of taste and smell are less frequent than those of the other senses, but occur occasionally.

The visual hallucinations develop from obscurity of vision, sparks, flames, shades, etc., to grinning faces, animals, etc.

The auditory hallucinations from hummings, buzzings, and whistlings to voices and tumultuous cries. They disappear in the same order and at first in the daytime then at night, and a prolonged and quiet sleep terminates the attack. In some individuals amelioration is less rapid and regular, sleep is disturbed, there is irritability, and there are vague ideas of persecution. In yet other cases delirious conceptions giving rise to suspicions, jealousy, and misery remain after the acute symptoms have subsided, and these cases furnish many examples of suicide and homicide (Magnan, *op. cit.*, pp. 125-128).

Two or three cases under my care (not in asylums) have been free from actual visual hallucinations, but have suffered from visual illusions, such as taking lamposts for policemen, and red letter-boxes for soldiers. In these cases, auditory hallucinations and illusions were well marked, the patients hearing whisperings, shoutings, and abusive and threatening language when there were no sounds whatever, and construing nearly every real noise into a reproach or a threat.

One patient, a medical man, a passenger on board ship, gave me the first indication of his condition a few hours after joining the ship by taking me down to hear what he called the "death watch" ticking; afterwards he became quite delirious, and, on his recovery, told me he fancied he was going to be killed during his excitement, and that the engines kept constantly saying, "Waiting for the doctor," "Waiting for the doctor," "Waiting for the captain," "Waiting for the captain" (auditory illusions). A coloured man, who passed his cabin door when it was open, he took to be the devil. He had other visual illusions, but no actual visual hallucinations.

Tremor may be general or localised, transitory or persistent, slight or intense. If local, transitory, and slight, the disease is mild; if intense, general, and persistent, not disappearing during sleep, accompanied by shudderings, slight muscular shocks and muscular undulations, on the second or third day nervous exhaustion sets in. These muscular twitchings and undulations are more frequently observed in

(*b*,) Pyretic delirium tremens: and if not perceptible visually, should be sought for digitally.

Fever and tremor ordinarily co-exist, but in some cases the former is very high, with very little of the latter (Magnan, "Influence de l'Alcoolisme sur les Maladies Mentales," p. 7). In mild cases, owing to agitation, the temperature may rise to 101° or 101·4° F. (taken in rectum), sinking during temporary quietude to 100·2°. In grave cases the temperature will oscillate round 102·1° for two or three days, and then rise to 105·8° (*op. cit*, p. 6). Febrile or pyretic delirium tremens is marked therefore by fever, severe muscular tremor, and muscular feebleness, and occurs almost always in consequence of recent and numerous alcoholic excesses.

(*c*,) Complicating Delirium Tremens manifests itself in the chronic alcoholic who is accidentally attacked by some disease, or who suffers some mechanical injury. The fever follows the course pertaining to the intercurrent affection, whether pneumonia, erysipelas, pericarditis, or other disease, or accidental or therapeutical traumatism (Magnan, *op. cit.*, p. 7).

The other alcoholic phenomena present very different degrees of intensity. Thus, the delirium may betray itself by nocturnal hallucinations and nightmares only. The motor troubles are very frequently limited to tremor, more or less extensive, of the hands, or of the arms and legs, of the arms alone, of the legs and face. There is an absence of the deep-seated general muscular tremor of febrile delirium tremens (Magnan, *op. cit.*, p. 8).

In severe attacks of delirium tremens, the nervous exhaustion and muscular feebleness may proceed to incomplete paralysis of the arms and legs; there may be embarrassment of speech, simulating that of general paralysis; profuse sweats, exhaling an alcoholic odour; the urine is scanty, the pupils are sometimes unequally contracted; the pulse is small, often accelerated (Bra, *op. cit.*, pp. 143–144).

C.—Chronic Alcoholic Insanity.

The passage to the chronic state does not in general follow a regular progression, but is made by leaps, and is accompanied by exacerbations, which Lasègue described under the name of "sub-acute alcoholism." Between the transient phenomena of frequently-repeated intoxication and the establishment of insanity there is an intermediate period of irritability, inquietude, and impressionability.

Precursory Symptoms.—Insomnia more and more pronounced, tremor of the hands, dreams, and, in a more advanced stage, hallucinations (Bra, *op. cit.*, p. 148). The hallucinations have the three characters described under the head of "Delirium Tremens," *i.e.*, they are nocturnal, mobile, painful.

Visual Hallucinations.—The patient fancies he sees unclean animals, thieves, assassins, etc.; he witnesses heartrending scenes; sees himself in the midst of flames, at the scaffold, etc.

Auditory Hallucinations.—He believes he hears abusive epithets, threats, accusations against his honour and morality; he hears groanings, lamentations, cries, the clash of arms, etc.

Olfactory Hallucinations.—He fancies he perceives most disagreeable stenches and suffocating odours; that he breathes a pestilent atmosphere, etc.

Gustatory Hallucinations.—He fancies he tastes all sorts of nauseating substances, as well as poisons.

Tactile Hallucinations.—He believes he is suffering terrible punishments; he feels knife-blades entering his flesh and mutilating him frightfully; he feels the crawling of a serpent, which glides over his skin and enfolds him; he feels insects and worms gnawing his body, which he fancies he sees falling in shreds; swarms of flies appear to him to enter his mouth, nostrils, and eyes; or he even fancies he is drowned or thrown down a precipice.

Under the influence of these hallucinations, the patient reacts variously; he becomes excited and defends himself, threatens, or strikes; or he may even remain immovable, overwhelmed, crushed. Hence his different attitudes, maniacal, melancholic, or stuporous, all arising from the same cause, but varying according to the degree of intensity of this cause. There is a successive gradation in the development of these phenomena. The mere functional trouble passes into the illusion, and this into the confused hallucination—at first simple, then multiple (compound), and becoming more and more precise, neat, and distinct, simulating reality.

When amelioration is taking place, the phenomena disappear gradually, in accordance with an analogous decreasing order; the distinct hallucination fades into the confused one, the latter into the illusion, and this again into the mere functional disturbance. Such is the usual method of evolution and involution of alcoholic hallucinatory phenomena; although exceptionally they may at once reach their acme (Magnan, "Influence de l'Alcoolisme sur les Maladies Mentales," p. 9).

As the disease progresses, all the psycho-sensorial phenomena gradually lose any acuteness which they possess; the moral and affective faculties and the will of the patient become considerably weakened, and the individual abandons himself entirely to the caprice of his instincts. After a long series of excesses, the intelligence is absolutely annihilated, and the patient falls into a state of complete dementia (Bra, *op. cit.*, p. 151).

Spitzka, writing of Chronic Alcoholic Insanity ("Insanity," p. 252), observes that the alcoholic tremor is the most important and constant somatic sign, and that it is best seen in the hands, tongue, and lips; he also says it has the peculiarity that it *decreases* under the influence of alcoholic beverages, and is most marked when the patient is perfectly sober.

When intoxication is prolonged either slowly and progressively, or by means of several relapses with acute accidents, there remain certain delusions, hypochondriacal ideas, illusions, and sometimes even hallucinations, which reflect the general characters of the intellectual and sensory disturbances of the earlier periods, but without their acuteness or activity. Chronic alcoholics may have melancholic delusions and suicidal ideas, but these are only the outline of the initial phenomena. Some patients, with or without renewed excesses, become, at irregular periods, semi-maniacally excited, turbulent, and destructive, and act in every sense automatically (Magnan, "L'Alcoolisme," etc., p. 130).

In some cases, after the acute symptoms, certain delusive conceptions emanate from the hallucinations, and give rise to jealousy and suspicion. In other cases troubles of the general sensibility remain, accompanied by hypochondriacal ideas and fears of being poisoned (Magnan, "L'Alcoolisme," etc., p. 128).

"The persecutory delusions of alcoholism relate to the sexual organs, to the sexual relations, and to poisoning. This fact is so constant a one, that the combination of a delusion of mutilation of the sexual organs with the delusion that the patient's food is poisoned, and that his wife is unfaithful to him, may be considered to as nearly demonstrate the existence of alcoholic insanity as any one group of symptoms in mental pathology can prove anything. With this there are unpleasant hallucinations. . . . Delirious exacerbations are likely to occur in consequence of the patient's morbid fear, and in brutal fury he may hack the wife whom he suspects of infidelity to pieces" (Spitzka, "Insanity," p. 254).

Magnan ("L'Alcoolisme et son action sur l'Intelligence," pp. 131-132) says, in chronic alcoholism (chronic alcoholic insanity), the memory is enfeebled, the judgment is less sound and is incapable of discernment, the imagination is extinguished, the power of associating ideas is much lessened, giving rise to incoherence; the moral sensibility is blunted. Apathetic, indifferent, stupid, the chronic alcoholic has no care for his person, and no thought for his family. In the last periods intelligence is annihilated; insensibly all the delirious flights disappear, the hypochondriacal preoccupations and the sensory troubles fade away little by little. Sometimes a false sensibility (sensiblerie). analogous to that of apopletic dements, supervenes.

There are frequently also "stunnings," vertigo, apoplectiform and epileptiform attacks, and partial paralysis, corresponding to the lesions of the nervous centres, discovered post-mortem.

D.—ALCOHOLIC PSEUDO-GENERAL PARALYSIS.

This affection attacks by preference those in whom alcoholic excesses are rarely accompanied by well marked intellectual disturbances, but rather by profound somatic perturbations. It resembles real general paralysis so nearly as to be easily mistaken for that disease. It commences brusquely, and quickly attains its apogee, breaking out usually after an attack of sub-acute alcoholism. There is embarrassment of speech, which occupies a secondary place, and tends to disappear rapidly. The pupils are unequal, and very little sensible to the influence of light.

The tremor is general and massive. The paralysis is incomplete, and commences at the distal extremities of the limbs.

There are symptoms of alcoholic intoxication; disturbances of sensibility, anæsthesiæ, hyperæsthesiæ, *visual hallucinations.* Sometimes there are epileptiform attacks, gastric embarrassment, anorexia.

The tendency to ameliorate soon shows itself; the patient often recovers from his attack, but soon falls into excesses causing an almost fatal relapse (Bra. *op. cit.*, pp. 137–139).

E.—COMBINED ALCOHOLIC DEMENTIA AND ALCOHOLIC PARALYSIS.

A form of combined alcoholic dementia and alcoholic paralysis which occurs mostly in females—in them to a certain extent taking the place of delirium tremens in men (Blandford "Insanity," 1st ed., pp. 65 and 278–279).

After years of habitual drinking, quite suddenly, without illness, sleeplessness, or excitement, memory gives way. Muscular power and co-ordination are lost, but the weakness does not extend to the organs of articulation. "The delusions of such patients chiefly depend on the entire obliteration of memory. The patients want to see and visit people who have long since died; and when told of this, when the circumstances are brought back to their recollection, they make the same request five minutes after."

II.—INSANITY FROM ABSINTHE AND ALCOHOLIC PREPARATIONS OF ABSINTHE.

In absinthe drinkers the hallucinatory phenomena reach their acme suddenly (Magnan, Infl. de l'Alc., etc., p. 9). In a general way the hallucinatory phenomena are similar to those of alcohol, differing only in the suddenness of their onset. In addition to the tremor, there are epileptic seizures, which are seldom observed in uncomplicated alcoholic cases (Magnan, "Infl. de l'Alc.," etc., pp. 4–5).

A small dose of essence of absinthe injected into the veins of dogs causes vertigo and muscular twitchings; a large dose, epileptic attacks and delirium.

During the first stage of the absinthic attack (in dogs) the pupils are dilated and the papillæ and fundi oculorum are injected (Magnan, "Physiologie Pathologique et Recherches Cliniques," p. 113).

III.—INSANITY FROM OPIUM OR ITS ALKALOIDS.

Clouston (*op. cit.*, p. 445) says, "I have seen many cases of insanity resulting from opium-eating, and one from the hypodermic use of morphia. These were very like the insanity of chronic alcoholism, but not so suicidal, with greater weakness of the heart's action, and more sleeplessness, sickness, and intolerance of food for the first fortnight."

Savage ("Insanity," p. 430) believes that the opium crave is stronger than any other. He states that symptoms resembling delirium tremens may be set up by opium eating or the injection of morphia, and mentions the fact that it has been said that a morphia injection will quiet in morphismus, but alcohol will cause excitement. There are present the same tremor, want of appetite, refusal of food, ideas of poison, hallucinations, and tendency to erotic ideas. Chronic morphismus may also be set up with suspiciousness, auditory hallucinations, feelings of galvanic shocks.

Cocainomania is a form of excitement caused by the abuse of cocaine, generally in conjunction with morphia.

IV.—INSANITY FROM CHLORAL.

Clouston (*loc. cit.*) says, "I have seen two cases of insanity brought on by the use of chloral. They, too, were of the same generic type as the alcoholic cases, and demanded the same treatment."

Savage (*op. cit.*, p. 429) states that chloral will give rise to a crave, a sleepless habit, a feeling of deep depression, most marked on awaking in the morning; this feeling is associated with anxiety and a hypochondriacal feeling at the epigastrium. It may produce very great emotional disturbance and irritability, passing into deep melancholia, with suicidal tendencies. It produces loss of control and tendency to impulses, and thus causes suicide or homicide.

V.—INSANITY FROM CANNABIS INDICA.

This drug when taken habitually for a lengthened period produces a form of excitable mania with delusions, from which the patients for the most part recover. These cases are common in India (Blandford, "Insanity and its Treatment," 1st ed., p. 66).

VI.—INSANITY FROM LEAD. SATURNINE INSANITY.

A.—DELIRIOUS FORM.

Precursory Symptoms.—In the great majority of cases it is preceded by physical symptoms of saturnine intoxication, blue gingival line, colic, etc. The commencement of the attack is rarely sudden, but is generally announced by *cephalalgia, depression, acceleration of the pulse, insomnia;* sometimes by ocular troubles vertigo, tremor.

Developed Stage.—This form of insanity is sometimes devoid of special characteristics, at other times the state is one of (*a*,) *acute mania*, with furious and more or less continuous delirium, with incoherence, extreme agitation, *abusive vociferations*, and every sort of violence. In some cases the condition is one of (*b*,) *melancholia*, either continuous or interrupted by periods of exaltation. Visual and, more rarely, auditory hallucinations appear. The loss of memory is often complete. It sometimes happens that (*c*,) *dementia* is suddenly established at first, without the physical saturnine symptoms being such as to lead one to expect such an occurrence.

B.—Comatose Form.

It may show itself at first, or appear after the delirious form or after an epileptiform attack. In general the coma is not very pronounced. It is a sort of somnolence, disturbed by fits of agitation, automatic movements, and cries. The patient replies in a confused and evasive manner to questions, and then falls again into his torpid state. Little by little the patient comes to himself, without remembering what has happened.

C.—Convulsive Form.

Saturnine convulsions have an extreme analogy with epileptic attacks. There is, however, no aura. They begin by a sudden loss of consciousness, are of long duration, and terminate in an intellectual obtusion; a state of profound somnolence, which lasts till the outbreak of a fresh attack.

D.—Mixed Form.

It presents successively all the forms already described, melancholic, maniacal, convulsive, comatose.

E.—Saturnine Pseudo-General Paralysis.

It commences suddenly and noisily. It at once attains its acme. The embarrassment of speech is sometimes so marked at the commencement that the voice is unintelligible. It is a true stammering. There are symptoms of saturnine intoxication; blue line on gums, earthy hue of skin, acute cephalalgia, stunnings, cramps, formications, various neuralgiæ, anæsthesiæ, partial hyperæsthesiæ, arthropathies, paralyses, epileptic or eclampsic troubles. The patients are subject to insomnia and nightmare. Visual hallucinations and ideas of persecution and poisoning intermingle with the delirium.

The pupillary inequality is more often absent than in general paralysis.

The tremor is more intermittent, more pronounced, and more spasmodic than in general paralysis.

The patients are often dirty and completely paralysed on their entry into the asylum.

The dementia, which manifests itself suddenly in its greatest intensity, is much more apparent than real. It is a suspension rather than an abolition of the intellectual faculties. The tendency to amelioration makes itself rapidly felt, the intellect sometimes waking up at the end of a very short period (Bra, *op. cit.*, pp. 131-134).

VII.—INSANITY FROM MERCURY.

Generally preceded and accompanied by the mercurial cachexia.

The symptoms of chronic mercurial poisoning as it affects the central nervous system are, according to Naunyn, great mental irritability, extraordinary terror, perplexity, anxiety, sleeplessness, and tendency to suffer from Hallucinations ("erethismus mercurialis"). At the same time there are symptoms of mercurialismus (anæmia, gastro-intestinal catarrh, salivation, tremor). Mania, melancholia, or mental enfeeblement may develop out of this condition (Krafft-Ebing, *op. cit.*, p. 224).

Cocaine, salicylic acid, iodoform, and ergot sometimes give rise to hallucinatory delirium. Hyoscyamus, conium, stramonium, belladonna, poisonous fungi, chloroform, and paraldehyde occasionally cause mental disturbances. The prolonged use of large doses of the bromides may result in mental and muscular weakness, loss of throat reflex, amnesic aphasia, stammering speech, staggering gait, cachectic pallor, tremulousness, etc. Poisonous gases, *e.g.*, carbon monoxide and the vapour of carbon disulphide may cause, when habitually inhaled, headache and vertigo, followed by transitory mania or melancholia and then mania. Cerebral symptoms may be occasioned by autogenous poisoning (uræmia, acetonæmia, cholæmia). The cachexia strumipriva (anæmia, cachexia, and mental torpor following extirpation of the thyroid in young persons) is worthy of note (Krafft-Ebing, *loc. cit.*). Morselli in his classification includes pellagrous insanity under the head of the "toxic encephalopathies" (*op. cit.*, p. 437). He gives pellagrous pseudo-general paralysis, carbonic oxide pseudo-general paralysis, etheric insanity, lathyrism, and nicotism (smokers' dementia) in addition to many of the forms above-mentioned.

TRAUMATIC INSANITY.

Spitzka (p. 371) and Clouston (p. 414) say that insolation causes a form of insanity very much like that due to traumatism.

Spitzka further states that it is far more likely to lead to general paralysis than is traumatism. He also observes that radiant heat (artificial) acts in a similar way to sunstroke. Head injuries may originate an ordinary attack of insanity in a person predisposed to the disease (Clouston, p. 417).

Whilst fractures of the skull, with depression, are more likely to lead to serious mental mischief, yet even simple concussion may induce chronic incurable insanity, or the disposition to it.

Directly after an injury, and intercurrent with the stupor and coma following shock, delirium, hallucinations, and excitement often manifest themselves. Or there may be serious lacunæ of the memory, the patient either recovering, or passing into a condition very similar to primary mental deterioration. But the true traumatic insanity comes on some time after the injury to the head (Spitzka, " Insanity," p. 370).

Spitzka further writes (p. 371), "The subjects of this disorder are noticed to undergo a change of character, to exhibit a tendency to alcoholic excesses, to become morally perverse, suspicious, brutal, and quarrelsome, and to manifest murderous or other violent impulses, occasionally associated with fits of maniacal self-exaltation or furor, usually of short duration. This condition is remarkable for its long duration and its frequent and sudden changes, the occasional lucidity of the patients being accompanied at the time by hypochondriasis. As a rule, progressive deterioration sets in, and dementia terminates the history of the case. Tinnitus aurium, photopsia, scintillation before the eyes, headache of a pulsatory or grinding character, vertigo, paresis of various muscular groups, particularly of the eyeball, without fibrillary tremor, anæsthesias, and hyperæsthesias, as well as insomnia, are frequent accompaniments, and some of these enumerated signs are present in every case."

Clouston, speaking of the more characteristic type of traumatic insanity, says ("Ment. Dis.," p. 414), "It is accompanied by motor symptoms, either in the shape of speech difficulties, slight hemiplegia, general muscular weakness, or convulsions. Usually in such cases there are, in addition, sensory symptoms, such as cephalalgia, vertigo, hallucinations, a feeling of confusion and incapacity for exertion of any kind, mental or bodily. The mental symptoms are usually a form of melancholia at first, tending in time towards an irritable and sometimes impulsive and dangerous dementia or delusional insanity. In my experience such cases are all absolutely intolerant of alcoholic stimulants, a very little of which will always make them maniacal, and often very dangerous and even homicidal. Many of them have a craving for stimulants, too, which they indulge, and which aggravates all these symptoms." He further states that a few cases become ordinary epileptics. He also recommends that in all traumatic cases the condition of the urine as to sugar and albumen should be carefully tested.

Traumatism is one of the rare causes of general paralysis, and Spitzka observes (p. 370) that the prodromal period of cases so caused is apt to be marked by the "furious outbreaks and murderous impulses characteristic of what might be called the traumatic neurosis."

UTERINE, OR AMENORRHŒAL INSANTIY.

Of the cases arising from disordered or suspended menstruation, nearly two-thirds are melancholic in character, and nearly one-third maniacal, a few are stuporous, and a few delirious.

Occasionally, but very rarely, acute delirious mania may be induced in a young, full-blooded, healthy woman of nervous heredity by the suppression of menstruation.

Stupor is, however, more common than acute delirium as a mental result of suppressed menstruation in young women of nervous heredity.

In the melancholic cases, hallucinations of hearing of a disagreeable character may develop (Clouston, " Ment. Dis.," p. 473 et seq.).

YOUNG CHILDREN *(Delirium of)*.

"In most cases it is a pure delirium, without consciousness, attention, or memory, but in some instances there are frightful hallucinations; in others an excited melancholia of short duration, with violent screaming, tearless weeping, and all the usual signs of mental depression" (Clouston, p. 606).

GROUPING OF THE FOREGOING FORMS OF MENTAL ABERRATION ACCORDING TO ONE OR TWO OF THE MOST PROMINENT SYMPTOMS.

I.—*Mental pain, or hindrance (hampering) of mental action, or both.*—Abdominal diseases, insanity of; anæmic insanity; cataleptic insanity; choreic insanity, some cases; circular insanity, melancholy phase; climacteric insanity; insanity from deprivation of senses; diabetic insanity; gestational insanity, most cases; hysterical insanity, some cases; katatonia, one phase in most cases; lactational insanity, most cases; melancholia, simple, agitated, suicidal; masturbational insanity, second stage; metastatic insanity; neurasthenic insanity; oxaluria and phosphaturia, insanity of; paralysis agitans, insanity of; pellagrous insanity;

periodical melancholia; podagrous insanity, some cases; post-connubial insanity; pubescent insanity; puerperal melancholia; saturnine and mercurial insanity, some cases; uterine insanity, nearly two-thirds of cases; delirium of young children, some cases; rheumatic insanity, most cases.

II.—*Emotional exaltation, or excitement (mental and motor), or both.*—Adolescent insanity; choreic insanity, some cases; circular insanity (maniacal phase); post-febrile consecutive insanity, some cases; inflammatory consecutive insanity; epileptic insanity; insanity of exophthalmic goître; gestational insanity, a few cases; acute hysterical insanity, most cases; chronic hysterical insanity, some cases; lactational insanity, a few cases; mania, acute, simple, and transitory; partial exaltation; periodical mania; podagrous insanity, many cases; pubescent insanity, most cases; puerperal mania; rheumatic insanity, some cases; expansive syphilitic insanity; mania a potu; saturnine and mercurial insanity, some cases; uterine insanity, nearly one-third of cases; masturbational insanity, first and third stages; a few cases of rheumatic insanity.

III.—*Delusion, hallucination, fixed idea, morbid impulse, or extraordinary actions.*—Choreic insanity, most cases; confusional insanity, primary and chronic; post-febrile consecutive insanity, some cases; delusional insanity (monomania, paranoia); folie à deux; folie du doute; impulsive insanity; katatonia, excited stage; mania, delusional; melancholia, delusional, hypochondriacal, and religious; moral insanity; ovarian insanity; partial emotional aberration; phthisical insanity; somnambulism pseudo-insanity of; insanity from cannabis indica; traumatic insanity.

IV.—*Acquired mental weakness.*—Severe and advanced cataleptic insanity; advanced consecutive insanity; terminal dementia; chronic epileptic insanity; advanced katatonia; masturbational insanity, fourth stage; primary mental deterioration; alcoholic dementia; advanced pubescent insanity; puerperal insanity, some cases; advanced saturnine and mercurial insanity; senile insanity; depressive syphilitic insanity; advanced traumatic insanity.

V.—*Acquired mental weakness with delusions and hallucinations.*—Chronic mania; insanity of myxœdema; delirium tremens; chronic alcoholic insanity; insanity from absinthe; insanity from opium.

VI.—*Acquired mental weakness with paresis.*—General paralysis; alcoholic pseudo-general paralysis; senile insanity, some cases; severe delirium tremens; chronic alcoholic insanity.

VII.—*Acquired mental weakness with paralysis.*—Insanity from coarse brain disease; alcoholic dementia in females; syphilitic pseudo-general paralysis; advanced general paralysis; traumatic insanity; saturnine pseudo-general paralysis; senile dementia.

VIII.—*Stupor.*—Anergic stupor; melancholic stupor; general paralytic, epileptic, and cataleptic stupor; rheumatic insanity, most cases; uterine and anæmic insanity, some cases; sometimes after acute mania; some cases of periodical melancholia and of melancholy phase of circular insanity.

IX.—*Delirium with unconsciousness.*—Acute delirium; delirium of young children, most cases; insanity of Bright's disease; insanity of cyanosis from Bright's disease, cardiac disease and asthma.

X.—*Congenital mental or moral weakness.*—Idiocy; imbecility; cretinism; moral imbecility.

CHAPTER IV.
ETIOLOGY.
A.—GENERAL.

CAUSES OF INSANITY IN ENGLAND AND WALES FROM 1878 TO 1887 INCLUSIVE, AS SHOWN BY THE REPORT OF THE COMMISSIONERS IN LUNACY, JUNE, 1889.

MORAL.	Proportion per cent. on Total Number admitted.		
	M.	F.	T.
Domestic trouble (including loss of relatives and friends)	4·2	9·7	7·
Adverse circumstances (including business anxieties and pecuniary difficulties)	8·2	3·7	5·9
Mental anxiety and "worry" (not included under the above two heads) and overwork	6·6	5·5	6·
Religious excitement	2·5	2·9	2·7
Love affairs (including seduction)	·7	2·5	1·6
Fright and nervous shock	·9	1·9	1·4
PHYSICAL.			
Intemperance, in drink	19·8	7·2	13·4
" sexual	1·0	·6	·7
Venereal disease	·8	·2	·5
Self-abuse (sexual)	2·1	·2	1·2
Over-exertion	·7	·4	·5
Sunstroke	2·3	·2	1·2
Accident or injury	5·2	1·0	3·0
Pregnancy		1·0	·5
Parturition and the puerperal state		6·7	3·4
Lactation		2·2	1·1
Uterine and ovarian disorders		2·3	1·2
Puberty		·6	·4
Change of life	·2	4·0	2·0
Fevers	·7	·5	·6
Privation and starvation	1·7	2·1	1·9
Old age	3·8	4·6	4·2
Other bodily diseases or disorders	11·1	10·5	10·8
Previous attacks	14·3	18·9	16·6
Hereditary influence ascertained	19·0	22·1	20·5
Congenital defect ascertained	5·1	3·5	4·3
Other ascertained causes	2·3	1·0	1·7
Unknown	21·3	20·1	20·7

PREDISPOSING CAUSES.

Under this head the International Congress in 1867 included hereditary influence, pure consanguinity, great difference of age between parents, influence of soil and of surroundings; convulsions or emotions of the mother during gestation; epilepsy; other nervous affections; pregnancy; lactation; menstrual period, critical age; puberty; intemperance (habitual excess, dating far back); venereal excess, and onanism.

Under this head Bucknill and Tuke (p 57, *et seq.*) include heredity; constitution, or diathesis; consanguineous marriages where parents are both neurotic; male sex (to a slight extent); age between 30 and 40, then 40 and 50, 20 and 30, 50 and 60, 60 and 70, 70 and 80, 80 and 90, 10 and 20; the summer season, especially in June (in France, at least); certain professions and trades; (1st,) soldiers; (2nd,) merchants, bankers, etc.; (3rd,) unoccupied proprietors; (4th,) professional men, artists, scientists, etc.; (5th,) innkeepers, servants, etc.; (6th,) carriers, porters, messengers, etc.; (7th,) agriculturists, fishermen, etc.; (8th,) mechanics, etc.; (9th,) butchers, bakers, grocers, greengrocers, etc.; civilisation pauperism, celibacy.

EXCITING CAUSES.

Bucknill and Tuke (page 91) give as the order of frequency :—

PHYSICAL.

Intemperance.
Epilepsy—six per cent.
Affections of the head and spine—six per cent.
Uterine disorders, viz., those of menstruation, pregnancy, parturition, lactation—five per cent.
Sexual vice.
Fever and febrile diseases—two to three per cent.

MORAL.

Domestic troubles and grief—fourteen per cent.
Religious anxiety and excitement—three per cent.
Disappointed affections—three per cent.
Fear and fright—two to three per cent.
Intense study.
Political and other excitement (joy, etc.).
Wounded feelings.
Or, without reference to the division into physical and moral :—
Domestic trouble and domestic grief.
Intemperance.
Epilepsy.
Affections of head and spine.

Uterine disorders.
Religious anxiety and excitement.
Disappointed affections.
Sexual vice.
Fever and febrile diseases.
Fear and fright.
Intense study.
Political and other excitement.
Wounded feelings.

Under the head of "Exciting Causes" the International Congress of 1867 included:—

I. *Physical Causes.*—Artificial deformities of cranium; convulsions of infancy and dentition; cerebral congestion (primary, not that which arises in the course of certain forms of insanity); organic affections of the brain; senility; pellagra; anæmia; constitutional syphilis; intermittent fevers; acute rheumatism; gout and chronic rheumatism; organic affections of the heart; pulmonary phthisis; intestinal worms; other acute diseases; other chronic diseases; suppression of hæmorrhoidal flux; menstrual disorders; metastasis; alcoholic drinks; abuse of tobacco; other vegetable poisons; mineral poisons (lead, mercury, copper, etc.); insolation; intense heat; intense cold; blows and falls upon the head; other traumatic causes.

II. *Moral Causes.*—Religion; education; love (love thwarted, jealousy); family affections; fluctuations of fortune; domestic troubles; pride; disappointed ambition; fright; irritation; anger; wounded modesty; political events; nostalgia; ennui; misanthropy; sudden joy; simple imprisonment; solitary confinement.

III. *Mixed Causes.*—Excess of intellectual work; prolonged vigils; evil habits of libertinism; onanism (sometimes simply predisposing); disorders of the reproductive system; destitution and want; bad treatment; sudden change from a life of activity to idleness and *vice versa*; loss of one or more of the senses (Bucknill and Tuke, pp. 107–108).

The insane diathesis, constitution, or predisposition is generally hereditary, but it may be acquired by the influence of certain somatic and psychical causes; amongst the former are traumatism, insolation, radiant heat, syphilis, alcohol and other narcotics, etc. Amongst the latter are faulty educational systems, early emotional overstrain, harsh treatment, sensational reading, and ambitious rivalry (Spitzka).

Batty Tuke remarks ("Brit. Med. Journ.," May 30, 1891), that in nearly 90 per cent. of the cases insanity is traceable to eight great classes of causes: (1,) Idiopathic morbid processes; (2,) Traumatic injury; (3,) Adventitious products; (4,) Secondary effects of other neuroses; (5,) Concurrent effects on the brain of diseases of the general system; (6,) Toxic Agents; (7,) Concurrent effects on the brain of evolutional and involutional conditions; (8,) Heredity.

B.—SPECIAL.

Forms of Insanity.	Causes.
ABDOMINAL DISORDERS, INSANITY FROM, INCLUDING SIBBALD'S GASTRO-ENTERIC INSANITY.	Irritation and catarrh of the gastric or intestinal mucous membrane. Constipation. Stricture or other causes of distension of the viscera. Epigastric tumours. Catarrh of bladder. Hepatic derangement (Sibbald, "Quain's Dictionary"; Bucknill and Tuke).
ADOLESCENT INSANITY.	Neurotic heredity, in a very large percentage of cases. Male sex. Mental weakness; intellectual overstrain; masturbation. Age 18–25, notably 20–25 (Clouston).
ANÆMIC INSANITY.	Starvation; chlorosis; prolonged indigestion, or other cause of anæmia.
BRIGHT'S DISEASE—INSANITY OF.	Chronic Bright's disease, with contracted kidneys and enlarged heart.
CATALEPTIC INSANITY. CAUSES OF CATALEPSY.	Neurotic heredity; female sex; puberty; hysteria; epilepsy; chronic cerebral disease; gastric and intestinal irritation. Great emotional disturbance (Ross).
CHOREIC INSANITY. CAUSES OF CHOREA.	Neurotic heredity; rheumatic diathesis; rheumatism; early life; female sex; onanism; pregnancy; menstrual disorders; chlorosis; emotional disturbances (fright, sorrow, and discontent) (Ross).
CIRCULAR INSANITY.	Neurotic heredity; youth; education; ancient lineage; diurnal, menstrual, sexual, and seasonal periodicities of the brain (Clouston).
CLIMACTERIC INSANITY.	Heredity, female sex. Age, 40–50 in females; 55–65 in males (Clouston).
COARSE BRAIN DISEASE, INSANITY FROM.	Cardiac disease; arterial atheroma; cerebral hæmorrhage; cerebral softening from thrombosis or embolism. Cerebral tumour. Cerebral atrophy. Chronic cerebral degeneration.

Forms of Insanity.	Causes.
CONFUSIONAL INSANITY, PRIMARY OR ACUTE.	Cerebral exhaustion; emotional shock; cerebral overstrain; exhausting diseases; excesses (Spitzka).
CONFUSIONAL INSANITY, CHRONIC.	Mania; melancholia; primary confusional insanity (Spitzka).
CONSECUTIVE INSANITY.	Small-pox; typhus, typhoid, intermittent fever; measles; erysipelas. The acute anginas; cholera; acute rheumatism.
CYANOSIS FROM BRONCHITIS, CARDIAC DISEASES, AND ASTHMA, INSANITY OF	Advanced age; hereditarily weak brain (Clouston).
DELIRIUM, ACUTE.	More common in females than males. Often hereditary predisposition. Insolation; privation; physical exhaustion; intellectual exhaustion or emotional strain; alcoholic excesses; business crises. Puerperal state (Spitzka, Bra).
DELUSIONAL INSANITY.	Neurotic heredity; congenital, cerebral, or bodily malformations; other somatic anomalies; exaggerated physiological mental states (pride, suspicion, sexual emotions, etc.) passing into delusions. Hallucinations slowly leading up to the development of fixed delusions. The delusion or delusions may simply arise out of some false idea, accepted without discussion. Typhus fever; head injuries; alcoholism; dreams; great emotional strain; monotonous ideation (Bra, Spitzka, Clouston, Savage).
DEMENTIA, TERMINAL. (1,) APATHETIC. (2,) AGITATED.	(1,) Anergic stupor; stuporous melancholia; violent outbreaks of maniacal furor. (2,) Chronic mania; agitated melancholia (Spitzka).
DEPRIVATION OF THE SENSES, INSANITY FROM.	The loss of one or more of the special senses, sight, hearing, etc. (Clouston).

Forms of Insanity.	Causes.
DETERIORATION, PRIMARY MENTAL.	Excitement; intellectual exhaustion; emotional strain; continuous mental worry (Spitzka).
DIABETIC INSANITY.	Diabetes Mellitus.
EPILEPTIC INSANITY.	Epilepsy.
EXOPHTHALMIC GÔITRE, INSANITY WITH.	Graves' disease.
FOLIE A DEUX.	See Chap. III.
FOLIE DU DOUTE.	Neuropathic or insane heredity in most cases. Age of puberty; female sex; high social position; inveterate onanism; eruptive fevers; fright; emotional excitement; excessive intellectual work. Neurasthenia, *q.v.*
FOLIE RAISONNANTE.	Heredity; neurasthenia; hysteria; uterine affections, especially infarcts and displacements (Krafft-Ebing).
GENERAL PARALYSIS OF THE INSANE.	Age between 20 and 60, especially 35 to 45; male sex; mental overwork, especially attempts to do work above the capacity; deception; ambition; jealousy; disappointment and vexation; sexual excess; alcoholic excess; syphilis; erysipelas of the scalp; insolation; traumatisms, especially falls on the head. Combination of intellectual over-work, emotional strain, and alcoholic excess most potent. Mickle gives as predisponents: male sex; age 30–50; energetic mental life with ardent imagination; neurotic heredity; married life; military and naval life; exposure to great heat or alternate heat and cold; prostitution; professional and literary occupations, as they entail strain, worry, and overwork; ambitious projects; prolonged and violent or sudden and frequent feeling or passion, as worry, indignation, rage, or lust, chagrins; forced erethism of the intellectual faculties; intellectual overwork, especially if sustained by stimulants;

Forms of Insanity.	Causes.
	cessation of discharges; cranial injuries; urban life.

As excitants, he gives: (1,) Alcoholic excess; (2,) Excessive and prolonged intellectual labour with undue emotional tension; (3,) Protracted painful emotional strain; (4,) Exhausting heavy physical labour; (5,) Sexual excess. Each of the last four being in many cases associated with undue alcoholic stimulation.

Savage ("Brit. Med. Journ.," May 4, 1890) says general paralysis is a degeneration "most commonly met with in middle-aged married men, inhabitants of cities, flesh eaters, and drinkers of alcohol." It frequently follows syphilis, especially if the disease has affected the higher nervous organs or their envelopes. Head injuries not uncommonly originate it, as do causes of nerve tissue change, such as lead. "It is not common among the congenitally deficient, or among epileptics."

In addition to several of the causes already mentioned, Voisin ("Traité de la Paralysie Générale des Aliénés," page 309 *et seq.*) gives: Bad moral hygiene; political events; intestine dissensions; domestic troubles; abuse of tobacco; dietetic abuses; pellagra; diabetes; epilepsy; simple insanity.

Gestational Insanity.	Generally after third, most frequently after sixth month of pregnancy (Clouston). Formed one per cent. of cases in Royal Edinburgh Asylum during nine years.
Hysterical Insanity.	Of hysteria, Ross ("Diseases of the Nervous System," p. 865) gives neurotic heredity; female sex; puberty; depressing passions, as fear, anxiety, jealousy, and remorse; exhaustion from overwork combined with anxiety; uterine derangements, structural or functional; imitation.

ETIOLOGY.

Forms of Insanity.	Causes.
(1,) IDIOCY AND IMBECILITY.	Neurotic heredity; advanced age of parents; consanguinity, especially when both parents are scrofulous or rachitic; acute alcoholism at time of conception; syphilis; diseases and emotions of the mother during pregnancy; difficult labour; maladroit obstetrical manipulations; accidental traumatisms; too forcible compression of the child's head; eruptive fevers in the child (Bra, p. 210).
(2,) CRETINISM.	Marriage of inhabitants of infected districts; consanguineous marriages; advanced age and bad constitution of parents; defective alimentation; misery; dirt; overcrowding; narrow valleys surrounded by high mountains; snow water; water of chalky districts, charged with carbonate or sulphate of lime; water containing too much magnesia, aluminum, or too little oxygen or iodine; miasmata. Although essentially endemic, it may be acquired by children sent to live in the infected districts (Bra, p. 223).
IMPULSIVE INSANITY.	Insane or neurotic heredity; critical periods of life; menstruation, pregnancy, and puerperium; acute moral suffering; overwork; sexual excess; alcoholic excesses on the part of patient or his parents.
KATATONIC INSANITY, OR KATATONIA.	Masturbatory excesses; disappointment in love (Spitzka, p. 150).
LACTATIONAL INSANITY.	Lactation in weak, overworked, and ill-fed women.
MANIA.	Heredity, similar or dissimilar; mental anxiety; adverse circumstances; overwork; alcoholic excess; sexual excess.
MASTURBATIONAL INSANITY.	Neurotic heredity; nervous temperament; masturbation.
MELANCHOLIA.	Heredity; puberty and the climacteric; female sex; certain occupations, viz., those of clergymen, artists, soldiers, doc-

Forms of Insanity.	Causes.
	tors, lawyers, journalists, actors, politicians; intellectual overwork; worry; adverse circumstances; faulty education; the puerperal state; masturbation; consanguineous marriages; civilisation; political, religious, or social commotions; solitary confinement; any debilitating causes.
METASTATIC INSANITY.	Metastasis; rheumatism, chronic ulcer, erysipelas, and other diseases; neurotic heredity and constitution.
MORAL INSANITY.	Any ordinary cause of insanity. Apoplectic or epileptic seizures; alcoholic excess; an attack of acute mania or melancholia (Blandford, "Quain's Dictionary," p. 727).
MYXŒDEMA, INSANITY OF.	Etiology of myxœdema (Ord, "Quain's Dict.," p. 1015); female sex; adult age.
NEURASTHENIC INSANITY.	Of neurasthenia cerebralis, mental overwork, especially in conjunction with emotional excitement. Of spinal neurasthenia, bodily overwork, severe illnesses, puerperia, sexual excesses, emotional strain. Of neurasthenia cordis (visceral neurasthenia), in asthenic individuals, emotional strain, too warm baths, excessive tobacco smoking. Of sexual neurasthenia, masturbation, sexual excesses, psychical onanism, gleet; in addition in women, puberty; the climacteric; incomplete or interrupted coitus; uterine tumours, infarcts, displacements, and erosions. The neurasthenic degenerative psychosis with fixed ideas is founded on a basis of neurasthenia which is sometimes, though rarely, *acquired* through mental overexertion, emotional strain, exhausting diseases, parturitions in quick succession, lactation, sexual excess, onanism; it is then curable (Krafft-Ebing).
OVARIAN OR OLD MAID'S INSANITY.	Age shortly before the climacteric.

Forms of Insanity.	Causes.
Oxaluria and Phosphaturia, Insanity of.	Middle age; oxalic diathesis; better class of society; too free indulgence in eating, especially in eating sweets.
Paralysis Agitans, Insanity of.	Long-continued paralysis agitans (Bra); paralysis agitans, etiology of; advanced life; male sex; exposure to cold and damp; great emotional disturbance; wounds and other injuries.
Partial Emotional Aberration.	Neurasthenia (see above).
Partial Exaltation, or Amenomania.	Neurotic heredity; exaggerated physiological emotions, pride, vanity, ambition, etc.; epidemic religious influences.
Pellagrous Insanity.	Pellagra frequently occurs in Italy. Of Pellagra; heredity; poverty; insufficient and improper food and clothing; malaria; unwholesome maize; sporisorium maydis. Exciting cause, sun's rays, especially "vernal insolation" (Erasmus Wilson, "Quain's Dictionary," p. 1103).
Periodical Insanity.	Neurotic heredity; cranial injuries; alcoholic excess; menstruation; puberty; climacteric.
Phthisical Insanity.	Latent phthisis (Sibbald). Insane heredity 7 per cent. more than in the insane generally (Clouston).
Podagrous or Gouty Insanity.	Gout, especially where there is strong gouty heredity (Clouston).
Post-Connubial Insanity.	Mental excitement and sexual excess.
Pubescent Insanity.	*Neurotic heredity always.* Irregularity in coming on of reproductive or menstrual function (Clouston, p. 525).
Puerperal Insanity.	Direct heredity; commonly old primiparæ; prolonged previous suckling; rapidly recurring pregnancies; twins; alcoholic excess during pregnancy; Ulceration or abscess of breast; eclampsia; pyæmia; chloroform. Grief; worry; anxiety; seduction; birth of natural children; desertion or

Forms of Insanity.	Causes.
	death of husband; loss of child (Savage, p. 372).
RHEUMATIC INSANITY.	Male sex; age 20–40; heredity; mental overwork; alcoholism; melancholy or nervous disposition. Hysteria, epilepsy, alcoholism, plumbism in the parents predispose; metastasis; hyperpyrexia; cardiopathy; therapeutic agents (Bra, p. 99.)
SENILE INSANITY.	Insane heredity; vascular diseases; previous mental overwork or disturbance of brain function. Generally over 60 years of age (Clouston, p. 566).
SOMNAMBULISM, PSEUDO-INSANITY OF.	Neurotic heredity; nervous temperament (Clouston, p. 608). Disorders of digestion.
STUPOR, ANERGIC (ACUTE DEMENTIA; ACUTE PRIMARY DEMENTIA; STUPOROUS INSANITY).	Adolescence; masturbation; excessive sexual intercourse, with or without mental and emotional exaltation; starvation; exhausting discharges; mental overwork during adolescence; mental and moral shocks; profuse hæmorrhage; alcoholic excess (Spitzka, p. 159; Clouston, p. 307).
SYPHILITIC INSANITY.	Syphilis; neurotic predisposition, most commonly hereditary; traumatism; mental fatigue; fright. Excesses may act as determining causes. Symptoms generally appear between the third and the tenth year of the disease, but may develop as early as the end of the secondary or beginning of the tertiary stage, or may be postponed for 30 years (Bra, p. 105). Cerebral symptoms may appear in a case of syphilis of any degree of severity, treated or not treated, and are more liable to appear in cases where the symptoms have been mild or only moderately severe (Fournier).
TOXIC INSANITY.	Neurotic heredity; alcohol; absinthe; opium or its alkaloids; chloral; cannabis indica; lead; mercury, etc. (See "Toxic Insanity," Chap. III.)

Forms of Insanity.	Causes.
TRAUMATIC INSANITY.	Blows on head; falls on head; other traumatic injuries to brain; sunstroke; radiant heat (Clouston, p. 414; Spitzka, p. 371).
UTERINE OR AMENORRHŒAL INSANITY.	Disordered or suspended menstruation.
YOUNG CHILDREN, DELIRIUM OF	Bodily ailments; pyrexia.

No two cases of insanity are exactly alike, and when endeavouring to discover the cause, or more correctly speaking, the causes (as there are generally several) of any given case, the past history of the patient and his immediate relatives (the anamnesia or anamnestic symptoms of the case) should, if possible, be ascertained. In the patient's history care should be taken to discriminate between causes and symptoms or prodromata. When enquiring about the relatives, it is best to ask what diseases they have suffered from or died of, and not to put the direct question whether they have ever suffered from insanity or any other nervous disease, or any chronic intoxication, or phthisis, etc. Although statistics only show heredity in about 20 per cent. of total admissions, 50 per cent. would be nearer the truth. Structure and environment bear an inverse ratio to each other in the causation of the insanities as of other diseases (Campbell). Given a perfectly healthy cerebral structure, only the most powerful causes will produce insanity, if it can be produced at all. Given, on the other hand, strong hereditary tendency with somatic stigmata (these are often accompanied by cerebral structural anomalies), as in some of the psychical degenerative forms, then a very slight disturbance will be sufficient to upset the unstable mental equilibrium and complete the causation.

CHAPTER V.
DIAGNOSIS.
A.—DIAGNOSIS OF INSANITY FROM OTHER CONDITIONS.

Of insanity from (1,) Eccentricity; (2,) Feigned insanity; (3,) The delirium of fevers and inflammations; (4,) Intoxication, alcoholic or other; (5,) Cerebral meningitis; and (6) Aphasia.

I.—ECCENTRICITY.

In mere eccentricity the intellectual faculties are not perverted, and, with the exception of the judgment, are not even defective. In one form of eccentricity the judgment is strong and there is an excess of individuality, with great independence of thought and action. In another or weak form of eccentricity (often premonitory to or a sequel of insanity), "The practical judgment is invariably weak; the character is marked by obstinacy or fickleness; unaccountable states of emotion often present themselves, but they are remarkable for their strangeness rather than their force. The perverted emotions of the eccentric man are feeble in comparison with those of the lunatic, and it is seldom that they result in offences against the law. The propensities of the eccentric man are normal, and his countenance, demeanour, and state of muscular activity are devoid of the signs of insanity" (Bucknill and Tuke, p. 444). The eccentric man of the latter type is influenced by the love of applause and the desire to attract attention, and his conduct is ill-regulated, vacillating, and capricious, varying with the emotion of the moment.

II.—FEIGNED INSANITY.

(1,) There is always a *motive* for feigning insanity, usually to escape punishment, often capital; but it may be to render some contract void; to obtain admission into an asylum, or to be enabled to remain in one; to escape from or shorten the term of military service.

(2,) There is no history of insanity previous to the commission of the crime.

(3,) The simulator always over-acts his part, especially if mania is chosen.

(4,) Acts are committed which are incompatible with the form of insanity assumed, *e.g.*, a melancholic eating filth, etc.

(5,) There is total inability to maintain motor excitement and loquacity for hours and days together without fatigue.
(6,) Inability to do without sleep.
(7,) The simulator alters his demeanour and conduct when he thinks he is unobserved.
(8,) The emotional state belonging to the form of insanity assumed is not (in some cases cannot be) correctly imitated.
(9,) When dementia is assumed the facial expression does not tally.
(10,) The simulator adopts suggestions thrown out in his hearing.
(11,) There is a pretended total loss of memory, though this occurs in but few forms of real insanity.
(12,) The simulator will sometimes say that he is suffering from "delusions," "hallucinations," "mania," "monomania," etc. An insane person never speaks in this way.
(13,) A dose of opium will act much more powerfully on the simulator than on the real lunatic.
(14,) Certain objective symptoms are absent, *e.g.*, frequent pulse, furred tongue, flushed or pallid countenance, injected conjunctivæ.

An insane person commits a crime without a motive, or with a trivial or outlandish motive. There is frequently no attempt at concealment: sometimes even boastfulness concerning the crime. Very often there is a history of insane or neurotic heredity, and of conversation and conduct indicating the existence of mental aberration before the commission of the crime.

III.—The Delirium of Fevers and Inflammations, (Especially Typhus and Pneumonia).

Fevers and inflammations may complicate insanity, and it may be consecutive to them. In the absence of the rash and the physical and other symptoms, sputum, pain, cough, etc., the elevated temperature distinguishes febrile and inflammatory delirium from all forms of insanity except pyretic delirium tremens; acute rheumatic insanity; some cases of puerperal insanity; the congestive phases of general paralysis; acute delirium (acute delirious mania).

(1,) Pyretic delirium tremens is distinguished by the character of the hallucinations (painful, mobile, and nocturnal); by the tremor; by the history of alcoholic excess.

(2,) Acute rheumatic insanity generally commences during an attack of acute rheumatism, most frequently from the fifth to the twentieth day.

(3,) Puerperal insanity is known by the conditions under which it occurs, by the noisiness, sleeplessness, destructiveness, dirtiness, jocularity, and erotic tendency; by the dislike of child, husband, and other relations; by the tendency to commit homicide and suicide; by the diminished or suppressed lacteal flow; by the suppressed, or altered and offensive, lochial discharge; by the enlarged, perhaps tender, uterus, and the patent rigid, irregular os, if the circumstance of parturition is concealed from the examiner.

(4,) The congestive phases of general paralysis are known by the previous history of the patient; by the somatic signs (lingual, labial, and facial tremor, pupillary inequality, cutaneous anæsthesia, etc.), and by the grandiose delusions.

(5,) Acute delirium (acute delirious mania). The prevalence of pneumonia and the presence of an epidemic of typhus or other fevers must be taken into account. The delirium of typhus is generally low and muttering, and the strength is prostrated from the first. (Typhoid, as a rule, develops slowly.) After a few days, in the absence of all physical signs, or of characteristic expectoration, or of an eruption, or of diarrhœa, headache, or vomiting, acute delirium may be diagnosed.

The temperature may rise in some cases of lactational insanity, or severe periodical mania, but it rarely exceeds 100°.

It should be borne in mind that insanity may complicate fevers or inflammation (*e.g.*, complicating delirium tremens, febrile consecutive insanity), and that fevers and inflammations may complicate insanity.

IV.—ALCOHOLIC OR OTHER INTOXICATION.

This is distinguished by its transitory nature, generally ending in sleep after a few hours. In the case of alcohol, there is the smell, vomiting, and the presence of alcohol in the gastric contents, obtained by means of an emetic or the stomach-pump if vomiting is absent; in opium or morphia intoxication there are the strongly-contracted pupils; in belladonna, the fixed widely-dilated pupils; in Indian hemp, the dreamy-delusional state and hallucinations of an agreeable nature. Early insanity may be complicated by any of these intoxications.

V.—CEREBRAL MENINGITIS.

This is distinguished by the acute headache, aggravated at intervals; the persistent vomiting, especially immediately after taking food; the intolerance of light and sound; the contracted pupils; the hard pulse.

VI.—Aphasia.

Aphasia is generally accompanied by paralysis, which is usually confined to the right arm and the right side of the face; the patient understands what is said to him (with an exception to be mentioned afterwards), and he is able to reply by gestures, or (unless he is agraphic) in writing; when he does not understand what is said to him (word-deafness) he understands writing, unless he is word-blind, and then he understands gestures, and reacts in a sane manner.

In amnesic aphasia the patient knows when he has miscalled any object, and immediately says "yes," with a pleased expression, on the correct name being mentioned, and he repeats the name, although quite unable to recall it, when shown the object a few minutes afterwards. An insane female patient in Norfolk Asylum, in 1873, afforded a typical instance of this form of aphasia.

In all forms of mere aphasia the judgment is sound and the actions are well regulated. Aphasia may, however, complicate insanity, and the latter may supervene on the former.

It should be stated here that Clouston says he has never observed complete aphasia along with perfect mental sanity, and he is inclined to believe that the two are incompatible. Max Müller ("Science of Thought") adduces strong arguments in favour of his view that thought is impossible without language, or language without thought, that thought is impossible without words or the signs replacing them, that, in fact, language is thought. Darwin ("The Descent of Man," p. 89) referring to this aphorism says, "What a strange definition must here be given to the word thought!" However, he admits (*op. cit.*, p. 88) that a complex train of thought *absolutely requires words*, as long calculations do figures or letters; and that an ordinary train of thought almost requires, or is greatly facilitated by words. Ireland ("The Blot Upon the Brain," p. 272) says, "Speech is, if not an indispensable method of arriving at any high mental endowment, at least too direct and obvious a one ever to be dispensed with."

In acquired insanity there is always an alteration in the disposition, demeanour, habits, and conduct of the individual affected. This should be minutely inquired into. There is very frequently a history of heredity, sometimes of intemperance, cranial injuries, masturbation, or sexual excesses. There may be a more immediate history of overwork, domestic troubles, business anxieties, adversity, disappointments, injuries, illnesses, etc. Age and sex should be taken into account. As to the symptoms, the patient may evince emotional exaltation or depression; extreme loquacity, or reticence amounting to absolute mutism; loss of power of attention; loss of memory; loss of power of calculation; diminished will power; weakened judgment; morbid or exaggerated physiological impulses; partial or complete loss of consciousness; stupor; illusions; hallucinations; delusions.

Method of Examining a Patient.

Having learnt as much as possible (including the hereditary and personal history or anamnesia) of the patient from the relatives or friends, or, if necessary, from his or her acquaintances and neighbours, the medical man should be introduced to him or her as a doctor, and not, under any circumstances, as anything else. The patient can be examined physically, and whilst investigating the state of the tongue, pulse, temperature, bowels, urine, appetite, cardiac sounds, pulmonary sounds, and inquiring about pain, sleep, etc., etc., the doctor can generally form some idea as to the patient's temper, memory, and emotional state (simple melancholia—acute mania), and sometimes as to his delusions (in hypochondriacal melancholia, general paralysis, delusional mania, delusional melancholia, acute mania, puerperal insanity, chronic mania, dementia with delusions). At the same time the dress, attitude, actions, demeanour, gestures, action of facial and oral muscles, facial expression, and appearance of eyes may be noticed, and when the patient is absolutely mute, or when he refuses to speak to the doctor, these are the principal and often the only symptoms to constitute the statutory "facts observed by myself" (anergic stupor, stuporous melancholia, and some cases of delusional insanity; see "Differential Diagnosis," post). In these mute cases food is generally refused, or only swallowed if put well back into the mouth. This can be verified by the medical practitioner. Inquire as to habits, whether "wet," or "dirty," or both. Inquire as to sleep, rest at night, impulsive violence in word or deed.

Where the patient converses, or even answers questions, and it has been found in the preliminary physical examination that his temper is facile and his intellect more or less enfeebled, the extent of this enfeeblement may be ascertained by some simple questions, requiring the patient to name the year, the season of the year, the month, the day of the week; asking him where he is and how long he has been in such place, his age, the year in which he was born, his occupation, and how long he has been so employed, the number and names of his or her children, brothers and sisters, etc., etc. Where from these questions it is inferred that the power of calculation is weak, this may be still more accurately tested by a few easy arithmetical questions.

The defective memory and the weakness of the power of calculation suggest some form of dementia, idiocy or imbecility, or general paralysis. Therefore, where these mental symptoms are present, look carefully for somatic stigmata and symptoms; microcephalus, hydrocephalus, vaulted or cleft palate, decayed teeth (idiocy and imbecility); hemiplegia contractures, aphasia, hemi-anæsthesia (organic dementia); tremulous, stammering, or stuttering speech; lingual, labial, and facial fibrillary tremor; slight paralysis; cutaneous anæsthesia; tripping-stumbling gait; awkwardness in turning (general paralysis).

When the patient's memory is good, propounding simple questions, such as those indicated, is liable to cause irritation and prevent the delusions being ascertained at first hand.

In such cases the conversation may be carefully led from the patient's physical ailments to such topics as his friends or relations, his enemies, his occupation and employer, his social position, his past life, his worldly prospects, the future state. In this way most delusions may be brought out, and even cunning chronic patients will generally, by their manner of answering questions, betray the fact that they still retain their delusions. Hallucinations should be sought for. Their presence will sometimes be betrayed by the facial expression, gestures, or actions of the patient. Voice-hearing patients often assume an attitude of listening, or turn their eyes or head suddenly round without any cause apparent to the onlooker, or even reply aloud to the voice or voices. These "voices" are frequently found in persecutory or suspicious delusional insanity, and are often the danger-signal of violence, homicide, or suicide. Patients suffering from visual hallucinations often suddenly look at a blank wall or empty corner with a surprised, frightened, angry, or amused expression; this is often observed in delirium tremens and other forms of alcoholic mental derangement. Hallucinations and illusions of taste and smell are generally disagreeable, and cause the patient to assume a facial expression of disgust (delusional insanity, masturbatory insanity, delusional melancholia). Hallucinations and illusions of the general sensibility, formication, itching, pricking, stinging, blows, etc., may be betrayed by the patient rubbing or scratching the part, no vermin, skin disease, or bruises being present (delusional insanity, general paralysis, melancholia, chronic mania). If possible, examine fundus oculi, test grip or hand-grasp, on both sides, with dynamometer, and investigate the condition of the special senses as to their acuteness, dulness, etc.

To Investigate the Acuteness, etc., of the various Senses.

(1,) The *cutaneous sensibility* is best tested according to Gowers' suggestion, by means of a quill, the sharp end being used for the sense of pain, the feathered end for that of touch. The æsthesiometer may be applicable in a few cases.

(2,) The sense of *smell* may be tested by means of such simple well-known articles as pepper, tobacco, onions, eau-de-cologne, the patient's eyes being kept closed during the examination, and each nostril tried separately. For testing this sense the French have invented a "boîte aux odeurs."

(3,) *Taste* may be investigated by means of camel-hair brushes dipped in solutions of sugar, salt, citric acid, and quinine, both sides of the tongue being separately tested.

(4,) The sense of *hearing* may be fairly well tested by means of a watch held at various distances, commencing with actual contact at the vertex of the head, or behind the ear, and withdrawing it gradually until the ticking is no longer heard, this distance being noted for each ear. Musical notes, high and low, may also be used.

(5,) *Vision* may be roughly tested (one eye at a time) by means of the ordinary tests: types at various distances for acuteness; by coloured wools for the colour sense; and by the hand of the examiner held above, below, and at each side of the eye under examination, at various distances, the eye being fixed on that of the examiner, for the field of vision.

(6,) The sense of *muscular resistance* ought to be tested.

(7,) The *electric sensibility* should be tested.

(8,) The state of the *muscular sense* may be ascertained by placing the limb in any position and asking the patient (whose eyes are closed) to put the other limb in a similar posture.

(9,) The *cutaneous thermic sensibility* should be tested.

The *pupils* should be examined as to their size, equality, regularity, light reflex, skin reflex, and accommodation reaction.

The *reflexes*, superficial and deep, especially the latter (ankle clonus and knee-jerk), ought to be carefully investigated, and their presence or absence, extent and duration, noted.

The *reactions* of the muscular and nervous apparatus to faradism and galvanism are to be ascertained. When there is paralysis the presence of R. D. is important.

The patient's *height* and *weight* should be noted, and his *head* measured as to the length of its antero-posterior and transverse curves, its horizontal circumference (most important), its diameters antero-posterior and transverse; by multiplying the latter diameter by 100 and dividing by the former the *cephalic index* is obtained (Morselli). The condition of the *general sensibility*, of the *sexual organs*, and of the *masticatory apparatus* should be ascertained. In mental diseases accompanied by hypokinesis, tremor, or ataxy, Morselli's *dynamograph* gives useful diagnostic indications. The patient's *"reaction time"* should when possible be ascertained. According to H. Münsterberg ("Beiträge zur experimentellen Psychologie") coffee and (when taken slowly in small sips) alcohol shorten reaction time. Alcohol in larger quantity delays it, though to the subject of experiment it always appears to hasten it. It is delayed also by fatigue, epilepsy, and mental disease ("Brit. Med. Journ.," July 11, 1891, p. 81). The *blood* should be examined chemically as well as by means of Gowers' hæmoglobinometer and hæmacytometer, and the microscope.

DESCRIPTION OF MENTAL AND BODILY CONDITION ON ADMISSION.

The Commissioners in Lunacy direct that the following particulars should be entered in the Medical "Case Book" kept in every asylum, licensed house, and lunatic hospital. This will be a guide to the first entry after admission, to be made in a blank page at the beginning of the Medical Visitation Book required to be kept where there is a single patient. It will also be a guide as to the method of "taking" a mental case.

"1st.—A statement to be entered of the name, age, sex, and previous occupation of the patient, and whether married, single, or widowed.

"2nd.—An accurate description to be given of the external appearance of the patient upon admission; of the habit of body and temperament, and appearance of eyes, expression of countenance, and any peculiarity in form of head; physical state of the vascular and respiratory organs, and of the abdominal viscera, and their respective functions; state of the pulse, tongue, skin, etc.; and the presence or absence on admission of bruises or other injuries, to be noted.

"3rd.—A description to be given of the phenomena of mental disorder; the manner and period of the attack; with a minute account of the symptoms and the changes produced in the patient's temper or disposition, specifying whether the malady displays itself by any, and what, illusions, or irrational conduct, or morbid, or dangerous habits or propensities; whether it has occasioned any failure of memory or understanding, or is connected with epilepsy, or ordinary paralysis, or symptoms of general paralysis, such as tremulous movements of the tongue, defect of articulation, or weakness or unsteadiness of gait.

"4th.—Every particular to be entered which can be obtained respecting the previous history of the patient; what are believed to have been the predisposing and exciting causes of the attack; what the previous habits, active or sedentary, temperate or otherwise; whether the patient has experienced any former attacks, and.

if so, at what periods; whether any relatives have been subject to insanity; and whether the present attack has been preceded by any premonitory symptoms, such as restlessness, unusual elevation or depression of spirits, or any remarkable deviation from ordinary habits and conduct; and whether the patient has undergone any, and what, previous treatment, or has been subjected to personal restraint."

B.—DIFFERENTIAL DIAGNOSIS OF THE FORMS OF INSANITY.

The operations of the nervous centres obtain expression immediately or remotely through the medium of the muscular system. The methods by which they do so are various, *e.g.*, facial expression; gestures; attitudes; actions, constituting habits and conduct; language, written or spoken. The latter (spoken language or speech), as by means of it we most readily, accurately, and fully gauge the thoughts and feelings (including their extent, aberration, or absence), especially in the insane, will be here used as the leading test in formulating a rough scheme of differential diagnosis.

In making a diagnosis the more common forms should be borne in mind: statistics of their relative frequency will be found at the end of this chapter.

The possibility of the co-existence of one or more forms should also be remembered. One or more symptoms of any form may be slightly marked or absent.

When the patient is appropriately questioned, he either answers the questions or he does not (either through incapacity or unwillingness, mostly the former).

I.—THE PATIENT ANSWERS QUESTIONS.

Under this head will come:—Delusional insanity (monomania, paranoia); impulsive insanity; moral insanity; folie raisonnante; simple mania; most cases of hysterical insanity; phthisical insanity; insanity with paralysis agitans; mild melancholia; mania in all its forms (not, of course, including acute delirium); neurasthenia and most cases of neurasthenic insanity; masturbational insanity; imbeciles and higher class of idiots; gastro-enteric insanity; insanity of abdominal diseases; insanity of Bright's disease; anæmic insanity; insanity from deprivation of senses; partial emotional aberration; partial exaltation; folie du doute; gouty insanity; general paralysis, except near termination; ovarian or old maid's insanity. Most cases of acute, religious, delusional, and suicidal melancholia; pellagrous insanity; diabetic insanity; climacteric insanity; delirium tremens and acute opium and morphia insanity; katatonia, except cataleptic phases; hypochondriacal melancholia (hypochondriacal insanity); insanity of phosphaturia and oxaluria; insanity from cannabis indica; gestational, lactational, and puerperal insanity; some cases of delirium of young children; traumatic insanity; post-connubial insanity; periodical insanity; circular insanity; pubescent and adolescent insanity; choreic insanity; insanity with exophthalmic goître; primary mental deterioration; chronic alcoholic insanity; alcoholic dementia; alcoholic dementia with paraplegia (mostly occurring in females); senile insanity; insanity from coarse brain disease; alcoholic, syphilitic, and saturnine pseudo-general paralysis; epileptic dementia; insanity of myxœdema; some cases of rheumatic insanity; insanity from chloral hydrate; saturnine insanity; primary and secondary confusional insanity.

1.

The patient answers questions and can sustain a conversation. The memory is not very defective, and the power of calculation is little, if at all, diminished.

A.—There are systematised, fixed, and permanent delusions, and no marked emotional aberration.

The forms of insanity coming under this head are *delusional insanity* (monomania, paranoia); *phthisical insanity*; some cases of insanity with *paralysis agitans*; some cases of *gouty insanity*; some cases of *chronic morphismus*; *ovarian* or *old maid's insanity*; and some cases of *consecutive insanity*.

In Phthisical Insanity there are delusions of suspicion. Physically, there are pallor, sallowness, emaciation, and frequently thoracic signs and tuberculous heredity. Sometimes the signs of phthisis, or frequent pulse and fever, with evening exacerbations.

In some cases of Insanity of Paralysis Agitans there are delusions of suspicion and persecution. It may be distinguished from monomania by the presence of the tremors, which occur in advanced life (over 40), continue during rest, and do not affect the head.

Cases of Gouty Insanity resembling persecutory monomania may be diagnosed by the existence of the gouty diathesis, the history of gout, and, as a rule, by the gouty heredity. Advanced life. Male sex.

The persecutory delusional insanity of Chronic Morphismus may be distinguished by the history of prolonged and persistent morphia intoxication. It is accompanied by auditory hallucinations and sensations of "galvanic shocks."

Ovarian, or Old Maid's Insanity resembles erotomania, or delusional insanity with erotic delusions, and may be included in that form of insanity, but it always breaks out shortly before the climacteric in a person who has been previously exceedingly modest and discreet, some clergyman being generally the object of affection.

In delusional insanity, or monomania, the delusions may refer to :—

(1,) Persecution or suspicion. This form has to be distinguished from those just mentioned and from melancholia with delusions of persecution. In the latter there are the inability to sustain a conversation and the overwhelming emotional depression. In monomania (paranoia) there are frequently somatic stigmata, such as cranial anomalies, etc., and there is generally a strong neurotic heredity. The delusions of neurasthenic paranoia arise out of the neurasthenic sensations.

(2,) Pride or ambition.

(3,) Religion. These forms (2 and 3) may be differentiated from partial exaltation by the emotional exaltation in the latter being more prominent than the delusions.

(4,) Jealousy. In alcoholic jealousy there is an alcoholic history. There are delusions of mutilation of the sexual organs and of poisoning (Spitzka).

(5,) Erotism (erotomania). To be distinguished from nymphomania and satyriasis, which are morbid impulses, not necessarily connected with delusions. To be distinguished also from the mere strong erotic tendency of acute mania, early general paralysis, etc.

B.—There are unsystematised, fixed, and permanent delusions, motor excitement, and, at times, emotional exaltation. The mental condition is secondary to a more acute one. There is a constant and gradually increasing tendency towards dementia. This includes the better and earlier cases of *chronic mania* and some cases of *chronic hysterical insanity*.

C.—There are morbid impulses without delusions and without any continuous morbid emotional state. This group constitutes *impulsive insanity*, including destructive mania, dipsomania, homicidal mania, kleptomania, lycanthropia, necrophilism, nymphomania, planomania, pyromania, satyriasis, suicidal mania.

As forms of insanity these should be distinguished from the mere symptomatic actions of mania, melancholia, etc. As forms of insanity they exist without any other apparent symptoms.

D.—Change of habits, disposition, and demeanour. The committal of vicious and cruel deeds : viciousness and wanton cruelty are also sometimes congenital. There are no delusions or hallucinations. The disease is moral insanity: distinguished from monomania by the absence of delusions; from impulsive insanity by the nature of the acts, and by their being committed when the opportunity rises, and not when the patient is seized with a sudden impulse; from simple mania by the absence of exaltation, and by not being preceded or succeeded by acute mania.

E.—Change of habits, with exaltation and ill-regulated conduct. There are no delusions or hallucinations. Sometimes precedes, sometimes follows acute mania; occasionally remains unaltered many years, having commenced as a primary affection. The form of insanity is *simple mania*. For differential diagnosis, see preceding paragraphs. Some cases of *chronic hysterical insanity* and some cases of *gouty insanity* resemble simple mania.

F.—Change of habits, etc., with depression. There are no delusions or hallucinations.

(*a*,) Without any essential bodily defect or disorder. *Mild melancholia; mild chloral insanity;* some cases of *pubescent* and *adolescent* insanity; *neurasthenia*, distinguished by tremor, nosophobia, feeling of prostration, reactivity, etc.

(*b*,) With disease of abdominal organs. *Gastro-enteric insanity; insanity* of *abdominal diseases;* and early stage of *insanity* of *Bright's disease*.

(*c*,) With anæmia. Most cases of *anæmic insanity*. In some cases of anæmic insanity there is slight stupor; in others, an alternating maniacal condition.

(*d*,) With catalepsy. Mild cases of *cataleptic insanity*.

(*e*,) With deprivation of the senses. Mild cases of *insanity* from *deprivation* of the *senses*.

(*f*,) Some cases of *podagrous* or *gouty insanity*. The bodily disorders and defects must be the cause and not the effect, or mere concomitant of the emotional depression.

(*g*,) Alternating with periods of *exaltation*. Mild cases of *folie circulaire*. According to Falret persons may suffer from a mild form unknown to their *entourage*.

G.—There is congenital abulia (weakness of will) combined with abnormally strong propensities. As a rule there are somatic stigmata. The patient is *weak-minded*, or one of the higher class of *imbeciles*.

H.—There is depression, preceded by restlessness, and succeeded by acute maniacal excitement, and occurring in youth. There are disagreeable hallucinations of smell and sometimes of taste, these symptoms supervening on the habit of masturbation. *Masturbational insanity* in its early stages.

I.—The patient is given to doubting, and is perplexed and vacillating. He is introspective, probably apprehensive, and frequently asks the same questions. He is gloomy, dreamy, and punctilious, and pays extreme attention to all his words and acts. The patient suffers from *folie du doute* (*doubting* insanity, doubting mania). These patients may be metaphysicians (sufferers from grübelsucht), realists, scrupulous, timid, counters, or touchers.

J.—There is morbid feeling, usually fear or apprehension, some abulia (weakness of will), and no delusions or hallucinations. The patient is suffering from *partial emotional aberration*, some of the forms of which are agoraphobia, claustrophobia, mysophobia.

K.—There is a morbid state of emotional exaltation. This is partial, and is one of pride, vanity, or religious feeling. There may be delusions, but they are secondary to, and less prominent than, the emotional state. The patient suffers from *partial exaltation* or *amenomania*.

L.—There are unsystematised, fixed, and permanent delusions, mostly resembling those of melancholia, sometimes those of mania. There is an absence of emotional exaltation and motor excitement. The state is secondary to melancholia, mania, or primary confusional insanity, and there is a constant and gradually increasing tendency towards dementia. It is *chronic confusional insanity* (Spitzka); generally included in chronic mania by other alienists.

M.—There are change of habits, readily induced muscular fatigue, increased susceptibility to the effects of alcohol, fine fibrillary tremor of tongue, and, perhaps, of lips and face, various neuralgiæ, some depression, sometimes pupillary inequality. *Prodromal stage* of *general paralysis*; *remission* of *general paralysis*; in latter, often mendacity and irritability, and there may be slight exaltation instead of depression.

N.—The patient may be able to sustain a conversation, and be free from delusions and hallucinations, but with a history of attacks of exaltation, depression, alternating exaltation and depression with short intervening lucid intervals, or epilepsy or epileptic vertigo with motor excitement and unconsciousness or semi-consciousness. He is in one of the lucid intervals of *periodical insanity, folie alternante, folie à double forme,* or *epileptic insanity*.

O.—There are perversions, *délire des actes*, etc., yet the patient is able to talk sensibly and to excuse his foolish acts with shrewdness. There is marked periodicity. There are degenerative signs. *Maniacal folie raisonnante* (mild periodical mania).

P.—There are depression, ill-humour, quarrelsomeness, aggravated at the menstrual periods. *Melancholy folie raisonnante.* Most of these cases are neurasthenic and degenerative. (For signs of degenerative taint, see "Periodical Insanity," Chap. III.)

Q.—The patient has no initiative, no originating power, no active desires, no power of self guidance or resisting capacity; or he may have complete loss of memory of recent events, with good memory of long past events. He is suffering from *partial dementia*, succeeding mania, or melancholia. Some of these cases belong rather to the next group.

2.

The patient answers questions, but cannot sustain a conversation. The memory is markedly defective, the power of calculation is much diminished, and the ideas of time and locality are vague.

A.—Absent-mindedness, general inertia, insomnia, forgetfulness of recent, and afterwards, if recovery does not take place, of both recent and long past events; the symptoms occurring in *young* or *middle-aged persons* as a *primary affection*. The patient suffers from *primary mental deterioration* (*simple* primary dementia), the primary partial dementia of Savage. In some cases of *consecutive* insanity the symptoms are similar to those of primary mental deterioration. Mild cases and insipient ones belong rather to group I., 1.

B.—History of prolonged addiction to drink, tremors, painful, mobile, nocturnal hallucinations, or (especially in women) weakness of lower extremities. *Chronic alcoholic insanity, alcoholic dementia, alcoholic dementia with paraplegia* (last form occurs mostly in women).

C.—Occurring as a primary affection in middle or old age, with a long past history of syphilis. With or without incoherence. *Depressive syphilitic insanity.*

D.—Occurring as a terminal affection, succeeding mania, melancholia, etc. Old faces and long past events are remembered better than new or recent ones during the earlier stages of the affection, but all memory is finally lost. The form of insanity is *terminal dementia.*

E.—Occurring in old age (generally after 60), and being accompanied by egotism, ideas of suspicion, parsimony, senile tremor, etc. The loss of memory of recent events, side by side with a remarkable recollection of long past occurrences, is most marked in this form of insanity, especially in the early stages; later on the memory ceases to recall even long past events. *Senile dementia* is the form of insanity.

F.—The mental state frequently follows a paralytic or apoplectic attack. It is generally accompanied by paralysis, or some muscular weakness, and sometimes by aphasia (organic dementia). Occasionally the paralytic or apoplectic seizure is preceded and followed by depression, the motor symptoms supervening after the seizure (organic melancholia). In insanity from coarse brain disease the speech is thick, but equally so from beginning to end of a sentence, whilst the defect increases at the end in general paralysis. Speech is not accompanied by labial and facial tremor as in general paralysis. In the form caused by tumour (which much resembles general paralysis) there is very intense cephalalgia. Insanity from *coarse brain disease (organic dementia, organic melancholia)*.

G.—Tremor of tongue, lips, and face; speech thick, stammering, or stuttering; defective gait; cutaneous anæsthesia; pupils frequently unequal; very often grandiose delusions. The defect of memory applies both to long past and recent events, and increases, as does the defect in the power of calculation and the vagueness of the ideas of time and space. Found in patients aged from 20 to 60, especially 35 to 45. Under this head come *confirmed general paralysis, alcoholic pseudo-general paralysis, syphilitic pseudo-general paralysis,* and *saturnine pseudo-general paralysis*. *Alcoholic pseudo-general paralysis* is distinguished by the *anorexia*, by the tremor being general and massive, by the paralysis commencing at the extremities of the members, by the history and symptoms of alcoholic intoxication, by the visual hallucinations, and, finally, by the tendency to amelioration, which soon shows itself.

Syphilitic pseudo-general paralysis is distinguished by the mental symptoms frequently following one or several apoplectiform or epileptiform attacks; by the frequent absence of tremor; by the paralysis being more accentuated than in general paralysis, and partial paralysis, such as ptosis, etc., being more frequent; by the evolution being less regular than in general paralysis, by the frequent presence of a very pronounced syphilitic cachexia, by the mental state being rather brutish or stupid (though sometimes extravagant) than expansive, and finally by the possibility of cure.

Saturnine pseudo-general paralysis is distinguished by the symptoms of saturnine intoxication, blue line on gums, colic, cramps, earthy hue of skin, cephalalgia, etc., etc.; by the presence of insomnia and nightmare, by the visual hallucinations and ideas of persecution and poisoning, by the dementia manifesting itself suddenly in its greatest intensity, and by the tendency to amelioration making itself rapidly felt.

H.—Succeeding repeated attacks of acute epileptic insanity, or as a consequence of epileptic fits continued through a succession of years. The patient is childish and addicted to onanism. *Chronic epileptic insanity* or *epileptic dementia.*

I.—Memory, power of calculation, facility for acquiring new facts and accomplishments below normal average from childhood. Noticing, speaking, walking below normal average from infancy. Hand reflex absent in early infancy. There are somatic stigmata, and generally insane, neurotic, or alcoholic heredity. There may be delusions. There may be strong propensities, and a tendency to emotional excitement on slight provocation. *Imbecility* and higher cases of *idiocy.*

J.—There are motor excitement and loquacity with absence of emotion. The patient is childish. The state is a terminal one, mania being, as a rule, the primary form of insanity. *Agitated terminal dementia.*

K.—Following depression, agitation, delusions, and peculiar pathos with cataleptic phases. The third or demented stage of *katatonia.*

L.—Solitary, unsocial, and impulsive. Succeeding acute maniacal excitement. Olfactory hallucinations occurring in youth. Final stage of *Masturbational insanity.*

M.—There are delusions of persecution and physical signs of myxœdema. *Insanity* of *myxœdema.*

N.—With acute maniacal excitement, *visual* hallucinations and delusions of suspicion. Occurring usually at the end of an attack of acute rheumatism, or during convalescence. A few cases of *rheumatic insanity* (properly so called).

O.—Maniacal exaltation, with insomnia, nocturnal motor restlessness, great egotism, suspiciousness. Occurring in old age. *Senile mania.*

P.—Acute depression, with egotism, delusions of suspicion of an ever-varying yet possible nature (things being stolen, etc., etc.), and insomnia. Occurring in old age. *Senile Melancholia.*

Q.—Depression, with or without fits of excitement; visual hallucinations. At other times mild or furious delirium. There are, or have been, blue line on gums, colic, constipation, cephalalgia, frequent pulse, and insomnia. No pyrexia. Most cases of *saturnine insanity.*

R.—There are unsystematised fixed delusions, with motor excitement, and episodical emotional exaltation. The state is secondary to acute mania, and the delusions resemble those found in that form of insanity, but finally become a mere parrot-cry. *Advanced chronic mania* and some cases of *chronic hysterical insanity*.

S.—There are unsystematised fixed delusions, generally resembling those of melancholia, sometimes those of mania. There is frequently a change in the sense of identity, owing to the failure of memory and logical power; not any motor excitement or emotional exaltation. The affection is secondary to melancholia, mania, or primary confusional insanity. *Advanced chronic confusional insanity* (generally included under head of advanced chronic mania).

T.—Depression, with suicidal tendency. *Post connubial insanity*.

U.—The patient cannot fix the attention, and the ideas are mobile and futile. Occurring in individuals suffering from chorea. *Mild choreic insanity*.

V.—Visual and auditory hallucinations are frequent, especially the former, and when maniacal excitement is added we have *severe choreic insanity*.

3.

The patient answers questions, but answers them irrelevantly, confusedly, or evasively, and cannot sustain a conversation.

A.—The affection is primary, and the patient suffers from hallucinations and delusions (often of identity). He speaks incoherently and irrelevantly, and does not finish his sentences. There is an absence of emotion, but at first the memory is good, the patient being able to relate circumstances (though in a piecemeal, halting manner) occurring immediately before the attack; afterwards he speaks of himself in the third person, or manifests a confused double consciousness. The affection is eminently curable. It is *primary* or *acute confusional insanity*.

B.—The patient suffers from a form of stupor, with fits of agitation. There are automatic movements and cries. The patient, when urgently questioned, answers confusedly and evasively, and falls again into a state of torpor. There are blue line on gums, colic, etc. It is the *comatose* form of *saturnine insanity*.

C.—There are tremor, subnormal temperature, anxiety, dulness of perception, blurring of consciousness, isolated delusions giving rise to dreamy, odd actions. The psychosis is founded on a basis of acquired neurasthenia, generally cerebral. *Transitory neurasthenic psychoneurosis* (some cases).

4.

The patient answers questions, but cannot sustain a conversation. There is marked emotional depression or exaltation.

A.—There is emotional exaltation, with motor excitement and incoherence, with or without delusions. *Acute mania, maniacal* phase of *circular insanity, periodical mania, pubescent* and *adolescent insanity,* fully developed *masturbational insanity, expansive syphilitic insanity, puerperal mania ;* most cases of *acute hysterical insanity.* Insanity with *exophthalmic goître,* fully developed *insanity of Bright's disease,* nearly one third of the cases of *uterine* or *amenorrhœal insanity,* some cases of *consecutive insanity,* maniacal phase of *general paralysis.* During the course of the latter the patient is facile and childish ; it lasts from 10 to 30 days, and leaves the somatic signs of general paralysis. According to Rabbas the early loss of reading power in general paralysis is, like the alteration in writing, of great use in making a differential diagnosis. A few cases of *gestational insanity.* Some cases of *metastatic insanity.* These can generally be distinguished from each other by the history of the case, the sex and age of the patient, the circumstances under which the attack has occurred, the presence or absence of some neurotic or other disease, or of some physiological condition, or characteristic symptoms; *e.g.,* circular and periodical cases may be distinguished from the others by the coherent loquacity, the greater frequency of illusions than hallucinations, the rarity or absence of delusions; from each other by the anamnesia, by the *méchanceté* of the former and the perversions of the latter; periodical mania with long lucid or sub-lucid intervals is usually severe and marked by some elevation of temperature. (See Chap. III.)

B.—There is extreme emotional depression, with self-abasement, with more or less suicidal tendency, with or without delusions, and without motor excitement. *Acute melancholia, chronic melancholia,* melancholy phase of *folie circulaire* in some cases, *diabetic insanity,* some cases of *consecutive insanity,* some cases of *insanity* of *paralysis agitans, severe chloral insanity,* some cases of *metastatic insanity.* There are morbid impulses in severe chloral insanity. *Pellagrous insanity.* Some cases of *acute hysterical insanity.* Several of these forms are distinguished by the diseases with which they are associated.

C.—Emotional depression with predominent suicidal tendency. *Suicidal melancholia.* Some cases of *periodical melancholia.*

D.—Depression, with suicidal tendency and extreme insomnia, with or without delusions. Occurring in women from 40 to 50 years of age, and occasionally in men from 55 to 65. It succeeds a period of shyness and various nervous and dyspeptic symptoms and menstrual irregularities. It is *climacteric insanity*.

E.—Emotional depression, tending towards an irritable and sometimes dangerous and impulsive dementia or delusional insanity. There are cephalalgia, vertigo, hallucinations, etc. There are also motor or sensory symptoms. Alcohol, for which there is a craving, causes violent outbursts of excitement. *Traumatic insanity*.

F.—There are hallucinations, painful, mobile, nocturnal, the patient seeing animals, devils, etc. There are muscular tremors of varying intensity, increased by the withdrawal of alcohol. There may be pyrexia, and in that case the symptoms resemble those of acute delirious mania (acute delirium), but in acute delirium the patient cannot be got to answer questions. The patient is suffering from *delirium tremens*. The symptoms of *acute opium* and *morphia insanity* are somewhat similar, but the tremor is increased by the withdrawal of opium or morphia respectively.

G.—There are delusions and hallucinations referring to the patient's bodily organs. There is not necessarily any tremor. There is no self-abasement and no suicidal tendency. *Hypochondriacal melancholia* (hypochondriacal insanity). In *insanity* of *oxaluria* and *phosphaturia* oxalates and phosphates are found in the urine, and there is an irritable, nervous, or apprehensive state, with despondency, but no actual delusions or hallucinations.

H.—There are extreme emotional depression and self-abasement. There are predominating and painful delusions. There is often strong suicidal tendency. *Delusional melancholia*, most cases of *gestational* and *lactational insanity*, *puerperal melancholia*, severe insanity from *deprivation* of the *senses*, some cases of *syphilitic depression* with *incoherence*; nearly two-thirds of the cases of *uterine* or *amenorrhœal insanity*. In the latter there are disagreeable auditory hallucinations.

I.—There are great emotional depression, agitation, and motor excitement. *Excited* or *agitated melancholia*; some cases of *insanity* of *paralysis agitans*; some cases of *delirium* of *young children*.

J.—There is great emotional depression, and the predominating delusions refer to matters connected with religion. *Religious melancholia*.

K.—At first there is depression, with or without delusions and hallucinations, generally of a painful or persecutory nature. Then there are atony, peculiar excitement, confusion, and depression, combined with a peculiar pathos, and alternating almost cyclically. There is verbigeration, and in a large number of cases there are spasmodic conditions during the first stage. The first and second stages of *katatonia*.

L.—There are emotional exaltation, motor excitement, and predominating delusions and hallucinations. *Delusional mania.* Insanity from *cannabis indica* (Gunjah, Bang, Hachish).

M.—Hampered mental action rather than psychical pain; nosophobia in early stages; great self-depreciation in the developed disease; olfactory hallucinations; suicidal attempts; in severe forms, filthy habits, fixed ideas, and religious delusions. The disease develops on a basis of sexual neurasthenia. It is *masturbatory melancholia*, one of the neurasthenic psychoneuroses.

II.—The Patient does not Answer Questions.

1.

The patient, though he cannot be made to answer questions, or to understand what is said to him, is very loquacious, and his language is utterly incoherent.

A.—There are pyrexia, furred tongue, weak frequent pulse, and disturbances of speech and locomotion, the bodily organs being healthy. *Acute delirium* (acute delirious mania). A few cases of *uterine* or *amenorrhœal insanity;* some cases of *delirium* of *young children.*

B.—The patient is very loquacious, his language is incoherent, there is no pyrexia, and the attack is of extremely short duration. *Transitory mania (transitory frenzy).*

C.—The patient is very loquacious and talks incoherently. The attack follows immediately the imbibition of alcohol, and is of short duration. There is no pyrexia. *Mania a potu.*

D.—The patient is very loquacious, talks incoherently, and frequently repeats the same phrases. There are motor agitation and a blindly impulsive tendency to violence, homicide, or suicide. The attack occurs in connection with epilepsy.—*Acute epileptic insanity;* attacks of excitement in *chronic epileptic insanity.*

E.—There are frightful hallucinations, and the attack is of short duration, and terminates abruptly. Most liable to occur episodically in cases of agitated melancholia developing in alcoholic subjects. *Melancholic frenzy.*

F.—Occurring mostly in old persons affected with cyanosis, from bronchitis, cardiac disease, and asthma, and often passing into torpor and coma. Insanity of *cyanosis*, from *bronchitis, cardiac disease,* and *asthma.*

2.

The patient mutters or uses isolated words or phrases habitually, but cannot answer questions.

A.—The defect is congenital, and there are somatic signs. Some cases of *idiocy*.

B.—There is emotional depression. The affection is primary. Some cases of *melancholia*.

C.—There is mental weakness. The affection occurs in old age, and may be primary or secondary, generally the former. There is senile tremor. *Advanced senile dementia*.

D.—There is mental weakness, secondary or tertiary. Generally occurring in middle age, but may be prolonged into advanced life. History of primary exaltation or depression, secondary mental weakness or confusion, epilepsy or general paralysis. *Advanced terminal dementia, advanced epileptic dementia*. Some cases of *general paralysis* in third stage.

3.

The patient, although he understands what is said to him, and is able to speak, refuses to answer, through obstinacy, ill-temper, or some delusion or delusions.

A.—The affection is congenital. Some cases of *imbecility*, especially *imbecility* with *epilepsy*.

B.—The affection is acquired. Some cases of *delusional insanity* (monomania, paranoia). Some cases of *hysterical insanity*. In these cases (coming under head 3) the facial expression must be noticed when annoying or ludicrous propositions are made. The patient's dress, demeanour, and actions must be studied. The history of the case should be brought to bear on the diagnosis.

4.

The patient is absolutely mute.

A.—There is an absence of facial expression. The pupils are dilated. The patient is quite passive. The limbs remain for a long time, without extraneous support, in any position in which they are placed. *Cataleptic phase* of *katatonia. Severe* form of *cataleptic insanity*. In the former case the condition alternates with excitement, etc. In the latter it supervenes on a cataleptic attack.

B.—There is an absence of facial expression. The pupils are dilated. The patient is quite passive. The limbs when raised and then let go obey immediately the laws of gravity, and fall like the limbs of a person in a state of syncope. *Anergic stupor* (*acute* primary dementia). A few cases of *uterine* or *amenorrhœal insanity*.

C.—There is an absence of facial expression. The pupils are dilated. The patient is *not* passive, but inclined to resist feeding, etc. He clenches his teeth, compresses his lips, etc. He resists being taken out of doors, etc., etc. Many cases of *stuporous melancholia* (melancholia attonita). *Periodical melancholia*. Most cases of *rheumatic insanity*, properly so called. Melancholic phase of *folie circulaire*. The latter when severe more nearly approaches group B than the others, and it alternates with maniacal attacks. Unlike group B consciousness is retained; this is shown by the patients, in the maniacal phase, talking about what they have heard during the melancholy period. The rheumatic form follows an attack of rheumatism. The periodical melancholia alternates with lucid or sub-lucid intervals.

D.—The facial expression is intelligent or ecstatic. Some few cases of *delusional insanity*. Some cases of *partial insanity* or *amenomania*.

E.—The facial expression is anxious or terrified. Some cases of *stuporous melancholia* (melancholia attonita). These cases constitute, according to some authorities, a distinct group, which they name *delusional stuporous melancholia*.

F.—There is loss of perception, consciousness, memory, and movement, the pupils are wide and sluggish; the pulse is small and wiry; the temperature subnormal. *Transitory neurasthenic psychoneurosis* (some cases).

5.

The patient is unable to speak, but makes inarticulate noises.

A.—The affection is congenital, and accompanied by various bodily diseases and defects. Pronounced cases of *idiocy*. It is sub-divided into genetous, epileptic, paralytic, hydrocephalic, microcephalic, eclampsic, traumatic, and inflammatory idiocy, idiocy by deprivation, and cretinism. (See "Idiocy," chap. III.)

B.—The patient suffers from acquired insanity; makes inarticulate noises, but does not speak, and does not in any way answer questions. Last stage of *terminal dementia*. The ultimate condition also in some cases of epileptic, senile, syphilitic, and toxic insanity, and of general paralysis.

REFERENCES TO PRECEDING PARAGRAPHS OF THIS CHAPTER.

Abdominal Disorders, Insanity from. I. 1.F.
Adolescent Insanity. I. 1.F.—I. 4.A.
Anæmic Insanity. I. 1.F.
Bright's Disease, Insanity of. I. 1.F.—I. 4.A.
Cataleptic Insanity. I. 1.F.—II. 4.A.
Choreic Insanity. I. 2.U.
Circular Insanity. II. 4.C.—I. 4.A.—I. 4.B.—I. 1.F.
Climacteric Insanity. I. 4.D.
Coarse Brain Disease, Insanity from. I. 2.F.
Confusional Insanity, primary or acute. I. 3.A.
Confusional Insanity, Chronic. I. 1.L.—I. 2.S.
Consecutive Insanity. I. 4.A.—I. 1.A.—I. 2.A.
Cyanosis from Bronchitis, Cardiac Disease, and Asthma, Insanity of. II. 1.F.
Delirium, Acute. II. 1.A.
Delusional Insanity (Monomania, Paranoia). I. 1.A.—II. 4.D.
Dementia (Terminal). I. 1.O—I. 2.D.—II. 5.B.
Deprivation of Senses, Insanity from. I. 1.F.—I. 4.H.
Deterioration, Primary Mental. I. 2.A.
Diabetic Insanity. I. 4.B.
Epileptic Insanity. II. 1.D.—I. 2.H.—II. 5.B.—II. 2.D.
Exophthalmic Goître, Insanity with. I. 4.A.
Folie du doute. I. 1.I.
Folie Raisonnante. I. 1.O.—I. 1.P.
General Paralysis. I. 1.M.—I. 2.G.—II. 2.D.—II. 5.B.—I. 4.A.
Gestational Insanity. I. 4.H.—I. 4.A.
Hysterical Insanity. I. 1.E.—I. 4.B.—I. 1.B.—I. 4.A.—I. 2.R.—II. 3.B.
Idiocy, Imbecility, and Cretinism. I. 1.G.—II. 3.A.—I. 2.I.—II. 2.A.—II. 5.A.
Impulsive Insanity. I. 1.C.
Katatonia. I. 4.K.—II. 4.A.—I. 2.K.
Lactational Insanity. I. 4.H.
Mania. I. 1.E.—I. 4.A.—I. 4.L.—II. 1.B.—I. 1.B.—I. 2.R.
Masturbational Insanity. I. 1.H.—I. 4.A.—I. 2.L.
Melancholia. I. 1.F.—I. 4.B.—I. 4.C.—I. 4.G.—II. 4.C.—II. 4.E.—I. 4.H.—I. 4.I.—I. 4.J.—II. 1.E.—II. 2.B.
Metastatic Insanity. I. 4.A.—I. 4.B. (See "Rheumatic Insanity.")
Myxœdema, Insanity of. I. 2.M.
Neurasthenic Insanity. I. 1.F.—I. 3.C.—II. 4.F.—I. 4.M. (See "Anergic Stupor," "Acute Confusional Insanity," "Folie du doute," "Partial Emotional Aberration," "Delusional Insanity" (Paranoia), "Melancholy Folie Raisonnante" (I. 1.P.) See also "Neurasthenia" and "Neurasthenic Insanity," Chap. III.)
Ovarian or Old Maid's Insanity. I. 1.A.
Oxaluria and Phosphaturia, Insanity of. I. 4.G.

Paralysis Agitans, Insanity of. I. 4.B.—I. 4.I.—I. 1.A.
Partial Dementia. I. 1.O.
Partial Emotional Aberration. I. 1.J.
Partial Exaltation. I. 1.K.—II. 4.D.
Pellagrous Insanity. I. 4.B.
Periodical Insanity. I. 4.A.—II. 4.C.—I. 1.N.—I. 4.C.—I. 1.P.
Phthisical Insanity. I. 1.A.
Podagrous or Gouty Insanity. I. 1.E.—I. 1.F.
Post-Connubial Insanity. I. 2.T.
Pubescent Insanity. I. 1.F.—I. 4.A.
Puerperal Insanity. I. 4.A.—I. 4.H.
Rheumatic Insanity. II. 4.C.—I. 2.N.
Senile Insanity. I. 2.O.—I. 2.P.—I. 2.E.—II. 5.B.—II. 2.C.
Stupor, Anergic. II. 4.B.
Syphilitic Insanity. I. 2.C.—I. 2.G.—I. 4.A.—II. 5.B.—I. 4.H.
Toxic Insanity. I. 4.F.—I. 2.B.—I. 2.G.—II. 1.C.—II. 5.B.—I. 1.F.—
 I. 4.B.—I. 4.L.—I. 2.O.—I. 2.G.—I. 3.B.
Traumatic Insanity. I. 4.E.
Uterine or Amenorrhœal Insanity. I. 4.H.—I. 4.A.—II. 1.A.—II. 4.B.
Young Children, Delirium of. II. 1.A.—I. 4.I.

FREQUENCY OF SOME OF THE PRINCIPAL FORMS OF INSANITY.

I.—ON ADMISSION.

Bucknill and Tuke, quoting Boyd, give :—

Mania—42·9 per cent.
Melancholia—18·4 per cent.
Dementia—10·6 per cent.
Monomania—5·3 per cent.
General Paralysis—5·1 per cent.
Idiocy—4·3 per cent.
Delirium Tremens—1·4 per cent.
Moral Insanity—1·1 per cent.
Epilepsy—10·9 per cent.

MALES.
Mania—39·4 per cent.
Melancholia—15·2 per cent.
Dementia—9·2 per cent.
Monomania—6·2 per cent.
General Paralysis—8·3 per cent.
Delirium Tremens—1·4 per cent.
Moral Insanity—1·4 per cent.
Epilepsy—12·2 per cent.

FEMALES.
Mania (including Puerperal Mania 6·0 per cent.)—46·3 per cent.
Melancholia—21·6 per cent.
Dementia—12·1 per cent.
Monomania—4·5 per cent.
Idiocy—3·1 per cent.
General Paralysis—2·0 per cent.
Moral Insanity—0·8 per cent.
Delirium Tremens—0·1 per cent.
Epilepsy—9·5 per cent.

Clouston gives at Royal Edinburgh Asylum :—

Pubescent Insanity—4·6 per cent.
Adolescent Insanity—8·7 per cent.
Climacteric Insanity—7·2 per cent., of which : Women—6·2 per cent. ; Men—1· per cent.

Senile Insanity—6·4 per cent., of which : Women—4·2 per cent. ; Men—2·2.
Paralytic Insanity (Insanity from Coarse Brain Disease)—2·89 per cent.
Phthisical Insanity—2·0 per cent.
Hysterical Insanity—1·08 per cent.
Masturbational Insanity—1·46 per cent.
Puerperal Insanity—1·9 per cent.
Lactational Insanity—1·2 per cent.
Gestational Insanity—·47 per cent.
Traumatic Insanity—·33 per cent.
Post-Febrile Insanity—1· per cent., of which ·4 per cent. after Scarlatina.
Folie Circulaire, less than—·5 per cent.
Acute Mania—12·49 per cent.

II.—Of 2,297 patients at the New York Pauper Asylum, over 14 per cent. were terminal dements (Spitzka), and 12·3 per cent. were paretic dements (General Paralytics), or dements with organic brain disease.

Spitzka found 2 per cent. of over 2,000 male lunatics suffering from Katatonic Insanity.

III.—ORDER OF FREQUENCY AT DEATH.
(Bucknill and Tuke.)

Dementia (frequently General Paralytic).—Mania (usually Chronic).—Melancholia.—Monomania.

IV.—Several forms, such as Folie du Doute, Partial Emotional Aberration, etc., are seldom met with in asylums. Many cases of Mild Melancholia, Senile Insanity, Insanity from Coarse Brain disease, Toxic, and Consecutive Insanity do not find their way to asylums. So that in private practice, Melancholia is perhaps met with more frequently than any other form of insanity.

CHAPTER VI.

PROGNOSIS.

A.—GENERAL PROGNOSIS.

I.—As to danger to life : —

(1,) Unfavourable when serious diseases of other organs, such as tuberculosis, heart disease, etc., are present.

(2,) Unfavourable when general paralysis is present. As a rule fatal in from one to three years, and very often sooner (Griesinger).

(3,) Unfavourable when there are extensive and intense hyperæmias of the brain, which very rapidly advance to softening, and prove almost immediately fatal by causing serous effusions, extravasation of blood, etc.

(4,) Œdema of the brain, especially if it comes on acutely ; unfavourable.

(5,) Long continued refusal of food unfavourable.

(6,) Much greater tendency to death in acute melancholia and acute mania than in the chronic forms, or monomania, imbecility, etc.

(7,) Tendency to death above the average in puerperal or senile insanity.

(8,) Strong suicidal tendency unfavourable.

(9,) High temperature unfavourable.

II.—As to recovery from mental derangement. The considerations are :—

(1,) The *form* of *insanity*. See forms, prognosis of (postea).

(2,) The *duration*, which bears an inverse ratio to the recoverability. Clouston gives the recovery rate of patients free from organic brain diseases, and who had been less than a year insane before admission, as 70 per cent.

(3,) The *causes*, exciting and predisposing ; age and sex. In hereditary insanity there is great danger of relapse. Prognosis unfavourable when insanity develops gradually in a person always eccentric (Griesinger). Also unfavourable when it occurs after years of vexation, suspense, etc. Much more favourable when arising from sudden mental shock, fright, etc. Youth and the female sex influence the prognosis favourably. See etiological forms of insanity, *e.g.*, anæmic, traumatic, alcoholic, etc.

(4,) The *number* of attacks. Probability of permanent cure diminishes with each attack, yet many patients who relapse repeatedly recover.

(5,) The *external circumstances* and *relations* of life of the patient. Poverty and inability to have change of surroundings unfavourable.

(6,) The *physical health* and *condition*. Return of bodily health unaccompanied by return of mental sanity unfavourable. Stoutness then frequently an indication of tendency to dementia.

(7,) The *secretions*. Return of suspended (especially menstruation), favourable. Profuse secretions sometimes critical (Bucknill and Tuke).

(8,) The *course* of the disease. Periodicity: early, with intervals longer than paroxysms, but becoming shorter each time, or regular periodicity, unfavourable. Short lucid intervals, or remissions gradually becoming longer, favourable. Gradual recovery more likely to be permanent than sudden return to sanity.

(9,) The *presence* or *absence* of certain signs: Somatic stigmata, paralysis, paresis, convulsions, onset or exacerbation at physiological periods, morbid impulses with one exception, morbid propensities occurring alone, moral perversions without other symptoms, affective perversions occurring alone, fixed ideas, systematised delusions, fixed delusions, grandiose delusions, very defective memory, unconscious delirium, are generally unfavourable. Maudsley says that the homicidal patient with persecutory ideas seldom recovers, "while the suicidal patient generally does recover, particularly after some serious and all but successful suicidal attempt." With regard to fixed delusions Maudsley says the prognosis is unfavourable in melancholics who believe their troubles are caused by external agency, but more favourable in those who attribute their suffering to faults of their own.

(10,) *Return* of *natural affection* and of former habits, tastes, manner, and modes of speech, and the patient's recognition of his state, past and present, favourable.

For further indications for prognosis see below, under heads "Mania" and "Melancholia."

B.—SPECIAL PROGNOSIS (INCLUDING DURATION).

ADOLESCENT INSANITY.

Out of 180 cases, Clouston only knew of 26 who became incurable. He therefore looks on adolescent insanity as very curable compared with many mental disorders, though not so curable as others, e.g., puerperal insanity. An inveterate habit of masturbation renders the prognosis more serious, especially in males. Clouston is inclined to give a favourable prognosis when he sees, accompanied by mental improvement, signs of physiological manhood appearing: the beard growing, the form expanding, the weight increasing. Only three of his 180 cases died.

Of B. Lewis's cases nearly three-fourths of the females and nearly three-fifths of the males recovered, about a thirteenth of the former and a seventh of the latter were discharged relieved, nearly an eighteenth of the women and nearly a tenth of the men died, and about three-tenths of the whole number remained in the asylum as chronic patients. Half the female cases that were cured, after a relapse or two, had recovered by the seventh, and nearly three-fourths by the tenth, month.

ANÆMIC INSANITY.

80 per cent. of Clouston's cases recovered, most of them within three months.

BRIGHT'S DISEASE (*Insanity of*).

In a case quoted by Clouston as typical, the patient only lived about two months after the appearance of his mental symptoms.

CATALEPTIC INSANITY.

Speaking of catalepsy (not cataleptic insanity) Ross states that the prognosis is unfavourable with respect to complete recovery. Cases arising from malaria, injuries, or mental shocks are more favourable than those arising from other causes.

CHOREIC INSANITY.

In the mild cases, which constitute the majority, the prognosis is fairly favourable, but the tendency to dementia should not be overlooked. In the extremely rare cases of very acute maniacal excitement the prognosis is unfavourable, half the cases proving rapidly fatal and the remainder often suffering from divers intellectual troubles of variable duration.

CIRCULAR INSANITY.

(Including Folie circulaire, Folie alternante, and Folie à double forme.)

Clouston says three things are sure about the prognosis:—
(1,) Its utter uncertainty; (2,) Recovery cannot be looked for at the climacteric period in many cases; (3,) About 20 per cent. may be expected to settle down into a sort of comfortable, slightly enfeebled condition, in the senile period of life. In another place (p. 216) Clouston says it is mostly an incurable disease.

Each phase may last a few days, a few weeks, one, two, or six months, or even longer.

Bra says this form of insanity inspires no great hope of recovery. It lasts as long as the individual, and eventuates, often after a very long *duration*, in dementia. Some individuals retain considerable mental lucidity all their lives. Falret pointed out the frequent occurrence of cerebral congestion in these cases.

"Alternations of excitement and depression, whether amounting to well marked circular insanity or not, constitute a bad sign" (Bucknill and Tuke, section "Prognosis").

CLIMACTERIC INSANITY.

Of Clouston's cases, 53 per cent. recovered; 31 per cent. of the men and 57 per cent. of the women.

Of those who recovered, 55 per cent. were discharged within three months, 65 per cent. within six months, and 91 per cent. within twelve months. A few recovered after two years of treatment. The maniacal and melancholic cases recovered in about equal proportion, but the maniacal in shorter time. Only 29 per cent. of women over 50 recovered. From 55 to 60 the men were most curable. Only 3, out of 11 over 60, recovered. Twelve per cent. of the whole number of men suffering from this form of insanity died, but only 9 per cent. of the women. Forty-seven per cent., or excluding cases complicated with organic brain diseases, 59·5 per cent. of Merson's cases recovered (Bucknill and Tuke).

COARSE BRAIN DISEASE (*Insanity from*).

Nearly 19 per cent. of Clouston's cases recovered mentally. Organic dementia arising from cerebral tumour may terminate fatally in a month, or may last twenty years from the onset of the symptoms (Clouston).

CONFUSIONAL INSANITY (PRIMARY).

Recovery takes place in all except a small proportion of cases, and in this small proportion the affection assumes the chronic form (Spitzka).

CONFUSIONAL INSANITY (CHRONIC).
(Chronic Mania.)
The tendency is towards dementia.

CONSECUTIVE INSANITY.
The cases resembling melancholia and mania, due to anæmia, rapidly recover. Some of the cases resembling persecutory monomania pass into dementia. Some cases pass into a permanent state of slight mental weakness. Duration: very variable.

CYANOSIS FROM BRONCHITIS, CARDIAC DISEASE, AND ASTHMA *(Insanity of)*.
Often passes into coma.

DELIRIUM (ACUTE).
In the majority of cases acute delirium terminates fatally. Nevertheless, if the disease is not of more than four or five days' duration, and the temperature has not attained 106° or 107° F. in the rectum, the symptoms may improve; in which case the agitation diminishes, the temperature lowers, the pulse becomes slower, and sleep re-appears, but the patients retain hallucinations and delusions for a long time afterwards (Bra).

DELUSIONAL INSANITY.
(Monomania, Paranoia.)

"The prognosis of monomania is very unfavourable. The chief feature to be consulted in reference thereto is the mental power of the patient. The more considerable this is, the more likely is a correction of the delusive beliefs, the delusive suspicions, or morbid fears to take place. Consequently the prognosis is best with those patients who suffer from simple delusions of persecution or of social ambition, worse with erotomania, and worst of all with religious monomania, for here, as has been already stated, a background of original weakmindedness is generally present. Bad as the prognosis is in this form, cases are reported where the hallucinations and delusions disappeared, and the patient, if not altogether recovering, showed nothing abnormal beyond an extravagant religious zeal, and a desire to convert mankind to what he happened to consider, in the excessive egotism of the fanatic, the right faith" (Spitzka, p. 318).

The jealous form is also unfavourable as to its prognosis, relapses frequently occurring after apparent recovery.

If not cut short by some bodily disease, or by suicide, monomania may last for many years, and then end in complete dementia (Clouston, Bra, etc.). There is a tendency to mental enfeeblement as time goes on, and in most cases the intensity of the conviction of the delusion and the aggressiveness with which it is put forward tend to diminish. A patient in Norfolk Asylum, who said he was the Lord God and had cured the cattle plague, became very industrious on the farm. Another, who called himself Lord Nelson, made himself generally useful about the building, doing, with evident pleasure, a good deal of hard work.

In another asylum a patient, who believed himself to be the rightful Earl of Derby, and who was well versed in the current topics of the day, and displayed considerable knowledge of the various systems of shorthand, worked contentedly in the dormitories. His delusion had to be sought, but when found was still stoutly defended.

"Most monomaniacs live long, all but the cases of morbid suspicion, who, as I said, mostly die of phthisis" (Clouston, p. 265).

DEMENTIA (TERMINAL).
(Secondary Dementia.)

"Dementia is, with rare exceptions, incurable." "Fever and acute maniacal paroxysms have, however, occasionally been the means of restoring to reason patients apparently sunk in hopeless dementia" (Bucknill and Tuke).

Patients suffering from severe apathetic terminal dementia consequent on anergic stupor or stuporous melancholia, as a rule, do not live more than a few years. Patients suffering from the milder forms of terminal dementia may, if well taken care of, live to a good old age (Spitzka).

DETERIORATION (PRIMARY MENTAL).

If the first signs are heeded, comparative mental health may be restored under treatment, "but if the exciting causes are kept in operation, actual dementia may be the result" (Spitzka).

DIABETIC INSANITY.

Prognosis unfavourable. The diabetes and the insanity sometimes alternate.

EPILEPTIC INSANITY.

Dementia is the usual termination. The prognosis is, therefore, very grave, especially where the epileptic seizures have commenced during childhood. When the attacks of acute epileptic insanity are separated by long intervals the intellectual faculties are sometimes very little weakened (Bra).

"Epileptic insanity offers little hope of really permanent recovery. The exceptions are so few that it may be regarded as almost incurable" (Bucknill and Tuke).

Duration of each attack of ante-convulsive epileptic insanity: ordinarily, several hours, occasionally, several days; of post-convulsive: usually short, not extending beyond a few days.

FOLIE DU DOUTE.

The prognosis is very unfavourable as regards recovery, the patient becoming more and more unsociable and inclined to seclude himself. The morbid ideas and emotions (obsessions) diminish in number and extent, but increase in intensity. The aberration, although lasting the lifetime of the patient, never ends in true dementia. In a few cases life has been cut short by suicide (Ball, Bra). Sometimes there are long intermissions. Some cases end in mental torpor. A few based on *acquired* neurasthenia recover (Krafft-Ebing).

FOLIE RAISONNANTE.

Is constitutional, and its course though stationary is prolonged for years or even a lifetime (Krafft-Ebing).

GENERAL PARALYSIS OF THE INSANE.

Voisin states that although the usual termination of general paralysis is death, yet it may pass into a chronic state, and last for years, or it may terminate in recovery, in proof of which he gives an account of thirteen cases drawn from various sources, and of several of his own, mostly treated by cold baths.

Death is caused most frequently by intercurrent diseases and complications, such as pneumonia. It may be caused by colliquative diarrhoea, cystitis, bed-sores and their consequences, mechanical asphyxia by food, drink, etc., or by exhaustion. Two-thirds of the patients are carried off by complications of cerebral origin, evinced by apoplectiform and epileptiform attacks, coma, contractures, etc., etc. Meningeal hæmorrhages, uræmia, or œdema of the ventricles may supervene.

Duration.—Mickle says the average duration of the disease in the soldiers under his care was more than two years, but in the gentlemen patients at the same asylum (Bow) it was very much longer, some of the cases lasting more than six or ten years.

Spitzka says, "The prodromal period may last only three months, usually a few years, and in rare cases nearly a lifetime. Dating from the explosion of the malady, the lethal termination may occur in six months, more commonly in three years; in not a small number of instances, in six, or ten, or even more years."

According to Voisin, the duration of the prodromal period is often many years, six, ten, or more; sometimes only three months.

Savage ("British Medical Journal," May 4, 1890) says the "warnings" may be present for years, and that they are almost certainly present for at least a year before even the specialist can diagnose general paralysis.

Bucknill and Tuke say the duration is generally estimated at thirteen months, but think this average quite too low. They say patients rarely live more than three years after the development of well-marked symptoms, but they quote a case lasting ten years, seven of which the patient was bedridden.

GESTATIONAL INSANITY.

Seventy-three per cent. of Clouston's cases recovered, nearly all the recoveries taking place within six months from the date of admission. None died of the disease *per se*.

HYSTERICAL INSANITY.

Favourable in acute form, but relapses are frequent. In the chronic form the ultimate prognosis is unfavourable. Sudden temporary recoveries are noted, but recurrence is very probable, and with each recurrence deterioration becomes more marked (Spitzka). Chronic hysterical insanity degenerates into dementia more frequently than might be supposed, yet much less rapidly and less frequently than epileptic insanity (Bra).

IDIOCY.

(Including Imbecility and Cretinism.)

Unfavourable as to recovery, but the condition may be much ameliorated by treatment.

IMPULSIVE INSANITY.

Unfavourable. *Nymphomania*, rather favourable in young persons, rarely admits of cure when it appears at the climacteric (Griesinger).

KATATONIA.

Favourable as regards life, but the tendency to pulmonary tuberculosis, in the depressed and atonic stages, must be kept in view. In the majority of cases recovery takes place after one or two cycles of the symptoms, but relapses are common. The progress to dementia is slow, and the dementia is rarely extreme (Spitzka).

LACTATIONAL INSANITY.

Very favourable when resulting simply from the anæmia induced by prolonged suckling (Bucknill and Tuke). Of Clouston's cases, $77\frac{1}{2}$ per cent. recovered; 80 per cent. of the recoveries occurring within six months, and all of them within eighteen months. There were few relapses.

MANIA.

Clouston gives the five usual terminations of mania as :—
(1,) Complete recovery in 54 per cent.
(2,) Partial recovery, leaving mental weakness, eccentricity, change of character, etc.
(3,) The substitution of fixed delusions or delusional states for the exaltation.
(4,) Dementia in about 30 per cent.
(5,) Death in 5 per cent.
Clouston includes acute delirium (acute delirious mania) under the head of mania, and this would tend to diminish the percentage of recoveries and increase the percentage of deaths. Spitzka says various authors estimate the recoveries at from 60 to 80 per cent., and that the latter figure accords with his experience. Under mania he does not include transitory mania (transitory frenzy, transitory insanity), but this is a very rare and very curable form.

Convalescence ordinarily takes place slowly, the occurrence of brief lucid intervals being the first indication of returning health.

Savage " would say that any case of acute mania in which the excitement was great, sleeplessness well marked, food either not taken at all or not assimilated, must be regarded with danger, especially in young and old cases." The prognosis in all cases of acute mania must be guarded.

Bucknill and Tuke say, " Acute mania is in a large proportion of cases recovered from." " A noisy boisterous mania is usually recovered from." They give the percentage of recoveries on admissions (for mania) as 57 in the Pennsylvania Hospital for the insane, in thirty-two years ; and 63·89 for acute mania, and 6·36 for chronic, at the York Asylum, in twenty-seven years.

Many mild cases of mania are treated at home or under private care successfully, and these would swell the percentage of recoveries.

Clouston gives as indications of prognosis in mania (including acute delirium) :—

1st.—Favourable : A sudden onset ; short duration ; youth of patient ; no fixed delusions or delusional conditions ; appetite for food not quite lost ; no positive revulsion against food or drink, or perversions of the food and drink appetites ; no indication of enfeeblement of mind ; no paralysis, or paresis, or marked affection of the pupils ; no epileptic tendency ; no complete obliteration or alteration of the natural expression of the face or eyes ; the instincts of delicacy and cleanliness not quite lost ; no unconsciousness to the calls of nature ; the articulation not affected ; the disease rising to an acme and then showing slow and steady signs of receding ; no former attacks, or only one or two that have recovered.

"Hereditary cases are often very curable, but relapses are more probable."

2nd.—Unfavourable indications : A gradual and slow onset ; great length of duration of attack, especially after twelve months' persistence of fixed delusions or delusional states ; extreme and increasing exhaustion of the patient in spite of proper treatment ; paralysis of the trophic power so that the nutrition cannot be restored ; persistent refusal of food ; requiring forcible feeding ; extreme failure of the cardiac action and circulation, so that the extremities are always blue and cold ; persistent affections of the pupils, especially extreme contraction ; persistently dirty habits ; a tendency towards dementia ; a tendency towards chronic mania ; an utter and persistent deterioration in the facial expression, especially if it be towards vacuity ; persistent and complete paralysis or perversion of the natural affections and tastes and appetites ; many former attacks ; convulsive, paretic, paralytic, or inco-ordinative symptoms ; such perverted sensations as cause patients to pick the skin, pull out the hair, bite off the nails into the quick ; a restoration of sleep and bodily nutrition without, in due time, an improvement mentally ; very persistent insane masturbation ; a tendency for the exaltation to pass off and fixed delusion to take its place ; excitation of the limbs and subsultus tendinum ; a typhoid condition.

Auditory hallucinations are unfavourable (Blandford).

Extreme and sometimes sudden exhaustion is a symptom always to be feared in acute mania, but loss of flesh need not occasion alarm either as to recovery of mind or body (Bucknill and Tuke).

Duration.—Spitzka states that the average duration of typical mania is about five months, including an initial stage of depression of six weeks, a maniacal period of about three months, and a period of convalescence of about a fortnight; but some intense cases may last only a few weeks, whilst some mild ones may last a year or more. The duration of transitory mania varies from an hour up to a few days (Clouston).

Of acute mania Clouston says 60 per cent. recover, 7½ per cent. die (he includes acute delirium), and 32½ per cent. become demented, or pass into chronic mania.

Of the deaths from mania generally, Clouston gives 5 per cent. as the average on the admissions, the immediate cause of death being maniacal exhaustion, or some cause directly traceable to the disease. Bucknill and Tuke give 4 per cent. and Spitzka gives 2 to 5 per cent. The three first mentioned evidently include acute delirium, and Spitzka, puerperal insanity (maniacal form). Excluding these diseases, and including many mild cases treated at home, the mortality would be somewhat lower.

MASTURBATIONAL INSANITY.

Of Clouston's cases, 34·78 per cent. (16 out of 46) recovered, and a somewhat larger percentage went home more or less improved, the remainder being "hopeless, incurable, and degraded." He states that there is no special risk of relapse, and no tendency to any special form of nervousness of lifelong duration. Spitzka says the prognosis is bad.

MELANCHOLIA.

Clouston gives 54 as the percentage of recoveries on admission. About one-third of the cases were relapses. He says the recovery rate would no doubt be much greater if the milder cases treated at home were included. He says simple melancholia is in most cases curable, except in very advanced life or when it accompanies organic brain disease. Clouston's experience is that hypochondriacal melancholia, "when it occurs at the more advanced ages, is apt to be permanent, or the prelude to senile dementia." He states that about 50 per cent. of the cases of melancholic stupor recover.

Maudsley says melancholia is the most curable of the forms of mental disease, acute mania coming next in order.

Spitzka states that about six cases of melancholia out of ten admissions recover completely, females recovering in greater proportion than males.

He says that the lighter forms are very common in general practice, and are often treated as neurasthenia and dyspepsia, and frequently cured. Apart from the tendency to suicide and the liability to recur, the prognosis of *melancholia sine delirio* is always favourable. Next come the typical and agitated forms, 80 per cent. of these patients recovering under early and proper treatment. "With stupid or atonic melancholia the outlook is bad."

Bucknill and Tuke are of opinion that "Acute melancholia, if cases which never reach an asylum are included, is at least as frequently cured (as acute mania), but certainly in asylum experience comes next in order." And that "Simple depression of mind is not unfavourable except in the decline of life."

Duration.—Spitzka places the average duration of melancholia at from three to eight months.

Clouston states that 50 per cent. of his recoveries occurred under three months, 75 per cent. under six, and 87 per cent. under twelve months. A few patients make sudden recoveries in a few days.

Clouston gives as *favourable signs or conditions:* Youth, sudden onset, a removable cause, want of fixed delusion, absence of hallucinations of hearing, taste, or smell; no visceral delusions; no strongly impulsive or epileptiform symptoms; no picking of the skin, or pulling out the hair, or such trophic symptoms; no long continued loss of body weight in spite of treatment; no long continued inattention to the calls of nature, and no dirty habits.

Bad signs: Slow gradual onset; fixed delusions, visceral or organic; gradual decay of bodily vigour; persistent loss of nutritive energy and body weight; convulsive attacks and motor affections generally, not ideomotor; persistent hallucinations, especially of hearing, smell, and feeling; picking the skin or hair; persistent refusal of food; an unalterable fixity of expression of emotional depression in face, or persistence of muscular expression of mental pain (wringing hands, groaning, etc.); persistent suicidal tendency of much intensity; arterial degeneration; senile degeneration of brain; no natural fatigue following persistent motor efforts in walking, standing, etc.; a mental enfeeblement like dementia. But the prognosis should be guarded in almost every case.

The prognosis is more favourable according as there is less stupor, less nutritive disturbance, more variation in the symptoms from day to day, and with youth and the female sex. It is of specially good import when the patient's condition shows a marked improvement from evening to evening (Spitzka).

Termination.—The tendency to dementia is much less marked than in mania (Bucknill and Tuke).

Clouston says a few (mostly middle-aged or old people) end in chronic melancholia and a few pass into dementia, which is never so complete a mental enfeeblement as when it follows mania. He says many of the cases pass into mania.

Spitzka states that the cases which do not recover or die pass into apathetic terminal dementia (a common sequel of *melancholia attonita*), or into chronic delusional insanity with deterioration (chronic confusional insanity), an occasional sequel of the typical, and a more frequent one of the agitated forms.

METASTATIC INSANITY.

Recovery often follows re-appearance of the original disease.

Prognosis favourable, "when the derangement has followed the suppression of an eruption or an accustomed discharge" (Maudsley, "The Physiology and Pathology of Mind," p. 489).

MORAL INSANITY.

Prognosis unfavourable (Bucknill and Tuke).

MYXŒDEMA *(Insanity of)*.

Unfavourable (Savage). The tendency is to pass slowly and steadily into dementia. The duration of life depends upon the bodily condition and upon the secondary degenerations which may occur. Presence of albumen unfavourable to longevity (Savage).

NEURASTHENIA AND NEURASTHENIC INSANITY.

Neurasthenia.—Favourable when acquired (post-febrile, post-puerperal, etc.); unfavourable when constitutional (commencement at puberty, traces in early childhood).

Neurasthenic Insanity.—Favourable in the psychoneurotic group; unfavourable in the degenerative, except in the rare cases founded on acquired neurasthenia.

OVARIAN OR "OLD MAID'S" INSANITY.

None of Clouston's cases recovered, but in two out of ten, as they passed into the senile period, the delusion became so theoretical that they almost ceased to allude to it.

OXALURIA AND PHOSPHATURIA *(Insanity of)*.

Few require asylum treatment. The nervous symptoms disappear under the treatment that cures the oxaluria (Clouston).

PARALYSIS AGITANS (*Insanity of*).

As the sensory troubles become increased or disappear, so do the mental.

PARTIAL EMOTIONAL ABERRATION.

(Pathophobia, Kovalewsky; Agoraphobia, Claustrophobia, Mysophobia, etc.)

These patients seldom or never find their way into asylums. Prognosis similar to that of *folie du doute* (q v.).

According to Kovalewsky the order of evolution is: (1,) Neurasthenia; (2,) Pathophobia; (3,) *Folie du doute*.

PARTIAL EXALTATION, OR AMENOMANIA.

Prognosis more favourable than in true ambitious or religious delusional insanity.

PELLAGROUS INSANITY.

Passes on to chronic dementia (Sibbald). Pellagra itself, if treated early, is curable in the proportion of 78 per cent. (Erasmus Wilson).

Unfavourable (Bucknill and Tuke), yet has been cured after lasting some years. This applies to pellagrous insanity as well as pellagra.

Duration of majority of cases of pellagra, about three years, but may be from one to fifty or sixty years (Bucknill and Tuke, Wilson, etc.).

PERIODICAL INSANITY.

Unfavourable. "Shares the bad prognosis as to recovery with other degenerative disorders, such as monomania (paranoia) and epileptic insanity" (Spitzka, p 267).

PHTHISICAL INSANITY.

Recovery rate 30 per cent. (Clouston).

PODAGROUS OR GOUTY INSANITY.

In the mild form, with matutinal dread, treatment is markedly successful (Bucknill and Tuke).

POST-CONNUBIAL INSANITY.

Curable and not prolonged (Clouston).

PUBESCENT INSANITY.

Spitzka says the prognosis is on the whole exceedingly unfavourable, and that the course is protracted. Imperfectly developed cases which appear to be only a pathological intensification or undue prolongation of the ordinary pubescent state present better prospects. The period he includes is from the fifteenth to the twenty-second year. The habit of masturbation increases the gravity of the prognosis.

PUERPERAL INSANITY.

Of Clouston's cases, 75 per cent. recovered, more than half of the recoveries occurring within three months and nine-tenths within six months.

Of Bevan Lewis's cases, four-fifths recovered, more than one half of the recoveries having occurred by the fifth month, and nearly nine-tenths of them by the ninth month from the commencement of the insanity.

"Very favourable unless it assumes an inflammatory or typhoid type" (Bucknill and Tuke). According to Savage's statistics, quoted by Bucknill and Tuke, the third month showed the majority of recoveries from puerperal mania and the sixth from puerperal melancholia, the average duration of the latter being longer than that of the former. Bra says the prognosis is more favourable in the maniacal than in the melancholic form. Bevan Lewis says the prognosis is influenced more especially by the duration before treatment and by the age of the patient; if the former be less than a week and the latter under thirty years, it is favourable. In spite of the generally favourable prognosis Savage advises caution, owing to the number of deaths and the frequency of cases passing into chronic insanity or weakmindedness.

Bevan Lewis and Clouston give their mortality as 8·5 per cent. and 8·3 per cent. respectively, four-fifths of the latter occurring within one month, and all within two months.

RHEUMATIC INSANITY.

The severe or supra-acute form of cerebral rheumatism is brief and fatal. The prognosis is less grave in the acute or less severe form, recovery sometimes taking place in the worst looking cases. In some cases the patient passes into a state of chronic insanity. The prognosis should always be guarded, and guided by the thermometric indications.

In the vesanic form, or rheumatic insanity (properly so called), recovery takes place in the majority of cases. It is gradual, and may be complete, but more commonly some weakness of mind remains. If rheumatic insanity continues more than three or four months dementia supervenes (Bra).

SENILE INSANITY.

Of Clouston's cases, admitted during a period of nine years, 35 per cent. recovered, half of them within three months of admission and 79 per cent. of them within six months; 33 per cent. of the cases died, 20 per cent. of the deaths making seven per cent. of the whole of the cases occurring within the first month, a month in which there is great danger of death in senile insanity. More than half the deaths occurred within the first six months of residence. Thirty-three per cent. of the cases were of senile melancholia, 30 per cent. of typical senile dementia (incurable), and the remainder were cases of senile mania. Only 30 per cent. of the melancholic cases recovered, but the recoveries were as a rule better than in the maniacal cases. Rather more than half the whole recoveries were mere gradual passings into manageable senility.

Bevan Lewis states that simple senile melancholia is a fairly recoverable form of alienation, but the strongly marked suicidal tendency must be kept in view. He says cases of senile mania, though often passing into permanent dementia, may completely recover, with scarcely any abnormal mental weakness remaining.

SOMNAMBULISM (*Pseudo-Insanity of*).

Few sleep-walkers ever become actually insane (Clouston).

STUPOR, ANERGIC.
(Acute Dementia.)

Typical anergic stupor in young persons is very curable (Clouston). Spitzka says the prognosis is highly favourable, probably 90 per cent. recovering. Youth and sudden onset, as from fright or profuse hæmorrhage, favour rapid recovery; masturbation unfavourable. Uncured cases sink into apathetic dementia or die of pulmonary disease.

Duration.—May be only a few weeks. Usually from one to three months (Spitzka).

SYPHILITIC INSANITY.

The cephalalgic form of cerebral syphilis is favourable, inasmuch as it arouses attention and leads to treatment.

The epileptic or convulsive form is, according to Fournier, one of the most favourable, for the same reasons. It yields readily to early specific treatment.

The congestive form is much less favourable. It is insidious, and leads to apoplectic and paralytic attacks, to intellectual disturbances, etc.

The paralytic form indicates disorganisation, often irreparable, of the nervous centres.

Of the mental form (syphilitic insanity properly so called) the prognosis is unfavourable when it assumes from the first the character of slow progressive mental weakness. When a state of complete hebetude is arrived at, the prognosis is hopeless.

Syphilitic pseudo-general paralysis is, of all forms of cerebral syphilis, the most unfavourable as to its prognosis. It is often refractory to treatment. Speaking generally, the intellectual disturbances in cerebral syphilis have a marked tendency to continuous aggravation and to terminate after a few months, or two or three years, in confirmed dementia.

Sometimes death supervenes rapidly, or sudden coma sets in, or a maniacal outburst takes place, or a hemiplegic attack occurs, etc., etc.

The intellectual troubles are less influenced by specific treatment than the other symptoms of cerebral syphilis. The prognosis is, therefore, always serious (Bra).

Hughlings Jackson, speaking of syphilitic cerebral neoplasms, calls attention to the fact that although the growths themselves may be removed by specific treatment, their destructive effects on the surrounding brain tissue are not amenable to any medication.

TOXIC INSANITY.

Acute Alcoholic Insanity.—Unless the brain has been weakened by repeated attacks, both forms (maniacal and melancholic) are curable, and generally of short duration (Sibbald).

Delirium Tremens.—If the temperature taken in the rectum, after oscillating two or three days round 102°1 F., rises to 105°8, the prognosis is unfavourable, and becomes more so if the temperature rises still further, or even persists at that elevation. When, on the contrary, the temperature rises somewhat rapidly to 102°1, or even 104°, and falls again at the end of 24 or 48 hours, the prognosis is favourable. In mild cases the temperature may rise to 101° F., or 101°4 F., but in such cases it sinks rapidly during temporary quietude and oscillates about 100°2 F. The extent and intensity of the tremor, the muscular weakness, the degree of chronicity of the intoxication, and, in the complicating form, the intercurrent disease have all to be considered. If the tremor is local, transitory, and slight, the case is mild; if it is intense, general, and persistent, continuing during sleep, and accompanied by shudderings, shocks, and undulations, nervous exhaustion sets in on the second or third day (Magnan). In the majority of cases of delirium tremens the patient recovers. Recovery may be delayed for eight, fifteen, or more days, and is announced by the return of sleep and the amelioration of all the symptoms. Yet the prognosis is always grave on account of the tendency to relapse (Bra). After two or three attacks the mind is sometimes hopelessly weakened with blunted moral feeling and considerable cunning (Maudsley).

Chronic Alcoholic Insanity.—"The prognosis of this form of insanity is very unfavourable, as there is a pronounced tendency to dementia. Complete cures are rare, and if the affection has lasted any length of time, impossible. The higher the mental status of the patient the better are his chances" (Spitzka).

As alcoholic intoxication advances, dementia, partial paralysis, and, in some cases, general paralysis, supervene (Magnan).

INSANITY FROM OPIUM.—"It may be cured, or, at least, the attack may be got over, by the total removal of the drug" (Savage). The prognosis (ultimate) is unfavourably influenced by the difficulty of breaking off the opium crave. Savage considers the opium crave the greatest of the craves, and the *chloral* crave the least, or among the least. Patients for the most part recover from the insanity set up by *cannabis indica* (Blandford).

SATURNINE INSANITY.—Favourable in saturnine mania and melancholia when the affection has not lasted more than four or five weeks. Recovery is announced by the return of sleep (Bra). The convulsive form most frequently terminates in death. In saturnine pseudo-general paralysis the tendency to amelioration is soon seen and sometimes very soon (Bra).

TRAUMATIC INSANITY.

Remarkable for its long duration. "As a rule progressive deterioration sets in, and dementia terminates the history of the case" (Spitzka).

In cases complicated with alcohol the mental symptoms are aggravated and there is a still stronger tendency to incurability (Clouston). Clouston admitted twelve cases in nine years, and he describes two cases which recovered, one after an operation, by which a depressed portion of the skull was removed. Only one of these cases belongs to the twelve admissions above mentioned.

UTERINE OR AMENORRHŒAL INSANITY.

The melancholic cases usually recover when the general health is restored and menstruation re-established (Clouston).

The maniacal cases do not show the same tendency to recover coincidently with the restoration of menstruation (Clouston).

YOUNG CHILDREN (*Delirium of*).

Favourable under treatment (Clouston).

CHAPTER VII.

PATHOLOGICAL ANATOMY, PATHOLOGY AND PATHOGENESIS.

MORBID ANATOMY.

A.—GENERAL.

I.—MACROSCOPICAL.

(*a*,) THE CRANIUM may be abnormally large or small, or it may be malformed. The cranial bones may be increased or diminished in thickness or in density, or in both.

(*b*,) THE DURA MATER may be adherent to the cranial bones. It may be redder than usual from increased vascularity. It may be abnormally thick, excessively tense or very lax (Griesinger).

(*c*,) THE ARACHNOID.—Opacity of the arachnoid is frequently met with in all forms of insanity, but is also seen in elderly sane persons. Opacity with thickening is especially frequent in general paralysis. This condition of the arachnoid with increase of the pacchionian granulations is found in drunkards and other persons subject to habitual cerebral congestion during life.

Osseous concretions may be found, or abnormal adhesions to pia and cortex. Frequently there are fine granulations on the external surface.

There may be ecchymoses. There may be inflammation of the parietal layer, pacchymeningitis interna, accompanied by firm pseudo-membranous or thin, quasi-gelatinous, structures. The serous fluid is frequently increased in quantity (Griesinger).

Durhæmatomata or arachnoid cysts are frequently found in the sac of the arachnoid, especially in general paralysis. They are also found after terminal dementia, chronic mania, and acute mania. Griesinger considers them to be of special importance, and speaks of them as spontaneous hæmorrhages. When the extravasation is considerable, the fluid at first thick and dark brown becomes clear with time. When the amount is small, the watery portion is absorbed, leaving a layer of fibrine at first brown, then yellow, then almost colourless. They should not be confounded with sub-arachnoid serous effusions, which, according to Griesinger, have no special significance.

The large ones cause compression, atrophy, and increased consistence of the hemisphere with infiltration of the meninges.

Hæmorrhage beneath the arachnoid is much less frequent and has no special significance.

(*d*,) THE PIA MATER.—There may be *active hyperæmia* appearing as a very intense uniform injection of the smallest vessels with small stellate ecchymoses. This condition is found after acute delirium (acute delirious mania) and violent mania (Griesinger).

There is also a form of *congestion* more connected with the large veins and due to varicosities, etc. This form (hyperæmia "ex vacuo") is chronic in its course and frequently associated with cerebral atrophy. It may also be the result of mechanical venous stases, and then always exerts an influence upon the form of the disease.

There may be *anæmia* of the pia mater, causing it to be remarkably pale. When chronic, it gives rise mostly to states of mental weakness and dementia (Griesinger).

As the result of chronic hyperæmia "ex vacuo," there may be opacity, thickness, and œdema of the pia mater. This œdema is frequent, especially after the secondary forms, with mental weakness well marked.

There may be *inflammation* of the pia, producing in combination with cortical inflammation adhesion of the membranes to each other and the cerebral cortex. This adhesion, the result of meningo-cerebritis, particularly affects the convexity and the contiguous internal surfaces of the hemispheres, the cornu ammonis being also often implicated. It occurs most frequently in general paralysis, and is the most constant lesion in that disease. It may occur in simple chronic insanity, especially in terminal dementia; and in chronic, alcoholic, and epileptic insanity (Griesinger).

It may occur in senile mania, and is frequently met with in traumatic insanity (Lewis).

(*e*,) THE CEREBRAL SUBSTANCE :—(*a*,) *Colour.*—The cortex may be markedly pale, as in anæmic insanity, phthisical insanity, and insanity after hæmorrhages, or in other forms of insanity accompanied by general anæmia. Sometimes in states of mental weakness and dementia without general anæmia (Griesinger).

The cortex may present violet or wine coloured softenings in acute mania when death has occurred suddenly during excitement. Pinel found redness of the middle layer in mania; Baillarger of the three gray strata. Most frequently the inflammation is of the most superficial layers; when the adherent pia is detached an eroded surface is left. This is the condition generally found after death from general paralysis.

In acute delirium there is redness of the cerebral cortex.

The medullary substance may be abnormally pale, may be mottled with red, or rose, violet or lilac colour, or on a hardened section may present a number of fine white points (miliary sclerosis, colloid bodies). Miliary sclerosis, described by Batty Tuke, is found in cases of chronic insanity. Spitzka says the so-called miliary sclerosis can be produced by the action of alcohol, the colloid bodies being tiny aggregations of *leucin*. The medullary substance may also be a dirty white colour, blue or gray, without puncta vasculosa. There may be hæmorrhagic foci in the ganglia, white substance or cortex. There may be dark spots with softening and loss of substance consequent on hæmorrhagic foci. There may be patches of yellow softening in the ganglia or cortex, or of white softening in the medullary substance. There may be widely spread but superficial yellow softening of many convolutions presenting itself, without loss of substance, on removing the soft membranes. Occasionally hæmorrhage may show darkly through the basilar cerebral tissue.

(β,) *Size and Form.*—Cerebral atrophy is frequently met with (Griesinger).

It is sometimes convolutional, sometimes general. It may be primary (senile or premature marasmus of the brain); or it may be secondary to inflammation, hyperæmia, or compression (Griesinger).

In the chronic forms of insanity atrophy is most frequently found in the fronto-parietal regions, especially in the frontal lobes (Lewis).

Atrophy is very common in general paralysis, terminal dementia, and chronic alcoholism. It is rare in recent insanity (Griesinger).

The cornu ammonis is atrophied in 50 per cent. of the epileptic insane (Meynert).

The brain is frequently very small in idiocy. In paralytic idiocy one cerebral hemisphere is generally larger than the other, the opposite half of the cerebellum being then frequently smaller than its fellow, and the olivary body on the side of the larger cerebral hemisphere of greater magnitude than the one on the side of the larger cerebellar hemisphere (Meynert, Schroeder van der Kolk, Fritsch, etc.).

Cases illustrating crossed atrophy of the cerebrum and cerebellum have been reported by me in the "Brit. Med. Jour." and the "Jour. of Ment. Science," 1882.

The brain may be abnormally large in idiocy, either from hypertrophy or from dilatation of the ventricles by fluid ; or one hemisphere may be abnormally large with dilated ventricles, whilst the other is excavated or fragmentary, being in great part replaced by a serous cyst (porencephalus). Such a case was reported by me in the "Brit. Med. Jour.," 1882. In insanity from coarse brain disease there may be more or less extensive porencephalous excavations of the cortex or cerebral ganglia, or both ; or tumours of the substance may protrude, or those of bone or membrane leave indentations in the brain substance.

According to Buchstab, the frontal lobes are normally larger in the male than the female, and the occipital larger in the latter than in the former ("Jour. Nerv. and Ment. Dis.").

(γ.) *Consistence.*—There may be softening, general or local, resulting from inflammation or œdema in the former case, and from thrombosis or embolism in the latter.

General softening is found in general paralysis, and is especially marked in the superficial cortical layer of the convexity of the hemisphere. It is also found in chronic insanity and in senile insanity. Local softenings are also found in these forms and in insanity from coarse brain disease.

The cortical area most prone to thrombotic and embolic softening is that supplied by the parieto-sphenoidal branch of the middle cerebral or Sylvian artery ; the left side being more liable to be affected than the right. Next in frequency comes the area supplied by the occipital branch of the posterior cerebral.

Localised softenings may also be due to fatty degeneration (Bucknill and Tuke).

There may be induration, general or local, the former in chronic insanity, the latter in idiocy.

(δ,) *Specific Gravity.*—The conditions which favour high sp. gr. are congestion and induration; those which favour a low one, œdema and fatty degeneration. A watery or œdematous condition of the brain is frequently met with in dementia and chronic insanity generally, and in such cases the sp. gr. is low. It is low in the softened condition of circumscribed portions, which the microscope shows to be one form of fatty degeneration (Bucknill and Tuke).

The same authors quote a case (p. 589) in which the sp. gr. of the cerebrum generally was 1041, that of the softened part 1035. According to them the normal average sp. gr. of the cerebellum is greater than that of the cerebrum, and that of the cerebral white substance greater than that of the gray.

Thudichum gives the mean sp. gr. of the gray tissue as 1032, of the white as 1041, of the whole brain as 1037.

(ϵ,) *Weight.*—Griesinger agrees with Parchappe that a diminution of brain weight takes place in insanity generally, and especially in chronic insanity. In some cases of insanity from coarse brain disease the weight is much diminished from local loss of cerebral substance. In general paralysis there is a diminution in weight varying from about one to about five ounces (Mickle).

In senile insanity the weight is also considerably diminished.

Luys ("L'Encéphale") says that whilst in healthy persons the left cerebral hemisphere is generally heavier than the right, the reverse holds good in cases of chronic insanity.

Bischoff has given the average normal weight of the brain as 1362 grammes in the male, and 1219 in the female; R. Boyd as 1325 and 1183 respectively (" Allg. Zeit. f. Psych.").

(*f*,) THE VENTRICLES.—They may be dilated (chronic hydrocephalus). Their surfaces may be granular as in general paralysis and hydrocephalus.

(*g*,) THE CEREBELLUM.—Diminished in consistence but not in weight; hyperæmic; membranes thickened and often adherent in general paralysis (Voisin, Mickle); asymmetrical, atrophic, and occasionally absent in idiocy. Sometimes softened locally or atrophied in insanity from coarse brain disease. Occasionally defectively developed in epileptic insanity.

II.—MICROSCOPICAL.

For microscopical morbid anatomy *vide infra* (special morbid anatomy).

B.—SPECIAL.

CHOREIC INSANITY.—Small clots have been observed in the nucleus lenticularis, internal capsule, and thalamus. In a case of mine there were lacunæ in the outer division of both lenticular nuclei, and in the left external capsule. These were seen in the fresh tissue, and were visible to the naked eye. Several authors believe chorea itself is caused by a bacillus, and they accordingly treat the disease antiseptically. Huntington, Jolly, Remak, etc., describe cases of "hereditary chorea" ("Centralblatt f. d. Med. Wiss.").

CLIMACTERIC INSANITY.—In a case observed by Clouston in a male there were atrophy and anæmia of brain convolutions, incipient arterial atheroma, dilatation of lateral ventricle, and a patch of white softening in centre of left hemisphere.

COARSE BRAIN DISEASE.—Insanity from softenings, hæmorrhages, atrophies, chronic degenerations, and occasionally tumours in various parts of the encephalon. *Vide supra* ("Consistence" and "Size").

Charcot ("Leçons sur les Localisations Cérébro-spinales," p. 93) says that one of the lenticulostriate branches of the Sylvian artery might with justice be called the cerebral hæmorrhagic artery on account of its size and its predominant *rôle* in intra-encephalic hæmorrhage. He also states (*op. cit.*, p. 76) that whilst hæmorrhage is common in the central ganglia and comparatively rare in the cerebral lobes, ischæmic softening is more frequent in the latter situation. The patches of softening are yellow in the central ganglia and cortex and white in the medullary substance. The cortical softening may be spread over many convolutions without loss of substance. Hæmorrhages according to their age may be evidenced by the presence of the blood itself, by dark spots with softening (with or without loss of substance) by cicatrices, by dark stainings, or by cysts. According to Bevan Lewis the brain is lighter in organic dementia due to softening than in any other form of acquired insanity.

Tumours.—"The most common intra-cranial growths are tubercular and syphilitic; next comes glioma, then sarcoma, and then cancer. Other forms are rare" (Gowers' "Diseases of the Brain," p. 199).

Clouston found tumours in seven of thirty-six cases of organic dementia. In most of these seven cases there was manifest secondary convolutional lesion through pressure or irritation. He found twenty-eight cases of brain tumour out of one thousand autopsies. Other authors say the proportion is about two in a thousand. In twelve out of thirty-six cases of organic dementia Clouston found very marked convolutional atrophy with or without softening. He concludes that gross brain lesions tend to cause mental disease in two ways: (1,) by reflex or other irritation, or excitement of morbid convolutional action; and (2,) by actual destruction, primary or secondary, of convolutional structure (p. 394).

CONFUSIONAL INSANITY, CHRONIC, with excitement.—The blood vessels are tortuous, twisted, and bent on themselves. They are irregularly dilated and constricted. Sometimes the dilatations resemble aneurisms (Spitzka, p. 107).

CONSECUTIVE INSANITY.—After fevers, protoplasm of nerve cells cloudy, their processes stain poorly, their nuclei are ill-marked (Spitzka).

DELIRIUM, ACUTE.—(Acute delirium, delirium grave, acute delirious mania.)

(1,) *Macroscopical.*—Hyperæmia of the cerebral membranes, sometimes nothing at all (Griesinger, Briand, Bra, etc.) Often a peculiar red coloration of the gray substance cortical and central. Sometimes a violet tinted mottling of the white substance, the gray showing nothing. The cord and its membranes are similarly altered, though to a less degree and after the cerebral change (Briand, Bra). Spitzka states that an intense hyperæmia of the brain and its meninges is constantly found in patients dying during the excited period, but that in those who die in the stuporous period this is "sometimes obliterated by a collateral œdema." "In all the brain appeared swollen." Layers of fibrine (newly formed) and white streaks consisting of leucocytes have been found by him around the vessels of the pia and cortex.

(2,) *Microscopical.*—"The cortical ganglionic elements are granular or opaque, stain poorly, and their periganglionic spaces, like the adventitial lymph sheaths, are literally crammed with the formed elements of the blood" (Spitzka). This author considers the disease to be "the result of a vaso-motor over-strain." Bra states that the blood corpuscles are diminished in number, and that Jehn has observed pigmentary granulations along the course of the blood vessels, and an increase of nuclei in the neuroglia, and the presence of cells attacked by fatty degeneration; further, that Briand has observed bacteria in the blood and a special red coloration of the internal tunic of the aorta at the arch.

DELUSIONAL INSANITY (Monomania, Paranoia).—Occasionally there are found anomalies of skull and brain, in part congenital, and in part acquired in early infancy, asymmetry of hemispheres or convolutions, or both, vascular abnormalities, anomalies of minute cerebral elements (Spitzka, p. 90). Frequently nothing is found.

DEMENTIA, TERMINAL.—Cases of chronic insanity in which no anatomical lesion is found are rare (Griesinger). Opacity and thickening of the membranes; atrophy of the brain, particularly of the convolutions; chronic hydrocephalus; subarachnoid effusions; cortical pigmentations; extended and profound cerebral sclerosis; granular ependyma are the lesions commonly met with (Griesinger, "Psych. Krank.," pp. 442-443). Considerable atrophy of the brain corresponds as a rule to a condition of profound mental weakness (*op. cit.*, p. 443). Griesinger says (*op. cit.*, p. 422) the morbid change is most frequently found in the anterior portion and superior part of the middle portion of the cerebral hemispheres. B. Lewis states that extreme atrophy of the frontal lobe is found associated with dementia accompanied by *extreme somnolence* and incapacity for the slightest mental effort. Spitzka states that the diminution of brain weight in dementia is considerable. He gives the macroscopical characters as diminished size of brain, enlargement of ventricles, shrinking of convolutions, widening and gaping of sulci, increased firmness on section, ready formation of gaps in white substance when being hardened, strands of fibres of a grayish or brownish tinge or dirty white colour.

Microscopical.—Spitzka also states that in chronic insanity the nerve-cells undergo passive atrophy, and still more frequently an intensification of the normal process of involution.. The final phase of the latter change, destruction of the cell, is noted in the larger pyramids in extreme dementia; the nucleus and cell processes disappearing and the cell being represented only by an irregular mass of granules and pigment. Pigmentation extends to the smaller pyramidal cells, and is not limited to the larger pyramidal nerve-cells as in healthy middle life. The nerve-cells are diminished in number. The neuroglia elements are increased, hence the greater firmness on section. Granular and yellowish, rarely brownish pigmentary matter in adventitia, especially at bifurcation of vessels. Proliferation of the nuclei in the vascular tissues and a fine granular or a colloid-like change of the muscular coat are common. General vascular sclerosis (a sequel of the nuclear proliferation) is common, and a whole vessel may degenerate into a fibrous filament, perhaps without a lumen (Spitzka).

EPILEPSY.—(1,) *Macroscopical.*—Tumours. Osseous spiculæ. Cortical erosions and softening. In 50 per cent. of the epileptic insane the cornu ammonis is found atrophied (Meynert). Clouston says that any irritation may cause epilepsy in a brain of a certain quality, but that irritation of the cortical motor area is much more apt to cause it. He calls attention to the fact that in nearly all cases one cerebral hemisphere is considerably heavier than the other, and that in children with one hemisphere better developed than the other, there is, in the majority of cases, epilepsy. In Jacksonian epilepsy there are found nearly always, erosions, adhesions, softenings, hæmorrhages, abscesses, or tumours in or near the cortical motor area.

(2,) *Microscopical.*—Atrophy of the cortical cells of the cornu ammonis (Meynert). Degeneration of the cells of the second layer (the small pyramids) of the general cortex cerebri. Abnormal excess of the neuroglia (Lewis).

GENERAL PARALYSIS.—(1,) *Macroscopical.*—Durhæmatomata (arachnoid cysts) between dura mater and arachnoid (or more correctly speaking, between the parietal and visceral portions of the latter membrane) occur more frequently than in other mental diseases. The arachnoid is very prone to be rough and granular on its outer surface; it is also frequently thick and opaque. Maudsley ("The Physiology and Pathology of Insanity," p. 456) says, "Great œdema of the membranes is one of the most frequent morbid changes." He also states that Sankey found effusion beneath the arachnoid (sub-arachnoid effusion) in eleven out of fifteen cases. Morbid adhesions between the cortex and its investing membranes constitute the most characteristic naked eye lesion, being only absent in a small percentage of cases. The walls of the ventricles have a roughened granular appearance (*condition sablée* of the French authors); this is well seen on the floor of the fourth ventricle. It is there rarely absent in general paralysis. The cortex is much thinned; it is often mottled with pink, often pale gray, or of a uniform dirty gray hue with but poorly defined lamination; the arterioles are frequently coarse and engorged; these changes are most marked in the frontal lobes, less advanced in the parietal, and seldom seen in the occipital. There may be redness of the medullary substance from increased vascularity. There is frequently general diminution of consistence. The average brain weight is diminished (Voisin, Mickle, Lewis).

(2,) *Microscopical.*—(*a*,) Vascular elements. The arterioles of the pia and cortex are distended and bulge at intervals; engorged, tortuous, thickened, and surrounded by nuclear proliferations; (*b*,) Neuroglia elements. They are greatly augmented. The cells are increased in number and size, and their radiating processes in number and thickness (Obersteiner, Luys, Mickle, Lewis). Auguste Voisin considers the neuroglia to be true nervous, and not connective, tissue (*op. cit.*, p. 459). B. Lewis attaches great importance to the action of the Deiters' cells (spider cells, lymph-connective elements), which, according to him, supplement the action of the blocked lymphatic channels, and, as above stated, increase in size, number, etc.; (*c*,) Lymphatic elements. The channels are blocked with *débris*, etc.); (*d*,) Nervous elements. There is degeneration of the cortical cells (Mickle, Voisin, etc.) Tuczek ("Allg. Zeit.") considers the wasting and disappearance of the cortical association fibres to be the most important alteration. The finest fibres of the frontal cortex are first affected, then the coarse fibres of the same lobe, then the fibres of the remainder of the cerebral cortex.

Of these changes the vascular are the first to show themselves.

Spinal cord, frequently affected. There are vascular alterations. Sometimes there is degeneration of the posterior columns.

IDIOCY.—(1,) *Macroscopical.*—The head is generally small. In some cases the cranial bones are very thick, the diplöe being of the consistence of ivory. This osseous condensation is often very unequally distributed (Bra).

Sometimes the cranial bones are very thin. There may be a prominence visible during life over the affected side when the patient is suffering from porencephalus (Griesinger).

The cranial capacity is usually much less than in normal individuals. This often arises from the premature ossification of the cranial sutures which may be uniform, giving rise to microcephalus, or may affect certain sutures and respect others, and thus produce the dolichocephalic cranium projecting either in the occipital or the frontal region.

In hydrocephalic idiocy the head on the contrary will be large, although the brain may be very small.

The average *weight* of the *brain* of male adult idiots is 1188 grammes, of female, 1057, the average physiological weight (according to Wagner and Broca) being respectively 1410 and 1262 grammes (Bra). This atrophy, or rather arrest of development may be caused either by premature ossification of the cranial bones or by trophic disturbances. There is often a great difference between the two cerebral hemispheres, the two halves of the cerebellum, and the olivary bodies.

The cerebrum may be markedly asymmetrical whilst the cerebellum is symmetrical, or nearly so.

Hypertrophy of the brain occurs occasionally in idiots. Griesinger says it is impossible to distinguish it during life from hydrocephalus. He quotes Baillarger as mentioning the case of a child of four years of age in whom the brain weighed 1305 grammes, and of another in whom the body weighed 46 pounds, and the brain 1160 grammes. Bricquet, Delasiauve etc., have reported analogous cases.

The *corpus callosum* may be atrophied or absent. The central ganglia may be atrophied on one or both sides, the pons may be atrophied; even the total absence of one of the cerebral hemispheres has been observed.

There may be *porencephalus*, the substance of the convolutions and centrum ovale being more or less destroyed, and the resulting cavity filled with serous fluid enclosed in a cyst.

Hydrocephalus is frequently observed; the fluid sometimes lying on the surface of the brain, sometimes accumulating in the ventricles only, forming two pouches which push the cerebral substance forward and cause the frontal region to project during life.

The *convolutions* are generally thin, narrow, attenuated, the *fissures* appearing consequently wider and deeper. The *frontal convolutions* are those most frequently irregular, the marginal coming next (Bra). According to Luys the *frontal ascendant* is the convolution which most frequently evinces traces of arrest of development. Of the fissures, the fissure of Sylvius displays widening most markedly. The occipital lobe is often shortened. It frequently presents abnormally small convolutions (microgyria) (Spitzka).

Griesinger states that various observers (Stahl, Röseby, Niépce) have remarked an unusual abundance of gray substance (in the ordinary localities) in the brains of certain idiots. Occasionally new formations of gray substance are found in parts where gray matter is not normally found (Virchow in an epileptic idiot, Griesinger in an epileptic whose mental state he did not know).

(2,) *Microscopical.*—Spitzka observed in imbeciles a disproportionate thickness of the outer or barren layer of the cortex cerebri. This was general. There was also a relative sparseness of the cortical ganglionic elements, "particularly noticeable in the granular layers." In one case the blood-vessels were sclerotic, and there was a general preponderance of connective tissue elements over the nervous structures throughout. He also found large numbers " of nuclear bodies surrounded by a little granular protoplasm, and contained in clear round spaces of the neuroglia."

In epileptic idiocy the nerve cells may be arrested in development (Bevan Lewis).

Bra says the most characteristic lesion is the arrest in development of the capillary net-work of the cortex, the meshes being imperfect, the vessels incompletely formed, varicose, and attacked by granular fatty degeneration. Sometimes instead of vessels little lacunæ filled with blood globules are observed. The nerve cells are slashed, irregular, infiltrated with calcareous salts.

H. Major ("Jour. Ment. Sci.," 1883) found, in the sclerosed and atrophied cerebellum of an epileptic imbecile, the molecular and granular layers diminished in thickness with increase of connective tissue elements in former, the Deiters' cells being numerous and prominent, disappearance of the Purkinjean corpuscles in the sclerosed leaflets, connective tissue in place of the white nerve fibres of the leaflets; he found similar, though less advanced, changes in the cerebellum of a paralytic idiot.

KATATONIA.—(1,) *Macroscopical.*—According to Meynert, there is ventricular dropsy; he says there is also, perhaps, premature ossification of the sutures (Verity, " Jour. Nerv. Ment. Dis."). There are indications of basal meningitis (Meynert, "Umfang, etc., der klin. Psych."). Julius Mickle found ("Brain," 1891, Part liii., p. 100) in one case patches of adhesion and decortication at the anterior part of the inferior and mesial surfaces of the cerebral hemispheres, "slight thickening and opacity of the pia-arachnoid over the brain base," and "great thickening, toughness and opacity over the supero-lateral, fronto-parietal region, with corresponding pial œdema, over-easy removal of meninges and wasting of gyri. Evidence of old hyperæmia. Meninges firm over inferior surface of cerebellum, resisting firmly when being cut. Cerebellar substance pale, relatively to cerebral, and slightly firmish. Pons, and medulla pink.

(2,) *Microscopical.*—Mickle's case, back part of left F^1; vacuolation, increase of nuclei in and around cell-walls, slight tortuosity of some of the blood-vessels.

Meynert says there are microscopic exudations (Verity).

MANIA.—In acute mania very frequently nothing whatever is found. The most constant appearance is hyperæmia more or less extensive of the membranes and of the brain substance, cortical and medullary, peripheral and central (Griesinger, Bucknill and Tuke, Spitzka, Bra). But this is seen in other mental diseases and even in sane persons (Spitzka). The microscope reveals nothing characteristic. In the *sub-acute* forms Luys says the membranes are thickened, and the cortical substance is thinned and sometimes indurated. The deep layers of the cortex may be hyperæmic. Several of the above-quoted authors include acute delirium under the head of Mania. In the *chronic* forms Luys states that more or less general atrophy of the cerebral lobes is found. The nerve-cells according to him are irregularly shaped, yellowish, and in some places have altogether disappeared (Bra). (See "Dementia, Terminal.")

MELANCHOLIA.—Nothing characteristic. Sometimes hyperæmia of cerebral membranes and substance, much more frequently anæmia, occasionally (especially in the stuporous cases) œdema. But all these appearances are found in sane persons dying from fever, phthisis, and other diseases (Spitzka).

Meynert ("Bau der Gross-Hirnrinde," etc.) mentions the presence of free nuclei in acute melancholia with trophic disturbances and proliferation of the nuclei of the ganglion cells in excited (crethische) melancholia.

For the morbid conditions in chronic cases tending towards dementia see "Dementia, Terminal."

PERIODICAL INSANITY.—In advanced stages the arteries are found twisted, tortuous, reduplicated, sacculated (Spitzka, p. 107).

PUERPERAL INSANITY.—If occurring after severe post-partum hæmorrhage, marked pallor of brain substance.

SENILE INSANITY.—(1,) *Macroscopical.*—Thickening of cranial bones. Opacity of arachnoid with thickening. Morbid adhesions between cortex and its investing membranes in senile mania. Atheroma and dilatation of arteries. Atrophy of cerebral substance with softening and loss of weight.

Lacunæ in white substance. The lesions may be localised in the frontal lobes or general.

(2,) *Microscopical.*—Meynert ("Bau der Gross-Hirnrinde," etc.) describes and depicts a large cortical pyramidal cell with a sharply oval nucleus containing a tripartite nucleolus; the latter (the tripartite nucleolus) he considers distinctly pathological; he found it in the cortex of a man aged 70, in whose brain there were also nuclear proliferations and masses of pigment along the course of the vessels with colloidal heaps in the white substance. Spitzka states that a destruction of the large pyramidal cells is noted in advanced senility as in extreme dementia; the nucleus becomes obscured or disappears, and the cell processes separate and also disappear; finally nothing but an irregular mass of granules and pigment is left to represent the cell. He says micro-chemistry has demonstrated that this is not a *fatty* degeneration. In *senile dementia* the capillaries are dilated. The vascular sheaths are loaded with pigment. The peri-vascular spaces are dilated. Neuroglia broken down in patches within which only molecules and nuclei can be detected. Granular degeneration going on to molecular disintegration of nerve cells, especially large pyramids. The nerve fibres are coarse, twisted, irregular, and sometimes broken up (Bucknill and Tuke).

SYPHILITIC INSANITY.—There may be gummata or gummatous infiltration (disseminated miliary nodules of Engelstadt and Lancereaux). The former occur most frequently in the dura mater and the sub-arachnoid space. They are of two varieties: (1,) Somewhat hard, white, dry, circumscribed tumours of varying size; (2,) Grayish red, moist, gelatinous, semi-transparent masses. They are formed from the round and stellar cells and the nuclei of the meshes of the normal intercellular tissue, and at their periphery the cellular infiltration fades gradually into the healthy tissue. There may be sclerosis (sclerotic hyperplastic encephalitis of Hayem) diffuse, limited to one convolution, or scattered in isolated patches; the blood-vessels are dilated and contain leucocytes in their peri-vascular sheaths; the neuroglia cells are swollen and contain several nuclei; atrophy and fatty degeneration of the nervous elements. The blood-vessels may only participate in the above lesions or they may be the sole structure affected; the lesion is situated under the endothelium; it is a sort of membrane formed of giant cells and connective elements (Bra). Syphilis (according to Heubner) attacks only the middle-sized and smaller arteries of the brain, whilst atheroma attacks any of the arteries of the body and only the greater and middle-sized arteries of the brain (Stretch Dowse, "Syphilis of the Brain"). There may be lesions which, according to Fournier and others, are only distinguishable from the lesions of diffuse interstitial meningo-encephalitis by a greater preponderance of the meningeal alterations (Bra). There are no specific lesions in cerebral syphilis. All is inflammation, proliferation, degeneration (Bra).

TOXIC INSANITY.—*Alcoholic Insanity.*—(1,) *Macroscopical.*—Arachnoid thickened and opaque. Fatty degeneration and varicose dilatation of the vessels of the pia; here and there extravasations of blood; occasionally adhesions of the pia to the convolutions, as in general paralysis; arterial atheroma; induration of encephalic tissue; localised softenings; hæmorrhagic foci; limited congestions.

(2,) *Microscopical.*—Numerous proliferating nuclei on the walls of the blood-vessels, the latter being coarse and tortuous. Fatty degeneration of large pyramidal and spindle cells. Increase of connective tissue elements (Griesinger, Magnan, Bra, Lewis).

The vessels, nerve-cells and connective elements of the spinal cord are also frequently affected.

Saturnine Insanity.—The brain contains lead. It is anæmic, yellowish resistant, sometimes œdematous. The convolutions are hard and seem crowded together. In saturnine pseudo-general paralysis adhesions have been observed between the meninges and the convolutions, and between the convolutions themselves (Bra).

TRAUMATIC INSANITY.—Adhesions between the cortex and its investing membranes, the result of chronic meningo-cerebritis, are very frequent (Bevan Lewis).

C.—MORBID ANATOMY OF SYMPTOMS.

DISORDERS AND DEFECTS OF CUTANEOUS SENSIBILITY.

Nearly all authors agree in the view that lesions of the posterior third of the hinder limb of the internal capsule cause anæsthesia, but there are differences of opinion as to the cortical centre for touch (that is, the centre in which tactile sensations become perceptions), Ferrier placing it in the hippocampal region, Luciani and Sepilli, Gowers, and others locating it in the "motor" area and the region behind it. Flechsig has traced the sensory fibres to the parietal and central region, "roughly speaking, the part of the cortex lying under the parietal bone" (Gowers).

Gowers quotes Demange's case of extensive cortical lesion confined to the outer surface of the right hemisphere, and symptomatised by left hemianæsthesia, hemianalgesia, loss of temperature sensation, and other sensory defects. In support of this view (and cases bearing it out are from time to time reported in the journals), is the case reported by Mason ("Lancet," May 30th, 1891), in which there was abscess of the lower part of the right central convolutions with paralysis of and loss of sensation in the left arm; sensation and motor power returned after trephining and evacuating. In another case ("Brit. Med. Jour." Supplement, Aug. 1, 1891, p. 35), a blow with a hammer on the left side of the head caused fracture over the fissure of Rolando. There ensued aphasia, right hemiplegia, and right hemianæsthesia. Postempski removed the depressed bone over an area of three square centimètres. Three hours afterwards the hemiplegia had almost disappeared, but there were still zones of anæsthesia on the outer aspect of the limbs. These zones disappeared in six days.

Of forty-four cases (organic dementia, senile dementia, idiocy, general paralysis, melancholia) in which I found well marked cerebral macroscopical lesions, nineteen had suffered from disturbances of cutaneous sensibility; sixteen of these had suffered from anæsthesia and analgesia, unilateral in ten cases; three had been troubled with cutaneous irritation, formication, etc.

In nine of the cases of anæsthesia the lesions affected the cortical motor area, seven being destructive, one (general anæsthesia) from pressure; and one (hemianæsthesia) from pressure and inflammation. Of these nine cases, four were unilateral with destructive lesions of the opposite motor area; and in one (the case of hemianæsthesia from pressure, etc., above-mentioned) there was a tumour in the opposite temporal lobe with membranous adhesions over the central gyri and base of the second frontal convolution.

There was softening of the opposite corpus striatum in two cases of hemianæsthesia, in one of which the loss of feeling was most marked in the foot; and in one case where the paralysis of sensation was at first unilateral and afterwards bilateral, both lenticular nuclei were softened. The anæsthesia was most marked on the face and arms.

In two cases the median surface of the cerebrum was affected, one unilateral as to the anæsthesia and opposite, there being a destructive lesion confined to the gyrus fornicatus (which is interesting in the light of Horsley and Schaefer's experiments on that region), the other bilateral, and in it the cerebellum was softened on the same side as the cerebrum. In one case a clot in the right lobe of the cerebellum caused (doubtless by pressure) left hemianæsthesia. In one case of general anæsthesia the right hippocampus and cornu ammonis, the tip of the left temporal lobe and the posterior parts of both thalami were soft and violet coloured, and the cerebellum was congested. In all the three cases with cutaneous irritation the temporal and occipital lobes were diseased; in one there was in addition superficial discoloration of the fronto-parietal region, and in another a durhæmatoma over the convexity of both hemispheres.

Auditory Disturbances.

Nearly all authorities are agreed that the cortical centre for hearing is in the temporo-sphenoidal lobe, Gowers, Ferrier and others locating it in the first convolution of that lobe. Meynert, however, gives it a more extensive area than this particular convolution. Luys places it in the occipital lobe, and especially in the cuneus. Luciani's experiments on dogs led him to the conclusion that the perceptive auditory area includes the whole of the temporal lobe, and radiates from there to the parietal and frontal regions, the hippocampus, and the cornu ammonis ("Brain," July, 1884). Schaefer ("Brain") found no deafness after removal of both temporal lobes, but says the animals appeared to be idiotic or stupid. Westphal and Gray have had cases of temporal lesion without word-deafness or any deafness, but with great loss of memory ("Jour. Nerv. and Ment. Dis."). Bevan Lewis has found temporal lesions where there have been auditory hallucinations, especially in alcoholic cases.

Of the forty-four above-mentioned cases of mine twelve had ascertained auditory disturbances, seven hallucinatory. In six of these seven the temporo-sphenoidal lobe was affected, in three bilaterally, in two unilaterally, and in all extending to other lobes; the lesions were either superficial softenings or membranous adhesions, and in one case a tumour in one temporal lobe with a small focus of ramollissement in the other. The seventh patient had suffered from hallucinations some time before admission, but not whilst in the asylum; the under surface of one lateral lobe of the cerebellum was grooved and indurated so that the auditory disturbances may have been caused by irritation of the nerve nuclei in the medulla. It is right, however, to mention that there was an old durhæmatoma, and that the occipital lobes were slightly softened The eighth case was one of word deafness. There was softening of the second and third temporo-sphenoidal convolutions on both sides; there was also arachnoid effusion over the left central and both parietal regions. Two other cases displayed slight deafness with slowness of mental reaction. In one the right supra-marginal gyrus was softened; in the other the right first temporal was so affected, and the left operculum was depressed on the island of Reil, by the abnormally large quantity of serous fluid, causing an indentation. In the two remaining cases there was extreme deafness. In one the membranes were adherent over the temporo-sphenoidal and occipital lobes on both sides; in the other there was softening of both lenticular nuclei, and in the right internal capsule a hæmorrhagic focus encroaching on the lenticular nucleus. A watch could not be heard when in contact with the left ear, and only at a distance of an inch and a half from the right. In one of the cases with hallucinations, a general paralytic in whom there were almost universal adhesions, there was a species of deafness which was rather word deafness than mere defect of hearing.

Visual Disturbances.

The visual sphere at first limited to the angular gyrus (*pli courbe*) by Ferrier, and to the occipital lobe by Munk, has been extended by Tamburini and Ferrier to both these regions, and by Luciani (in dogs) to the occipito-temporal region and parietal region of the vertex ("Brain"); and indeed, Goltz says removal of any portion of the hemispheres causes dimness of vision, but that this is greater when the occipital lobes are removed than when the frontal are so treated. According to Luciani, bilateral homonymous hemiopia is the most frequent symptom of occipital lesion. Bechterew says the visual area includes the whole of the occipital and part of the parietal lobes, and extends from the temporal to the median surface. He says there are two regions, one presiding over the retina of the same side, the other over that of the opposite side. Luciani and Goltz state that the blindness soon passes off, leaving mental or psychical blindness (Seelenblindheit). According to Munk each occipital lobe receives fibres from both eyes—from its own side of each eye, and removal of one lobe causes hemiopia, removal of both total blindness. This is of course denied by the above experimenters. Luciani and Sepilli limit the visual zone in monkeys to the occipito-parietal region ("Allg. Zeit.," 1887). Hysterical patients with hemianæsthesia are amblyopic, not hemiopic on the anæsthetic side. Gowers says he has seen a few instances of this "crossed amblyopia" in which there was certainly organic disease. He also states that several others have been recorded.

In Demange's case there was amblyopia with loss of colour vision in the left eye (*vide supra*). Seguin ("Jour. Nerv. Ment. Dis.") collected a series of forty-five cases of hemiopia in which there were lesions of the occipito-angular region, the tracts leading to it, or of the posterior part of the thalamus.

In a paper read at the Liverpool Medical Institution, Glynn described a case of word blindness with restriction of visual fields, caused by depressed fracture of the superior postero-parietal region of the skull an inch from the lambdoidal suture, and mostly on the right side.

Nine of my cases suffered from visual troubles. In two, vision was slightly defective; in one of these there was softening of the second and third temporal convolutions on both sides, with an arachnoid effusion extending over the convexity from the tip of the occipital lobes to the central gyri. In this case the defect was partly psychical; in the other the arachnoid was adherent over the occipital lobe, and there were superficial discoloration of parts of the frontal and parietal, and softening of part of the temporal convolutions on the right side, with a focus of softening implicating the cuneus, precuneus, superior parietal lobule, and gyrus angularis on the left. In another case with extensive cortical lesions the patient was nearly blind, but he had cataract of both eyes. Another patient could only distinguish light from darkness; in his case there was parieto-temporal softening on the right side. On the left there was atrophy of the lenticular nucleus, and the first temporal convolution was softened; the right lobe of the cerebellum was also soft and atrophied. In another case there was loss of vision of the left eye, and later on visual hallucinations appeared; the opposite gyrus fornicatus was softened, and there were small sclerotic portions scattered through the cortex of both hemispheres. Another patient had in addition to left hemiplegia, defective vision of the left eye. There were foci of softening in the right frontal and temporo-parietal regions, and the membranes were adherent over the right occipital lobe; the left hemisphere was macroscopically normal. A seventh case suffered from great restriction of the *left* visual field or from hemiopia of the *left* eye, affecting the right side of the retina; there were adhesions over both occipital lobes, and a destructive lesion of the left frontal lobe. In the two remaining cases there was no ascertained diminution of visual power, but in one there were visual hallucinations, and the left corpus striatum was softened. In the other there were delusions of identity based doubtless to a great extent on visual illusions; the cerebellum was congested with diseased vessels and probably softening, leading finally to hæmorrhage into its substance. There were lacunæ in both lenticular nuclei, and the columns of Burdach were slightly sclerosed on both sides.

OLFACTORY AND GUSTATORY DISTURBANCES.

Ferrier localises the centre for smell in the subiculum cornu ammonis, and that for taste in the lower part of the middle temporo-sphenoidal convolution. That there is a subordinate centre for the sense of smell in the subiculum is probable, as some of the fibres of the olfactory nerve pass directly to this convolution, and disease adjacent to these fibres has caused loss of smell on the same side (Gowers). That there is a centre in the opposite hemisphere is proved by the fact that disease of the posterior part of the internal capsule ("sensory crossway") gives rise to anosmia of the opposite side. That this olfactory area is, partly at least, on the outer surface of the hemisphere is shown by Demange's case already mentioned, in which there was, in addition to the other symptoms, loss of smell and taste on the side opposite to the lesion (Gowers). According to Luciani, it extends outwards and upwards from the cornu ammonis and hippocampal region to the parietal and frontal. He thinks the gustatory centre is close to the olfactory. More recently (1886) Luciani and Sepilli found that in animals, injury of the region in front of and near the Sylvian fossa, or (especially) of the cornu ammonis, caused decided bluntness of the sense of smell. Blows on the head, especially on the vertex or occiput, sometimes cause loss of smell and taste which may be permanent. That this is caused by *contrecoup* is not borne out by analogy, as blows over the lower part of the left motor area cause symptoms distinctly referable to the hemisphere injured and not to any other part of the encephalon. Broca believed the gyrus fornicatus to be functionally related to the sense of smell. Meynert is also of that opinion.

In only two of my cases were there decided and well ascertained olfactory troubles, and one of these was a case of hypochondriacal general paralysis with cortical adhesions everywhere, and an arachnoid effusion over the base of the first frontal encroaching on the ascending frontal convolution. There was also aneurism of the descending aorta. The patient from an early stage of his general paralysis suffered from an almost complete abolition of the sense of smell. He could distinguish strong snuff and pepper by sight, but not by smell; later he had disagreeable olfactory hallucinations and said he smelt offensively, and that he annoyed those about him. The other patient was completely anosmic. There was a large deep focus of softening affecting parts of the first, second, and third frontal, and, to a slight extent, the island of Reil. The arachnoid was slightly adherent over both occipital lobes. The anterior cerebral artery was exceedingly small, much smaller than the right, probably owing to embolism. The most feasible explanation of the anosmia is either that the frontal cortex has olfactory functions in man, or that the ischæmia, which caused the cortical necrosis, starved the head of the caudate nucleus, thus acting on the opposite nostril; and also deprived the olfactory bulb of blood, so acting on the nostril on the side of the lesion. Yet this hardly bears out Luciani's view that (contrary to the arrangement of the other sensory nerve tracts) the direct fasciculus is larger than the crossed one. The occipital lesions may have had some effect in both cases.

Meynert states that the head of the nucleus candatus is very large in animals with the sense of smell very keen; he also says the yellow nucleus of Luys is in close structural relationship with the olfactory lobe (Ranney, "Jour. Nerv. and Ment. Dis.");

and the recent researches of Steiner on bony fishes tend to prove by analogy that the human corpus striatum is a way station for olfactory impressions in their course from the sensory organ to the brain mantle. He states that the great brain of the shark is merely an olfactory centre, and that the vertebrate cerebrum has developed phylogenetically out of the olfactory organs ("Brain," 1890, p. 396).

The foregoing sensory areas (auditory, visual, cutaneous, and olfactory) overlap each other in the postero-parietal and hippocampal regions in animals. In man Luciani and Sepilli do not think there is any region common to them all ("Allg. Zeit.," 1887).

Motor Symptoms.

The part of the cerebral cortex concerned in voluntary motion in man is, roughly speaking, the superior middle region, or more accurately, the two ascending convolutions (frontal and parietal), the paracentral lobule, the superior parietal lobule, part of the precuneus, and the posterior part of the third or inferior frontal convolution. Most clinicians and pathologists seem to agree on this point, although this area is not quite co-extensive with that mapped out by the experimental physiologists. Schiff and Brown-Séquard deny the existence of a special cortical motor area. Goltz, formerly the strongest advocate of this negative view, now admits that ablation of the "motor zone" causes permanent motor defect. Charcot and Pitres limit the motor area to the central convolutions and the paracentral lobule. The cerebral cortex does not control individual muscles but muscular movements. Horsley and Schaefer localise in the gyrus marginatus the centre for the trunk muscles. The other "motor centres" are placed in the motor area in the following order from above downwards—leg, arm, face, lips, tongue. This order is not always borne out by clinico-pathological experience. Gowers thinks there is probably no sharp limitation between the "centres," at any rate between the centres for the extremities. Luciani found (and all his experiments are borne out by Sepilli's clinico-pathological observations) that removal of one centre affects others, but to a less degree. Von Gudden's experience was somewhat similar to that of the three last-named observers. In disease of the motor cortex and (according to Luciani) of the parietal region behind it, the patient loses the ability to localise the limbs in space. If his eyes are closed and one of his paralysed limbs held in any posture by the observer, he (the patient) cannot describe where the limb is, or imitate the posture with the sound limb. Bell called this phenomenon loss of the "muscular sense;" others believed that the ability of localising a portion of the body in space depended on the cutaneous sensibility, but the phenomenon above described is observed when the cutaneous sensibility is intact. Meynert believes the muscular sense is a function of the cerebellum, this organ receiving the impressions from the cerebral cortex. Cramer and others describe hallucinations of the muscular sense (in paranoia, etc.). Gowers found loss of the muscular sense where there was disease of the corpus striatum and internal capsule without cortical lesion, so that the symptom is not diagnostic of disease of the cortex.

Sharkey reported a series of six cases ("Lancet," 1883) strongly supporting the generally accepted localisation of the motor area. In his summary he states that he has never seen destructive lesion of that region unaccompanied by motor paralysis. This has also been my experience, though I cannot agree with the statement that destructive lesions confined to the regions outside the motor area are never accompanied by paralysis. This is certainly not the case so far as the right hemisphere is concerned. Cases supporting the view that the convolutions above named constitute the area or the greater part of the area for voluntary motion are constantly being reported, so that it will be unnecessary to go into the details of such cases.

In thirty-nine of my cases there was paralysis or paresis of voluntary motion, and in thirty-six of them on the side opposite the lesion. The other three were cases of right hemiparesis, and in each case the right half of the cerebellum was diseased. Nine of the thirty-six cases presented no lesions in the motor area, and in seven of these nine there was left-sided paralysis, one (brachio-crural monoplegia) arising from a glioma in the right temporal, another from an abscess of the same region (in both of these there was Jacksonian epilepsy), and a third from clot and softening in the base of the right second frontal. In another of these seven cases there was softening of part of the right supra-marginal gyrus with brachio-crural monoplegia of twenty years' duration, the arm being contracted and atrophied; in another of the right gyrus fornicatus; and in another of the right hippocampal region. The seventh was a case of general paralysis with cortical adhesions over the right temporal and orbital convolutions; there was facial monoplegia with epileptiform convulsions succeeding an apoplectiform attack. The preponderance of lesions outside the motor area in the right hemisphere as compared with the left (the effects of the two which were outside in the latter could be accounted for by pressure) shows that the former is less highly organised than the latter. Similar cases are occasionally seen in the journals, and in at least one case an operator, though using every care, failed to trephine over the seat of the lesion, which, being a right-sided one, gave rise to misleading left-sided symptoms. The two left-sided lesions outside the motor area, with dextral symptoms, were hæmorrhagic, and both in the prefrontal region. In one, besides the clot close to the white substance of the frontal ascendant gyrus, there were extensive lesions on both sides, but all outside the motor area. There was aphasia with dextral Jacksonian epilepsy. The other case was more interesting; there was only one lesion, a dark brown softening, 2 inches by $1\frac{1}{4}$ inches, situated in the anterior part of the upper and middle frontal convolutions. Six months before admission the patient had had an apoplectic attack followed by right hemiplegia, which passed off gradually and disappeared in four or five months. Afterwards her gait was staggering at times. Seven months after the first attack she had a slight apoplectiform seizure followed by right facial monoplegia and aphasia; the latter amounted at first to total loss of speech, but in a few days there was only some indistinctness of articulation (see "General Mental Weakness," *postea*).

Of the twenty-seven cases in which lesions were situated in the cortical motor area or the central ganglia, five of the cortical cases did not conform to the generally accepted limb-centre arrangement. None of these anomalies could be accounted for by pressure. In a sixth case in which there was a tumour as large as a walnut over the right ascending convolutions close to the falx, all the limbs were paralysed, and the left were slightly atrophied. In two of the five cases there was complete hemiplegia with the lower halves of the central gyri apparently free from disease; in another case there was complete hemiplegia with no lesion above the lower third of the ascending frontal; in a fourth there was brachio-crural monoplegia with brachial contracture, the precuneus being very small, the temporo-sphenoidal lobe small and sclerotic, and the central gyri apparently normal; in the fifth case there was right lingual monoplegia with lesion of the left superior parietal lobule and precuneus, and left facial monoplegia with softening of the first and second right temporal convolutions (not included in the seven right irregular cases, *vide supra*). Many such cases have been reported, notably one by J. Mickle, in which there was brachial paralysis (right) with atrophy of the leg centre. In two cases, when a part of the leg centre was removed by a distinguished operator to cure convulsions, the arm became paralysed immediately after the operation.

In very few of the twenty-seven were the lesions confined exclusively to the motor regions. In one there was extensive superficial softening with membranous adhesions over both prefrontal lobes. The right first temporal convolution was affected near its tip. There was a large quantity of serous fluid, and the operculum was depressed on the island of Reil, causing an indentation. There had been two apoplectic attacks followed by general muscular weakness, aphasia of the aphemic variety, and right unilateral convulsions. Between the fits, the patient complained of headache and giddiness, and after the second the power of co-ordination was diminished in both hands. There was also deafness which was at least partially word deafness. Another patient, mentioned under the head of "Olfactory Disturbances," had deep softening of part of the outer surface of the prefrontal lobe supplied by one of the branches of the anterior cerebral. This focus extended across the base of the third frontal to the surface of the insula. This patient suffered from a species of slight aphemia (indistinct jerking utterance) without any other localised paralysis, but he used to sit with his neck bent forward.

Of the five cases free from voluntary paralysis one had an extensive lesion in the prefrontal and insular region similar to that in the case last mentioned, but on the right side. He also kept his head or rather his neck always bent forward. In another case there was dilatation of the right pupil with a clot $\frac{1}{2}$ by $\frac{1}{8}$ inch in the left external capsule and opacity and adhesions of the arachnoid on both sides over the temporal and occipital lobes. In this case the patient (aged 62) was seized with the symptoms of cerebral hæmorrhage during a severe attack of diarrhœa, and never rallied. There was no hemiplegia as far as could be ascertained; had the coma passed off he would probably have been aphasic. In another case, chronic epileptic insanity, there was deviation of the head to the right during the fits. The cornu ammonis was sclerosed on both sides, more markedly on the right. In another case, suicidal melancholia, the arachnoid was slightly adherent over both occipital lobes; there were many dark points in the white substance; the aortic semilunar valves were thickened. The remaining case has been already described (eighth case under head of "Auditory Disturbances").

There was pupillary inequality in sixteen cases; the dilated pupil was on the side opposite in twelve cases of destructive lesion, and in one case of tumour. In one case of clot the dilated pupil was on the same side. Two cases were doubtful, there being in one a clot on the side of the large pupil and softening of the inferior parietal lobule on the opposite side, and in the other case softening of the median surface on the same side, and disseminated sclerosis in both hemispheres. In seven cases the so-called deep reflexes (knee-jerk and ankle clonus) were much more marked on the side opposite the lesion than on the same side. In three, tumour, disseminated sclerosis, and hippocampal softening, they were absent altogether, and in one case the knee-jerk was absent on the right side, but well marked on the other; the form of paralysis being right linguo-facial monoplegia with defective speech; there was also twitching. Some paraparesis, more marked on left side, existed. There was in the right hemisphere softening of the supra-marginal gyrus and adjoining part of the first temporal; on the left side there were softening of the first temporal and atrophy of the lenticular nucleus, accompanying which there was atrophy with softening of the right half of the cerebellum. This last lesion most probably caused the absence of the knee-jerk. In three cases there was ptosis, in two of them the supra-marginal gyrus was implicated, and in the third the nucleus lenticularis. The corpus striatum (candate or lenticular nucleus, sometimes the whole organ) was affected in six cases, and apparently this lesion caused the paralysis. Meynert says destructive lesions of the lenticular nucleus cause paralysis of the opposite side, especially marked in the face, tongue, and arm. Ziehen states that stimulation of the anterior part of the corpus striatum has the same effect as stimulation of the anterior part of the cortical motor area ("Brain.") In at least two cases I have found, with lesion of the lenticular nucleus, the paralysis most marked in some or all of the parts indicated by Meynert; one of these was mentioned above when speaking of the deep reflexes. In the other there was at first unilateral, afterwards bilateral, paralysis of the face, tongue, and arm, with lingual, labial, and facial tremor. There was anæsthesia of the paralysed parts. There was brachial contracture, but only slight crural weakness. Both lenticular nuclei were softened anteriorly. I was present at an autopsy performed by Luys on a case of aphasia at the Salpêtrière in 1878. The only lesion was a focus of softening in the lenticular nucleus. It was the third case of the kind Luys had observed. Meynert in support of his view quotes Nothnagel's experiments in which he destroyed the nuclei lenticulares by injecting a solution of chromic acid, and so caused loss of spontaneous movement. Gowers, on the contrary, says destruction of the nucleus lenticularis does not cause paralysis, unless the anterior part of the hinder limb of the internal capsule is implicated.

Wernicke states that the candate and lenticular nuclei are not connected with the cortex, and only indirectly, through the substantia nigra, with the pes pedunculi, and are therefore not way-stations for voluntary impulses as Meynert thinks, but more recent observers (Luciani, etc.) hold Meynert's view. Those who have seen Meynert's preparations (cleavages and sections) will not be surprised at this reversion. Wernicke agrees with Meynert that the tegmentum (directly connected with these nuclei) is the reflex path.

In several cases with cervical muscular weakness the insula was found diseased. There were fourteen cases of aphasia of various degrees, slight aphemia or dysphasia, amnesic aphasia, complete loss of speech, total inability to speak from early infancy, there being no congenital deafness. In these the posterior part of Broca's (the third or inferior left frontal) convolution, or (in one idiot) the left temporal lobe was either intrinsically diseased or subjected to pressure. In some cases the insula was also affected. In two cases of general paralysis the symptoms were apparently anomalous, there being left brachial or brachio-facial monoplegia, but the central gyri were found to be specially diseased on the right side, and the third frontal on the left. The right third frontal was affected in several cases, but in none of these was there aphasia. In one case there was a cyst over the lower part of the insula effacing the convolutions; there was marked weakness of the cervical muscles, *but no aphasia*.

There were nine cases in which Jacksonian epilepsy was present; the lesions were hæmorrhages, adhesions, softenings, abscess, tumour, cyst close to or in the motor area. In all except one the convulsions were on the side opposite to the lesion. The excepted case was doubtful as there was softening of the right gyrus fornicatus with disseminated sclerosis in both hemispheres; there were hemiplegia and hemianæsthesia of the left side with, at first, twitching and tremor; later there were convulsions confined to the right side. There was one case of dura matral tumour over the upper part of the right central gyri without any convulsions whatever. In nine cases displaying mental and motor excitement and restlessness, there were lesions of the fronto-parietal region or corpus striatum.

General Mental Weakness.

Defective attention, memory, judgment (general, not partial, as when there are delusions) and volition.—Griesinger pointed out long ago that the brain degeneration accompanying the mental weakness of the chronic forms of insanity affected especially the fronto-parietal convexity. Bevan Lewis says atrophy of the brain occurred in a little over two-thirds of a large number of cases, and that the segment of the brain most frequently atrophied was the fronto-parietal, particularly the frontal part of it. Ferrier, Ross, and many others think the prefrontal lobes are the seat of some of the higher cerebral functions, inhibition, attention, observation, etc. Hughlings-Jackson is inclined to think they are the principal seat of the re-representative cognitions (abstract ideas) and feelings (emotions). Sir J. Crichton-Browne considers it probable that, in addition to other faculties and acquisitions, the moral sense and religious emotions are localised in them. Harlow's crow-bar case is an instance of moral deterioration caused by prefrontal injury.

In three of four cases in which I found serious destructive lesions confined, or almost so, to the prefrontal region, one right, one left, and one bilateral, the phrase, "Takes notice of nothing," occurred in all the medical certificates. There was the usual general mental weakness displayed by organic dements, but a total absence of morbid "*sensiblerie.*" The patient with the left lesion, though generally apathetic, had explosive outbursts of anger at times, and the bilateral case had two apoplectic attacks followed by unilateral epilepsy affecting the whole of the right side, and aphasia. Some of these symptoms were however, doubtless, caused to a great extent by the pressure of the abnormally large quantity of serous fluid on the operculum, the insula being indented by this structure. To make a digression: an abnormally large quantity of serous fluid surrounding a diseased cerebral structure seems to endanger life in two ways: (1,) by pressure through its sudden increase on account of diminished excretion or other cause; (2,) by causing atheromatous arteries to rupture in consequence of lessened vascular support, when the serous fluid is suddenly diminished in quantity owing to increased excretion or other causes. In the fourth case there were no naked-eye lesions outside the left prefrontal lobe, and in it there was a large focus of brown hæmorrhagic softening anteriorly. This patient was also very weak mentally, and also free from morbid "*sensiblerie;*" but she was sleepless, restless, noisy, violent, and destructive. The latter symptoms resemble those of acute mania. In the beginning and at the end of her illness she suffered from right hemiplegia and aphasia, both the highest and middle centres of evolution of Hughlings-Jackson being then affected, whilst in the interim only the former suffered. In view of the connection between the prefrontal lobe and the cerebellum, it is noteworthy that two of these patients suffered from vertigo and staggering gait at times, and one from diminished power of co-ordination in the hands. The memory was most defective in patients with extensive temporal lesions.

Morbid "Sensiblerie."

Motiveless weeping was a marked symptom in nine cases, and in these the lesions were mostly confined to the occipito-temporal region or the central ganglia.

DELUSIONS.

There were depressive delusions in nine cases, and in six of these there were lesions (mostly membranous adhesions) of the occipital lobes; in one, lesions of the lenticular nuclei and cerebellum; in one of the lenticular nuclei alone; and in another of the *right* cortical motor area. There certainly seemed to be, as others have observed, a greater tendency to depression where there was right-sided cerebral disease. In one of the six cases, a hypochondriacal general paralytic, the lesions were very extensive, there being adhesions all over both hemispheres. Two general paralytics with expansive delusions had no adhesions over the occipital lobes, the macroscopical lesions being confined to the frontal, temporal, and anterior part of the parietal. It is right to state that the hypochondriacal general paralytic suffered from aortic aneurism.

SUICIDAL TENDENCY.

There were three cases with suicidal tendency. In one there was extensive right cortical disease with membranous adhesions over the occipital lobe, the left hemisphere being normal. A second case had disease of both lenticular nuclei and membranous adhesions on the inferior surface of the occipito-temporal regions (right and left) and cerebellum. The third in whom the tendency was very strong, had adhesions over both occipital lobes with numerous dark points in the white substance of these lobes, and no other lesion.

SITOPHOBIA AND ANOREXIA.

The last mentioned suicidal patient absolutely refused food. Another patient with adherent arachnoid over the occipital lobes and many dark points in their white substance, took food very unwillingly and sparingly, as did another with softening of the posterior extremity of both thalami. Ferrier observed refusal of food in most of the monkeys from which he had removed the occipital lobes, though defective vision or smell may have had something to do with this phenomenon. It is quite possible, however, that the occipital lobes subserve other functions besides those of vision or other sense, just as the so-called motor spheres subserve more than one function; and indeed this is very probable considering the eight-layered formation of the occipital cortex.

Erotic Tendency and Loss or Absence of Sexual Desire.

Ferrier noticed strong sexual desire in one of the monkeys from which he had removed both occipital lobes. Combette's patient with congenital absence of the cerebellum was a nymphomaniac. A patient of mine, a female epileptic, had a very strong erotic, or more correctly speaking, nymphomaniacal tendency; there was an extensive superficial brown discoloration of one hemisphere, and a destructive lesion of the inner and upper part of the opposite occipital lobe encroaching on the parietal. A paralytic (and epileptic) idiot, aged 19, with small cerebrum and large cerebellum showed no signs of puberty, and his genital organs were infantile. In the third stage of general paralysis, that is to say, when the cerebral cortical cells are extensively degenerated, sexual desire and sexual power are completely lost. The loss of sexual desire has been mentioned by some observers (Boeck, Le Bœuf, etc.) as the most marked or only phenomenon after the removal of the cerebrum in dogs and pigeons (especially the latter), if the animals can be kept alive through the drowsy and irritable stages. Impressions from the sensitive surfaces of the genital organs are probably received by the cerebral cortical cells at the posterior part of the cutaneous or tactile area, a region contiguous to the visual, auditory, and probably the olfactory and gustatory areas, and possibly overlapped by some of them. Hence the olfactory and gustatory hallucinations in masturbatory insanity, the olfactory and tactile in insanity caused by ovarian disease, the olfactory, auditory, and visual hallucinations and illusions in masturbatory paranoia. Sexual desire may be aroused from without by impressions made upon the special sense areas, or upon the sexual area.

The other symptoms accompanied lesions which varied exceedingly.

D.—LESIONS OF NON-NERVOUS ORGANS AND TISSUES.

HEART.—Valvular disease, with or without hypertrophy, is frequently found in organic dementia and organic melancholia. Obstructive valvular disease is often seen in simple and hypochondriacal melancholia; dilatation with irritability of the organ in chronic mania; fatty degeneration in dementia (Bucknill and Tuke). The muscular fibres of the heart are granular, and there may be chronic endocarditic lesions in chronic alcoholic insanity (Bra). The arterial system is atheromatous in the same disease.

LUNGS.—In phthisical insanity they become sooner or later affected if not already diseased. All chronic lunatics are prone to lung disease, especially pneumonia. Suspicious monomaniacs are peculiarly liable to phthisis (Clouston). In most cases these diseases are not revealed by subjective symptoms, cough being frequently absent, and there being frequently no complaint of pain or discomfort on the part of the patient. Often there is very little dyspnœa. Sometimes patients will cough in quiet intervals, but not during maniacal excitement. Some pallor, emaciation, weakness, singly or combined, may be the only indication of pulmonary disease. Chronic catarrh of bronchi in chronic alcoholic insanity.

STOMACH AND INTESTINES.—Dyspepsia is frequently met with. In chronic alcoholism the coats of the stomach are thickened and the glands hypertrophied. In chronic insanity the large intestines are often displaced. The most common displacement is that of the transverse colon to the lower part of the abdomen (Bucknill and Tuke, Spitzka, etc.). Foreign substances are frequently found in the stomach and intestines. Dysentery and diarrhœa are somewhat frequent in the chronic insane.

LIVER.—May be sclerotic or steatotic in chronic alcoholic insanity.

SPLEEN.—Usually small in chronic insanity (Bucknill and Tuke); hypertrophied and soft in chronic alcoholic insanity (Bra).

KIDNEYS.—Usually small contracted kidney (interstitial nephritis) in insanity of chronic Bright's disease. Sometimes cirrhosis and parenchymatous nephritis in chronic alcoholic insanity (Bra).

BLADDER.—Congested in chronic alcoholic insanity.

UTERUS AND OVARIES.—Congestion, displacements, false corpora lutea, cysts, atrophy of ovaries, tumours. In a case of mine (melancholia with *auditory* hallucinations), without visible cerebral lesion, there were several small tumours attached to these organs. (See "Morbid Anatomy of Symptoms.")

THYROID GLAND.—Frequently enlarged in cretinism and nearly always in insanity from exophthalmic goitre. Diminished in insanity from myxœdema.

BLOOD.—Diminution of red globules, and increase of white in anæmic insanity, chronic alcoholic insanity, lactational insanity. (See "General Paralysis.") Deficiency in hæmoglobine and hæmacytes progressing with age in dementia. Deterioration also in masturbatory insanity, general paralysis, and epileptic insanity, and after prolonged excitement. It improves with mental recovery (McPhail, "Jour. Ment. Sci.").

SKIN.—Bedsores or asthenic gangrene are liable to occur in general paralysis and organic dementia. The skin may be picked into sores by the patient. Bullous eruptions are sometimes present.

MUSCLES.—Fatty degeneration in chronic alcoholic insanity. Von Sass has found extensive atrophy of the fibres in paralysis agitans ("Jour. Nerv. and Ment. Dis.," Sept., 1891).

EARS.—One or both may be swollen from othæmatoma, or contracted or shrunken as a consequence of that affection. The auricle may be mal-formed in the degenerative insanities and in criminals, or present the Darwinian tubercle on its postero-superior border.

HAIR.—The chronic insane seem to be less prone to baldness than sane persons of the same age, but liable to become prematurely gray. The hair often becomes dry and rough.

BONES.—The bones are abnormally brittle, especially in general paralysis. They are liable to be affected with osteo-malacia. Walsh ("Lancet," July 25th, 1891) describes four cases of delusional and hereditary insanity with osteo-malacia and cardiac valvular disease. The osteo-malacia was symptomatised by kyphosis and tendency to fracture.

PATHOLOGY AND PATHOGENESIS.

The highest powers of the microscope and the latest discoveries in histological method do not enable the observer to find anything in the central nervous system but cells and their processes (A. Hill, "Brit. Med. Jour.," July 4th, 1891). Some of the processes are long, and are called fibres. Others "branch until they are lost in what seems to be a common homogeneous ground substance." These cells of the cerebral cortex are of various shapes and sizes, and in the adult are arranged in definite layers. Over the greater part of the cortex these layers are five in number (see Meynert's "Psychiatrie," I. Hälfte, Bau der Grosshirnrinde, etc.; Hugenin's "Allgemeine Pathologie der Krankheiten des Nerven Systems," I. Theil, p. 229, *et seq.*, etc.) Meynert, Bain, and most authors believe that these cells receive and register external impressions, and by intercommunication form idea-apparatus. Meynert ("Bau der Grosshirnrinde und seine örtliche Verschiedenheiten") has roughly estimated the number of the cerebral cortical cells at 612,112,000 or more; Bain ("Mind and Body") at about 1,200,000,000 with say four times as many association fibres; and as the most richly endowed brain does not possess 200,000 ideas, there are manifestly cells and fibres sufficient and to spare for the formation of idea-apparatus. For these purposes the cells are connected with the organs of special sense and other parts of the body by means of the fibres of the system of projection, and with other cells near or at a distance by the fibres of association (Meynert). In certain positions (to be mentioned below) some of the cells are connected with efferent fibres communicating ultimately with the fibres of muscles through the intervention of spinal motor cells. Besides the intercommunication between cells of different convolutions far and near by means of the arciform fibres or fibræ propriæ, the cerebral lobes are connected with each other by bundles of fibres called "association bundles." The central ganglia are connected by means of projection fibres with the cortex; the convolutions of the cortex of one hemisphere are connected by means of commissural fibres (corpus callosum) with the corresponding convolutions of the opposite hemisphere (though this is denied by Hamilton); and the cerebrum and cerebellum are connected by several bands (*vide infra*). These essential parts of the mind-tissue are supported by connective tissue (neuroglia), nourished, especially the cells, by a copious blood supply, and drained by wide lymphatic vessels (perivascular lymph spaces). The effects of the quality of the blood supplied to the cerebral cortex are seen in intoxications, slow mineral poisoning, chlorosis, etc. With regard to the quantity of blood Mendel, Fürstner, and Kousnezoff found, after the establishment of artificially produced hyperæmia, the chief changes to be "hypertrophy of connective tissue of the vessels and neuroglia, and degeneration of the cells and nerve fibres" (Batty Tuke, "A Plea for the Scientific Study of Insanity," "Brit. Med. Jour.," May 30th, 1891). In a general way these are the changes found in general paralysis, chronic alcoholism, chronic epilepsy, chronic mania, terminal dementia, etc. (see "Morbid Anatomy"). Occasionally in mania and melancholia terminating fatally at an early period, sub-inflammatory products (leucocytes, pigment, nuclear proliferations, etc.) are found between the hyaline sheath and the muscular coat of the vessels of the superior convolutions. Occasionally there are evidences of stasis with distension and blocking of the perivascular lymph spaces (Batty Tuke, *loc. cit.*). In acute mania the most constant appearance is hyperæmia, and in the sub-acute forms there are found the results of prolonged congestion (thickened membranes, etc.). Meynert found 47 per cent. of the total number of the brains of maniacs hyperæmic. In patients dying during the excited stage of acute delirium, an intense hyperæmia of the brain and its meninges is constantly found (Spitzka). Meynert in his classification divides prosencephalic diseases into anatomical, nutritional, and toxic; the nutritional diseases (cortical or sub-cortical) include the so-called functional forms; there being cortical anæmia with sub-cortical hyperæmia in depression, and the reverse conditions in exaltation. Transitory local hyperæmia is physiological and necessary to the performance of the functions of the cerebral cortex, as has been demonstrated by several observers who have had opportunities of studying cases in which a portion of the skull had been removed. Batty Tuke has had two such cases under observation, and he says (*loc. cit.*) that bulging occurred through the openings during mental action; that this bulging was proportional to the intensity of mental action, and steadily increased until a certain maximum point was gained, gradually disappearing when the stimulus to cerebral action was withdrawn. Others have noticed the same phenomena, and the experiments of Lombard on the regional temperature of the head during intellectual activity and emotional excitement, support the conclusions arrived at by these observers. (See "Temperature, Regional of Head," chap. II.) If this condition is long maintained through persistent intellectual, or emotional activity, the physiological line may be passed and pathological conditions induced. The readiness with which some emotions cause vascular dilatation is seen in flushing and blushing; whilst certain emotions such as sorrow (Lange) and fear, cause vaso-constriction and consequent anæmia (encephalic) accompanied possibly by venous hyperæmia, and probably active, local, cortical hyperæmia of the occipital lobes. The waste products of the action of the brain are acid, lactic acid predominating (Maudsley). Roy found that the injection of acid brain filtrates was rapidly followed by hyperæmia. Meynert ("Psychiatrie") ascribes two functions to the cerebral cortex, viz., mentation and vaso-constriction; when the first is active the second is inactive and *vice versa*. The process then is somewhat as follows: The cortex receives a stimulus, intellectual or emotional, through the special senses or the general sensibility ("vivid impressions" of Herbert Spencer), aided in all probability by stimuli from other portions of the cortex ("faint impressions" of Herbert Spencer), by vaso-dilating drugs (alcohol, ether, etc.), by sexual excesses, or a too stimulating dietary (see "Etiology). Hyperæmia, at first functional, is induced, but, either through the force of the cause or combination of causes, or through cortical structural defect (heredity), the physiological line is passed, and the intercommunicating pericellular and perivascular lymph spaces are loaded with effete matters which assist in maintaining the hyperæmic condition. The neuroglia elements become hypertrophied, nourished by the super-abundant refuse matters; and the nerve-cells, strangled by the increased connective tissue, degenerate, the condition being symptomatised at first by delirium or mania, and afterwards by incurable dementia. The cells may also degenerate intrinsically apart from the action of the neuroglia. In acute delirium in the latter stages œdema takes place, causing the stuporous condition then observed. Anæmia, caused by profuse hæmorrhages, exhausting diseases or discharges, or starvation (see "Etiology"), may give rise to mental disease through defective nutrition of the cortex. The nuclear elements of the cells also undergo pathological alterations, Meynert having discovered proliferating nuclei and also free nuclei in melancholia, and nerve-cell nuclei with tripartite nucleoli in senility. There may be primary physiological and pathological changes in the cell contents which the microscope is as yet not powerful enough to show, or which require untried histological methods for their discovery; just as there are chemical physiological processes, of which we know little, going on in the brain cells and fibres synchronously with function, and pathological chemical processes of which we know less, manifested outwardly by symptoms of mental disorder. With regard to these chemical alterations: (1,) Prolonged activity changes the reaction of the axis-cylinder from alkaline to acid (Gamgee); (2,) Cerebral activity increases the quantity of phosphates in the urine or, at all events, it increases the alkaline phosphates, though Hodges Ward says it at the same time diminishes the earthy (Cranstoun Charles), it increases the quantity of the urine and the amount of urea; (3,) The cerebral phosphorised bodies (kephalin, myelin, etc.) have strong affinities for the oxides of lead, copper, manganese, iron, etc. (Thudichum).

THE CELL IN ITS RELATION TO OTHER CELLS OF THE CEREBRUM, THE CEREBELLUM, AND THE CENTRAL GANGLIA.

When a child of a few months old perceives for the first time, say an orange, the colour cells of the cortex most probably in the occipital region (Meynert, Munk, Hitzig, Schäfer, etc.) receive the impression from the retina corresponding to the colour orange; other cells in the same lobe the impression of roundness, these cells being connected by the arciform fibres; ganglion cells in the fronto-parietal region (Luciani, Gowers, etc.) receive the impression of sphericity (touch); others in the same region of solidity; yet others of more or less roughness of surface. The smell of the fruit is impressed on the cells of some of the convolutions of the cerebral convexity (Gowers); or in or near the cornu ammonis and hippocampus (Luciani). If the child lifts and bites the orange the impression of weight and resistance will be received in or near the central convolutions (Meynert, Gowers, Luciani). An impression will also be produced (according to Ferrier) on the cells of the lowest part of the middle temporo-sphenoidal convolution, but more probably on some of the cells of the convexity or in the hippocampal region. When afterwards the child hears the word "orange" associated with the object which has produced in him pleasant or unpleasant sensations as the case may be, communication is made by means of the longer association fibres between the above-mentioned cortical regions and the temporal lobe (Munk, Meynert), and unused cells there are added to the network (Ferrier limits the perceptive centre for the sense of hearing to the upper part of the superior temporal convolution); and when the child himself begins to use the word, cells in the opposite wall of the fissure of Sylvius, viz., the posterior part of the inferior frontal convolution (Broca, Ferrier). Meynert believes that the centres for psychical speech are situated in the walls of the Sylvian fossa, and in the walls he includes the island of Reil. This whole network of cells and fibres represents the child's idea of an orange. As he grows older many more facts (habitat, varieties, botanical characters, etc., etc.) are added to the first simple idea apparatus, and abstract ideas (fragrance, acidity, natural order, etc.) are formed.

If after a lamb has been seen for the first time and heard to bleat, the sound of bleating is heard when the lamb is hidden from view, the appearance of the lamb will be re-called (Meynert). When the infant touches a part of his own body he feels the sensation of touching as well as that of being touched (Wundt, Meynert). When he moves his limbs he feels the effort and sees the effect; when he cries he has at the same time sensations of muscular effort. In this way he begins to distinguish impressions received from his own body from those received from his environment. A nucleus of the individuality, the primary ego, is formed (Meynert). But there is a secondary ego constituted by the most frequently repeated perceptions of the outer world, and the most often repeated memories, particularly those associated with the emotions. This secondary ego embraces perceptions of intimate persons, wealth, objects of ambition, patriotism, etc., etc. (Meynert).

The convolutions and lobes of opposite sides are intimately associated through the fibres of the corpus callosum (Meynert, "Psychiatrie"), the influence of the left hemisphere as a rule preponderating owing to the universal prevalence of right-handedness. Each so-called centre has a wide cortical area surrounding it to receive fresh impressions, or to replace in time disused portions. Goltz found that removal of any part of the hemisphere caused dimness of vision, but that the injury to sight was greater when the occipital lobes were removed than when the frontal lobes were taken away. The mere visual defect soon passed away; the animal could see objects, but it could not distinguish a stone from a piece of meat by sight alone. The cortical visual acquisitions or memories had entirely disappeared. The animal suffered in fact from mental blindness ("Brain" vol. vi., p. 153). Stimuli applied to the motor regions of the cortex of the new-born infant do not provoke muscular movements as in the adult (Soltmann, Meynert). The movements are at first reflex, but through the thalamus innervation-sensations are received by the cells of the motor area. Herbert Spencer ("Principles of Psychology") says the first dawnings of intelligence are developed through the multiplication and co-ordination of reflex actions. Sucking is at first a reflex act (as shown by the fact that anencephalous children can accomplish it). But when the child sucks during dreamy sleep it is a proof that the act has been registered in the cortex (Meynert). Every registered image depends upon impressions received primarily in the sub-cortical centres. That is to say, the representative cognitions (concrete ideas) and feelings (emotions), and the presentative-representative (or perceptions) depend upon the presentative feelings (or sensations) and the presentative cognitions. In the same way voluntary movements arise out of reflex actions. To use Meynert's illustration: let a sharp instrument touch the conjunctiva; the impression will be conveyed by the fifth nerve to its nucleus in the medulla; from this the sub-cortical bulbar centre of the seventh nerve will transmit the stimulus to the branch of the facial nerve ending in the sphincter palpebrarum. There will be three impressions registered in the cortex, viz.:—

(1,) The image of the sharp instrument, this image being conveyed by the second nerve and its continuation to the occipital lobe.

(2,) The sensation transmitted by the fifth nerve to its cortical centre.

(3,) Sensations of innervation conducted through the thalamus to the innervation-centre (in the cortex) of the facial nerve.

These cortical centres will be united by association fibres. If afterwards a needle be brought near the conjunctiva without touching it, the image of the needle will revive the sensation of pain, and also the sensation of innervation of the facial branch, and contraction of the orbicularis palpebrarum will again take place.

According to my own observations the muscular innervation-sensations are much more intimately connected with the sense of pain than is the sense of sight; at all events, it always takes the muscular movements to complete the association. If an infant of nine months or so touches the flame of a candle, it has the impulse to do the same thing if it sees the candle a few hours afterwards; it stretches out its arm, but the instant it has done so, it withdraws it quickly without touching the candle. This would seem to indicate that the pain sense and the muscular sense areas are in close proximity, and that inhibition is primarily a function of the motor area. It should be remembered, however, that a slight sensation of warmth from the flame may be the factor that completes the remembrance of the painful sensation. The child on the next occasion will not attempt to touch the flame, the visual image being then sufficient to warn him.

The optic thalamus has extensive connections with the cortex through centripetal (the cortex is here considered central) projection fibres, and, with the corpora quadrigemina, constitutes a way station for innervation sensations between the nerve nuclei and the cerebral cortex (Meynert).

The nucleus lenticularis is most highly developed in man and apes; it is also well developed in the bat and mole. According to Meynert it has special relations with the movements of the upper extremity, as well as with the hypoglossal and facial nerves. Hemiplegia results from destruction of the prosencephalic ganglion, especially of the nucleus lenticularis. The nucleus lenticularis and the nucleus caudatus are larger anteriorly than posteriorly, and therefore are in a position to receive more fibres from the anterior cortical regions (Meynert). Fibres connect the thalamus and corpora quadrigemina with the occipital and temporal lobes (Gratiolet, Meynert). The sensory fibres of the internal capsule also enter these lobes. The anterior commissure (supposed to be an olfactory chiasm) is connected, according to Arnold, with the cortex of the temporal lobe, and according to Burdach and Meynert, with that of the occipital also. In my two cases of anosmia there were adhesions over both occipital lobes (*vide supra*). In animals with a large olfactory lobe the nucleus caudatus is comparatively largely developed.

According to Mauthner, sleep is to be considered as an evidence of tiring of the central gray matter of the ventricles, leading to suspension of its functions and interruption in the conduction of centripetal and centrifugal impressions to and from the cerebral cortex, although the cortex and the senses are awake. The following seems to me to be a more satisfactory explanation of the production of sleep :—

(1,) Withdrawal of stimuli from the cortex.

(2,) Consequent functional anæmia of the cortex (*vide supra* as to mentation and vaso-constriction). This factor is aided by dilatation of the cutaneous capillaries by baths, etc., or of the gastro-intestinal by food, etc.

(3,) Passive congestion—hyperæmia ex vacuo—of the central ganglia, the venous sinuses at the base being very distensible, and according to Meynert, increasing or diminishing in size according to the amount of pressure in the intra-cranial cavity. The internal capsule is compressed, and in part grasped by the turgid ganglia.

(4,) The flow into the distended ventricles of the cerebro-spinal fluid, and also probably the pressure of the basal lymphatic cushions, still further increasing the pressure on the ganglia, capsules, and peduncles.

(5,) The acquisition of the habit of sleep.

This theory somewhat resembles that of Cappie (Mackenzie, "Jour. Ment. Sci.," Jan., 1891), but he does not limit the venous distension to the base. In support of the arterial contraction and venous distension theory of sleep, he adduces the fact that during sleep the arteries of the retina are extremely contracted and the veins very much dilated (Hughlings-Jackson, Gairdner). This phenomenon is, however, rather more in favour of the basal hypothesis than of the general. The fact that the cortex has been seen in repeated experiments to become pale and sink during natural sleep (Durham, Hammond, Carpenter) proves the absence of venous congestion in the cortex. That the contraction of the arterioles *per se* is not the cause of sleep is disproved by the frequent occurrence of insomnia in persons with contracted arteries and pulse of high tension. But it may be said that this disproves the vascular theory altogether. Not so, however, for it is not the contraction of the arteries that induces sleep, but this contraction followed by the compensatory basal venous in-flow, so that if the arteries are already contracted the latter process cannot take place. Vaso-dilating drugs, such as sulphonal, mercury, etc., enable this process to take place. The accumulation of lactic acid, etc., in the blood, as a result of mental or bodily exercise, has the same effect in the physiological condition. Small clots in or on the central ganglia, or at the base compressing them or the pons, cause coma more readily than larger ones on the vertex. Bristowe reports four cases of tumour compressing the central ganglia, and in three of them drowsiness was a prominent symptom. J. Hutchinson another in which the patient slept better than he used to do. Gowers ("Quain's Dict.") speaks of the frequent occurrence of rupture of the cerebral arteries during sleep, and in the same article he says the most frequent seat of cerebral hæmorrhage is "the corpus striatum and the region just outside it:" "nearly half the intra-cerebral hæmorrhages are in this situation." The effects of effusions varying with their position; the effects of habit in infancy; dreaming, somnambulism, stupor, hypnotism, insomnia and somnolence, are explicable on this hypothesis. (See below, "Connections of the Cerebellum," etc.).

Stupor is symptomatic of cerebral œdema and may vary from mere intellectual numbness (light stupor) to cessation of mental action (Gedankenstillstand, Meynert) for the time being (deep stupor). In partial coma and in light stupor the patient can be roused momentarily by shouting, etc., but the resulting cortical hyperæmia cannot of course cause removal of pressure from the capsules and peduncles as in sleep, and the patient therefore relapses into his comatose or stuporous condition. With greater intra-cranial pressure the only responses to stimulation are reflex (conjunctival reflex, etc.), and in profound coma the only reflex left may be the respiratory, and even this may succumb to the pressure encroaching on the lower regions of the medulla.

Morselli gives as the exciting causes of insomnia:—

(1,) *Cerebral.*—Hyperæmia, meningitis, encephalitis, tumours; psychical causes, grief or other emotions, fear, etc., etc.

(2,) *Peripheral.*—Sensory stimuli, physical pain, neuralgia, congestions, inflammations, etc.

(3,) *Hæmic.*—Cardiopathies, anæmia, plethora, chlorosis, intoxicants, poisons, medicaments, infections, septicæmia, etc.

In many cases of mania, in acute delirium, in the congestive stage of general paralysis, the insomnia really arises from cerebral hyperæmia; but as this is caused by vaso-motor disturbances it does not constitute a sufficient indication for general blood-letting. In fact, the cases are not less numerous in which insomnia is the effect of cerebral anæmia, as in some cases of furious mania, in agitated melancholia, in many cases of neurasthenia and hypochondriasis, in acute hallucinatory paranoia, in the third stage of general paralysis, in the delirium of typhus, in sitophobia, in protracted convalescences ("Malattie Mentali," pp. 208-209). Morselli terms dreamy sleep paragrypnia, and sleep troubled with nightmares agrypnia.

Haig ("Brain," 1891) says uric acid contracts the arterioles and causes high arterial tension. The insomnia thus produced is relieved by opium and mercury and *acids*, drugs which diminish the uric acid holding power of the blood.

It will be readily understood that increased or diminished activity of the cells, if carried to an abnormal extent, will affect the external manifestations of mind; that degeneration of the cells will cause hopeless dementia. In addition to the cellular and nuclear alterations already mentioned, Meynert has found divided cells, inflated cells, and nuclei with an abnormal tendency to gravitate to the lowest part of the cells. The cells of the fronto-parietal region, particularly the frontal, seem to be more concerned with the higher mental functions of the brain than those of other parts, those regions suffering most markedly from atrophy in chronic insanity. Kölliker and others ("Brit. Med. Jour." Supplement) have discovered by using Golgi's method of hardening that what are most probably centripetal fibres in the cerebellum, are not connected with cells, so that the intra-cellular substance is presumably not functionless. Kölliker's explanation, however, is that these fibres act on the cells by actual contact, and that these bodies then transmit the stimuli to the centrifugal fibres arising from them.

Unequal action of the cells and fibres of the various regions of the brain will also cause mental disturbances. This may be brought about, amongst other causes, by vascular and vaso-motor abnormalities. Any interference with the fibres connecting the thalamus with the cortex, or lesion of the thalamus, will cause, through lack of or disturbance of innervation-sensations, delusions as to the position of the extremities, forced positions, sensations of falling down abysses, etc. (Meynert).

CONNECTIONS OF THE CEREBELLUM WITH THE CEREBRUM.

As will be seen below, these are very intimate and would seem to indicate that the cerebellum has other functions besides assisting in co-ordinating muscular movements and in the maintenance of bodily equilibrium. Herbert Spencer's hypothesis is that the cerebellum is an organ of doubly-compound co-ordination in *space*, just as the cerebrum is in *time*; the former being concerned with co-existent, the latter with sequential impressions and acts. Langwieser, in a paper entitled "Zur Physiologischen Erklärung des Bewusstseins" ("Allgemeine Zeitschrift für Psychiatrie," Band 41, Heft 1, 1884), advanced the theory that the cerebellum is the organ of consciousness and of the ego; not alone, but in conjunction with the portion of the cerebrum with which it is co-operating for the time being. According to this view the cerebellum furthers (in the cerebral cortex) whatever subject interests it most, and other subjects must wait. Langwieser arrives at this conclusion because the cerebellum distinguishes our own movements from those of our environment. Consciousness commences with the ability to distinguish our own movements from those of our surroundings. Rotatory movements take place when one half of the cerebellum of an animal is injured. These movements take place because the animal is afraid one side will fall away, as it feels like a foreign body to it. In men with degeneration of the cerebellum the ground seems to be flying away from their feet, and there is staggering gait. An unconscious movement of the eyes, or an unaccustomed unnoticed movement of the body, will cause giddiness; there is confusion as to whether the movement is of one's own body or of the surroundings; the cerebellum is at fault.

Without subscribing to the theory that the cerebellum is absolutely necessary to consciousness (this being probably the potentiality of any group or groups of cerebral cortical cells plus a stimulus from the special senses, the general sensibility, or cortical cells already conscious), facts taken from comparative anatomy, embryology, human anatomy, physiology, pathological anatomy, and clinico-pathology, indicate that this organ (the cerebellum) takes an important place in the encephalic mechanism.

In fishes its form is simpler (consisting merely of two halves and having the appearance of a single, undivided organ) than in any of the vertebrata, and smaller except in the single instance of the shark, a cartilaginous fish which, as well as being one of the largest, is also one of the most active and intelligent of the whole class pisces. In birds it consists of a large median and two very small lateral lobes. In marsupials it still consists principally of the median lobe, the lateral lobes being mere small appendages. In the bat with active and varied movements, the cerebellum is very large, but this is owing to the size of the middle lobe. Rising in the scale the lateral lobes become more developed until in the cat and dog there is a marked increase in their size, and in the marine cetacea (porpoise, dolphin, etc.) this is still more marked. The dolphin evinces rare docility and intelligence, the porpoise, striking curiosity. In the quadrumana this development of the lateral lobes is very great, and in man the middle lobe becomes a relatively diminutive structure (Charlton Bastian, "The Brain as an Organ of Mind").

Meynert says the larger the animal the heavier the cerebellum in proportion to the rest of the encephalon. He mentions the elephant as an extreme instance of this relationship, the cerebellum here forming nearly a fourth, by weight, of the whole encephalon. But the elephant is noted for sagacity as well as for size, and is possessed of an efficient and unique prehensile organ.

In the intra-uterine period, according to Flechsig, the medullary sheaths appear first in the columns of Goll and Burdach; then in the anterior columns of the cord; next in the medulla oblongata, except the pyramids; then in the medullary substance of the central lobe of the cerebellum, *followed by that of the lateral lobes*. The white substance of the cerebral frontal lobes does not appear until some time after birth, and does not become as white as in the adult until the end of the fourth month (Flechsig, Meynert).

The cerebellum of the new-born child only weighs about half as much relatively to the cerebrum as does that of the adult, and the disproportion may be even greater (Meynert, Sharpey, Bastian).

Meynert gives the following ratios: (1,) Cerebral mantle; (2,) Islands of Reil to cord; (3,) Cerebellum, respectively; in the adult, $79:10\cdot5:10\cdot5$; in the new-born child, $83:11:5$.

According to Flechsig, quoted by Gowers in "Diagnosis of Brain Diseases," fibres pass from the prefrontal lobe to the lateral and posterior parts of the cerebellum, degenerating downwards; others pass chiefly from the upper surface of the middle lobe of the cerebellum, and running partly beneath the lenticular nucleus and partly between it and the corpus geniculatum externum enter the temporo-sphenoidal and occipital lobes without passing through the internal capsule. These degenerate upwards. Other fibres run from the caudate nucleus to the cerebellum, degenerating downwards; and yet others from the cerebellum to the lenticular nucleus, degenerating upwards (probably). These fibres connect the corpus striatum with the opposite half of the cerebellum through the pons. The caudate fibres enter the cerebellum by the middle peduncles, and the lenticular by the superior. The optic thalamus also receives fibres from the superior peduncle of the cerebellum. The fronto-cerebellar fibres only degenerate as far as the pons, the degeneration being there arrested, as usual, by the gray matter. These and the temporo-cerebellar and occipito-cerebellar fibres are absent in congenital absence of the cerebellum, and the corpus striatum is reduced to one third of its ordinary size. Gowers (*op. cit.*) considers that these extensive connections with regions of the cerebrum, which are not motor and only sensory in limited area, warrant the revival of the old theory that the cerebellum plays an important part in intellectual processes.

Visceral, as well as (special) sensory nerve fibres, have been traced to the cerebellum. According to Ribot, the former play an important *rôle* in the formation of the "*personnalité.*"

The removal of the cerebellum in animals has given rise to affections of equilibration, but no experiments have, I think, been recorded in which monkeys or dogs, previously intelligent and docile, were tried as to their capability of learning new tricks after the removal of the cerebellum, or of its lateral lobes. Nothnägel has proved that it is the middle lobe of the cerebellum that is connected with the maintenance of equilibrium, the lateral lobes having been removed experimentally and by disease without any consequent loss of motor co-ordination or equilibration (Gowers).

Weir Mitchell found that pigeons could walk steadily many months after the cerebellum had been destroyed, though at first they could not do so. He observed "only feebleness and incapacity for prolonged muscular exertion" when they had recovered steadiness in their movements (Ferrier, "The Functions of the Brain").

Luciani removed almost completely the cerebellum of a dog.

Immediately after and for two months there was inco-ordination of all the voluntary movements; the animal could neither stand, walk, swim, nor feed itself. Then the inco-ordination in swimming disappeared altogether. There was a special form of ataxy of the other voluntary movements; these lacked steadiness and force, and there was a constant clonic motion. This period lasted four months. The third and last stage was characterised by nutritional failure, and finally marasmus. The dog lived eight months after the operation. In another dog he removed the sigmoid gyri as well as the cerebellum. There was a greater amount of paresis of all the limbs than occurs when the sigmoid gyri are extirpated without the cerebellum. He says the cerebellum "is a central organ on which depend the tone and a great part of the disposable nervous energy of the motor elements of the muscles" ("Jour. Nerv. and Ment. Dis.," Oct., 1884).

To compare the above results with the remote effects of removing the cerebrum (*vide supra* "Erotic Tendency," etc.). Steiner found that osseous fishes caught and ate worms the next day after having the cerebrum removed and the aperture in the skull sealed up ("Brain"). Meynert says that animals (mammalian) from which the cerebral hemispheres have been taken, are able to preserve their equilibrium, but are stuporous, cataleptic and unconscious.

The cerebellum is sometimes altogether absent in idiocy (Hitzig, Krafft-Ebing).

Ferrier (*op. cit.*) mentions a case, as already stated, recorded by Combette, in which this organ was absent, and the only characteristic symptom was that the patient frequently fell; yet he says on another page that she suffered from nymphomania, a psychical symptom which is seldom solitary, and the presence of which is noteworthy, as the removal of the cerebral cortex in dogs and pigeons causes total loss of sexual desire (*vide supra* "Morbid Anatomy of Symptoms"). Ferrier also mentions the case of Guerin (reported by Bouillaud and referred to by Longet and Vulpian), "whose cerebellum was found after death to be almost completely destroyed by disease." He "still retained the power of co-ordination of movements and could walk, only he was observed to reel and totter when he walked."

Ross ("A Treatise on Nervous Diseases," vol. ii, p. 704) mentions cases of atrophy of the cerebellum reported by Combette, Meynert, Pierret, Fiedler, Clapton, Dugnet, Moreau, Lallement, and Otto. Seven of these authors observed motor disturbances, mostly ataxic, but in some cases there was slowness of gait with frequent falls, especially backwards. Speech was temporarily or permanently affected in all the cases with motor disturbance. Lallement and Otto noticed no motor disturbance. The latter's patient was impulsive in his movements. Analgesia and slight disturbances of sensibility are occasionally observed.

Foville found diminished sensibility, Renzi impairment of vision and hearing, Lusanna of vision only, resulting from destructive lesions of the cerebellum (Meynert, "Psychiatrie").

The patients of Combette, Clapton, Otto, and Fiedler were *weak-minded*, even *idiotic*, and Pierret's case suffered from *weakness of memory* (Ross, *loc. cit.*).

In reports of cases of adventitious products in the cerebellum, apathy and somnolence are often mentioned, but such cases should not have too much stress laid on them, as pressure on surrounding organs (pons, medulla, veins of Galen, aqueduct of Sylvius, etc.) may account for some of the symptoms. Alterations in colour, consistence, and size of the cerebellum are frequently found in the insane without loss of the power of equilibration or co-ordination, where the absence of cerebral cortical or ganglionic paralysis allows these functions to be tested.

In ten autopsies I found one cerebral hemisphere markedly lighter than the other, the difference in weight varying from 400 grains to 2102 grains, and in one case 5244 grains. In five of these, one hydrocephalic idiot with paralysis and epilepsy, two paralytic and epileptic idiots and two organic dements, one half of the cerebellum was lighter than the other, the difference in weight varying from 28 grains to 246 grains, the lighter half being on the side of the heavy cerebral hemisphere. In a sixth case (organic dementia), the heavy half (right) of the cerebellum was on the side of the heavy cerebral hemisphere and contained a dark hæmorrhagic focus, $1\frac{1}{2}$ inches long ; the left frontal cerebral cortex was apparently normal although there was a small hæmorrhagic focus with surrounding softening in the centre of the white substance of the lobe posteriorly. There were extensive lesions in the other parts of the cerebrum on both sides. The patient was very demented and feeble, the right side being the weaker, and the cutaneous sensibility more diminished on that side. Vision was nearly *nil*, but there was double cataract. The four remaining cases, two organic dements, one epileptic idiot, and one general paralytic, had the two halves of the cerebellum exactly equal. In one organic dement the destructive lesions were confined to the temporo-sphenoidal and occipital lobes, and in the other to the right gyrus fornicatus. In the latter case there were also isolated sclerotic portions in the cortex of both hemispheres. The former case was extremely demented, and there was general weakness with paraplegia. In the latter there were extreme dementia, complete left hemiplegia, left hemianæsthesia, twitching of muscles of left arm and leg, epileptic seizures, dilatation of right pupil, visual hallucinations, cerebral vomiting on one occasion, no aphasia, no contractions. The idiot's left hemisphere was shorter than the right, the left temporo-sphenoidal lobe smaller and harder than its fellow, and the left precuneus very small ; both occipital lobes had a constricted appearance, but there were no special lesions of the frontal lobe on either side. The right limbs (especially the arm) were shorter and thinner than the left. The right arm was paralysed and contracted, the right leg weak but not contracted. As to intellect the patient ranked last but one of the four idiots. In the general paralytic the right cerebral hemisphere was 800 grains lighter than the left; the arachnoid was opaque and adherent everywhere except at the posterior part of the right F^1 encroaching on F^a, where an effusion formed a puffy enlargement ; no softening or other focal lesion; floor of fourth ventricle *sablé*. The patient had suffered from complete left hemiplegia, and had had the typical general paralytic speech.

Reverting to the five cases showing crossed atrophy (or rather, perhaps, arrested development in the idiots), the hydrocephalic idiot (case reported, " Brit. Med. Jour."), who was also epileptic and paralytic (right side), displayed the maximum difference between the weights, both cerebral and cerebellar ; there was porencephalus, only shreds of the cortex of the left hemisphere remaining ; the right lateral ventricle was much distended. The right arm and leg were paralysed, contracted and atrophied. Speech was absent. Perception, memory, emotion, all present. The right cerebral hemisphere weighed 8,074 grains, and the left half of the cerebellum 996 grains.

Another idiot (epileptic), who also suffered from brachio-crural monoplegia with contraction, seemed totally devoid of intellect, and uttered nothing but howls and cries. His right cerebral hemisphere weighed 5,386 grains, and his left only 3,284 grains ; the left half of his cerebellum, 466 grains, and the right, 366 ; all very much below the normal weight ; all the convolutions of the left cerebral hemisphere were much smaller and paler than those of the right. The third epileptic idiot (case reported "Jour. Ment. Sci.) suffered from right brachio-crural monoplegia with contracture, and left crural monoplegia with contracture, absence of speech and nystagmus. He moved his right arm freely. He was the most intelligent of the four idiots. He was the only one of the four with paralysis and contracture of both lower extremities. The central convolutions on the left side were extensively, and on the right slightly atrophied or arrested in development. The weights were : left cerebrum, 4,598 grains ; right, 5640 ; left cerebellum, 1078 grains ; right, 1050 ; the cerebellum being larger relatively and absolutely than in any of the other idiots.

In one of the organic dements with crossed atrophy of the cerebellum, the left cerebral hemisphere was 560 grains heavier than the right, of which F^2, F^3, and I R were softened; the left half of the cerebellum was 60 grains lighter than the right. Prior to admission he had been alcoholic and suicidal. In the asylum he was demented, but there was no paralysis, and speech was unaffected.

In the other demented case the right cerebral hemisphere was 760 grains heavier than its fellow, and was affected by yellow softening of the supra-marginal gyrus and upper or posterior end of T^1; the light hemisphere also had softening of T^1, and a small contracted lenticular nucleus containing an old hæmorrhagic focus. The right half of the cerebellum was small; it was yellow inferiorly, and there was a depression on the inner part of the under surface (median lobe); it was 120 grains lighter than the other half. Patient was very demented and almost completely blind, but not deaf. There was right linguo-facial monoplegia with weakness, but no marked paralysis of the extremities; vaso motor paresis: the right knee-jerk was absent; the right pupil was larger than the left; there was twitching of the right facial muscles; speech was very defective (*vide supra* as to nucleus lenticularis and facial nerve).

In three cases of mine, not included in the ten above-mentioned, there was right hemiparesis with softening of the right lobe of the cerebellum.

From all these data concerning the cerebellum, and from other facts mentioned in the antecedent parts of this chapter, it may be inferred:—

(1,) That its lateral lobes increase in size and development with the intelligence of the animal; if also with the capability of performing varied and complex muscular movements, this inference is not invalidated. It is admitted by Sully ("Brain," 1889, Part II., p. 154) that "attention stands in a particularly close relation to the process of motor innervation," and that "there is a close affinity between muscular and mental exertion." "Each is a variety of the active phase of consciousness." Wundt and Ward "regard both modes of activity as at bottom one." Bain "seeks to account for the whole process of thought control by help of a motor process." Motor innervation sensations form a large proportion of the cerebral cortical acquisitions, and therefore of the intellectual processes. Lombard's experiments on the regional temperature of the head point to the motor area and the adjoining part of the pre-frontal lobe as the most active region, both in intellectual work and emotional excitement. (See chap. II.) The size also of the animal influences the relative weight of the cerebellum, and where great size and intelligence exist in combination, as in the elephant, the organ attains its maximum.

(2,) That each lateral lobe acts, in man at least, in conjunction with the opposite prefrontal lobe and central ganglia.

(3,) That its serious disease or removal entails permanently increased exertion on the part of the cerebrum, as well as in-co-ordination and defective equilibration which are more or less temporary.

(4,) That it assists in the process of attention (*vide supra*, function of the prefrontal lobe). During conscious effort, as in forming new acquisitions, whether mental or muscular, the cerebrum and cerebellum act in conjunction in opening new channels, the fronto-cerebellar fibres conveying impulses from the frontal lobe to the cerebellum where they are strengthened and transmitted to the occipital and temporal lobes or to the basal ganglia. When actions or processes of thought (calculation, memory, judgment, volition, etc.) have been so frequently rehearsed by the conjoined organs that the nervous paths of association are opened with facility, and without effort, they do not arouse consciousness; they become automatic (Ross, "Brain"); are, in fact, relegated to the lower centres, probably the central ganglia, or are carried on by the unaided motor or sensory cortex, whilst the combined frontal and cerebellar cortices attend to new ones, or to the revivification and completion of old ones. Muscular acts, which have been frequently performed and have become habitual, and are performed unconsciously or automatically, are carried out by the corpus striatum, olivary bodies, and possibly other ganglia; or by the motor cortex itself without the prefrontal lobe and the cerebellum. Equilibration is also accomplished (after it has been learned by the child) by these ganglia with the help of the thalamus and corpora quadrigemina, the afferent system being the fibres from the muscles and skin, those from the semi-circular canals, and those from the retina and ocular muscles; the other organs of sense are probably also in relation with the cerebellum. Recent researches would seem to indicate that it is the auditory fibres of the nerve and not the labyrinthine that are connected with equilibration. In the same way that the muscular actions are regulated, and over-action or under-action avoided, the impulses flowing from the sensory cells of the temporo-sphenoidal and occipital lobes are also regulated and co-ordinated.

(5,) That the cerebellum maintains the equilibrium of the ego with reference to the remaining cerebral acquisitions. False ideas as to self and surroundings arise in consequence of derangement of the cerebro-cerebellar mechanism (*vide supra* "Morbid Anatomy of Symptoms"). From its visceral, sensory, and cerebral connections it probably plays an important part in the formation of the personality.

(6,) That the cerebellar cortex is subordinate to the cerebral; the central and basilar ganglia to the combined cortices; the medulla and cord to these ganglia. This is not incompatible with Hughlings-Jackson's anatomico-physiological division into highest (prefrontal lobes, etc.), middle (motor and sensory regions), and lowest centres of evolution (medulla and cord).

Does this theory help to explain physiological and clinical facts?

First, to take such phenomena as dreaming, sleep-walking, and hypnotism (*vide supra* as to the theory of sleep-production). The fronto-cerebellar fibres run along the inner side of the crus; and the temporo- and occipito-cerebellar run along the outer; all superficially, and therefore more susceptible to the increased pressure of the swollen central ganglia and distended sinuses. In sound, deep sleep, the pressure is sufficient to cut off all stimuli from the cerebral cortex, but in light sleep the pressure is principally sustained by the superficial fibres; the cerebro-cerebellar cortical communications are cut off, but the deeper fibres convey stimuli to the cerebral cortex from the viscera, special senses, and surface of the body, causing dreams. The absence of the influence of the cerebellum would account for the difference between dreams and waking thoughts; for the altered relations of the ego and its environment; for the absurdity and extravagance, and partly at least for the feeling of helplessness, and of falling down precipices. Although the fibres from the cerebellum to the cerebral cortex are superficial, those from the cerebellum to the corpus striatum are somewhat deep, lying between the pyramids and the substance of Sömmering. The cerebellar cortex is awake during sleep, and its communications with the pons and motor tracts are open, and so the somnambulist is able to enact his dream; motion is added to ideation; the temporo- and occipito-cerebellar fibres pass between the nucleus lenticularis and the corpus geniculatum externum (*vide supra*).

When a person is exposed to a monotonous visual perception, *e.g.*, a bright light or passes, the visual cortex, though at first stimulated, may become in suitable weak-willed subjects quiescent, the impressions being relegated to the corpora quadrigemina and geniculata; these probably become congested and exert pressure on the above-mentioned fibres, thus cutting off the cerebral cortex from the regulating and invigorating influence of the cerebellum, and placing the subject under the control of the hypnotiser. Some authors describe three stages of the hypnotic state, lethargic, cataleptic, and somnambulic. Hypnotic or artificial somnambulism should be distinguished from natural somnambulism (sleep-walking or noctambulism). In the Nancy method the suggestions of the operator probably give rise to the circulatory alterations that occur in drowsiness, and if carried further, to those that occur in sleep. The first-mentioned method (Braidism) probably acts to some extent in the same way. The vertiginous sensations and feeling of falling from a height point to implication of the equilibration apparatus. Morselli (*op. cit.*, p. 210) remarks that these symptoms, along with a sense of ill-being, vague pains, anxiety, oppression, and accompanied by paralysis of volition, inability to cry out, together with the appearance of threatening phantoms, correspond exactly to those experienced by some anxious melancholiacs. Judging by analogy the dreamy confusional states (acute confusional insanity, etc.) may be the symptomatic expression of disturbance of some part of the cerebro-cerebellar mechanism; such states are transitory in sane persons after shock or excesses (see "Etiology of Acute Confusional Insanity.") Delusions as to the identity of the patient or those about him, also as to place and surroundings, may develop on the same basis; delusions also as to one half of the body being altered, fear of falling from heights, etc. It is also probable that the cerebellum is involved in neurasthenia and its derivatives. Hence the bodily weakness, the tremor after exertion, the nosophobia, the more or less unsteadiness, or rather uncertainty of gait, the hampered mental action. In choreic insanity the motor inco-ordination extends to the ideational apparatus, the middle lobe of the cerebellum, or the tracts connected with it, being first affected, and afterwards the lateral lobes with their tracts. As Maudsley remarks, the incoherence and automatic character of choreic delirium are striking. Hallucinations and illusions frequently accompany the delirium.

In acute delirium in which the whole encephalon is hyperæmic, there is absolute incoherence with unconsciousness, contrasting strongly with acute mania in which the cerebral cortex alone is congested.

The pallor of the optic disc found by Clifford Allbutt in mania does not necessarily indicate anæmia of the cerebral convexity, as, according to Edgar Browne, optic neuritis is caused by lesions, generally diffuse (cerebritis, meningitis), of the cerebral base, the pons, and the front of the cerebellum, but not by lesions of the cerebral convexity.

The cerebellum assists the cerebrum, even when the former is healthy in the organisation of systematised delusions.

According to Ferrier and others the afferent nerves from

the viscera have a close connection with the function of equilibration. They have also without doubt great influence on the cerebrum (even when the visceral derangements do not rise into consciousness); instanced in some cases of hypochondriasis, hypochondriacal and other forms of melancholia, and in gastro-enteric and other forms of abdominal insanity. Insanity of exophthalmic goître, insanity of myxœdema, insanity from chronic Bright's disease, are forms based on diseases of bodily organs, and although such diseases act in other ways than through the nerves, yet their method of action must be partly nervous. Cases of so-called melancholia arise from cardiac disease. There are frightful dreams in the same disease. Neurasthenia and mental aberration may arise from uterine and ovarian displacements and disorders (vide supra). Dreams and delusions about flying are caused by rapid respiration, about suffocation by laboured breathing; nightmares arise from indigestion or abdominal pressure.

The nutrition of the brain itself influences, it goes without saying, the sensational, ideational and motor functions of the organ. Disagreeable emotions and impairment and hampering of function are thus caused. This may be observed in some cases of melancholia, in primary mental deterioration, in senile insanity, and in neurasthenia and neurasthenic insanity.

In idiocy, in criminals (moral insanity, impulsive insanity), in the other degenerative forms, paranoia, etc. (very often), there are anomalies of the lobes, convolutions and fissures (according to Benedikt, the fissures are often confluent in criminals, the Rolandic and Sylvian, the occipital and inter-parietal, being those most frequently so arranged); in these degenerative cases there is hereditary neurotic taint, and the malformations are most frequently congenital.

With regard to the arrangement of the cerebral cells into simple and complex idea apparatus, the symptoms of insanity in children throw some light thereon. In children the ideas are simple, few and disconnected. The ideas are therefore incoherent by reason of an absence of organic association between the residua. The morbid phenomena are not systematised as in the adult, and the result is delirium rather than mania (Maudsley). The morbid idea in the child's mind, having little range of action on other ideas, acts downwards on the sensory ganglia, causing hallucinations, or on the movements, giving rise to morbid impulses (Maudsley). These impulses constitute impulsive insanity, called by Maudsley, monomania, and by Morselli, *paranoia rudimentaria impulsiva*. Under the head of monomania also, Maudsley places the epidemics of morbid ideas, which have from time to time, in the history of the world, affected children.

(Choreic Insanity is mentioned above, and described in Chap. III).

The ruling instinct in a child of 3 or 4 is self-gratification, involved in which is a tendency to destroy what it dislikes. Its insanity is manifested by perverse and unreasoning appropriation of whatever it sees, and by extreme destructiveness; it suffers from the instinctive variety (Maudsley) of affective or moral insanity. (For description of Moral Insanity and Moral Imbecility, see Chap. III.) Maudsley describes a case of moral insanity in a boy in which the cutaneous sensibility was completely lost during attacks of irresistible violence, returning in the docile intervals. This is interesting in the light of the present knowledge of the motor area. He also describes another case in which the cutaneous sensibility was deficient. More or less cutaneous anæsthesia is frequently observed in persons suffering from morbid or immoral impulses or propensities symptomatic of cerebral exhaustion. Lombroso and others find distinctive bodily as well as cerebral characteristics in criminals, adult or juvenile. Maudsley also describes a form in children called cataleptoid insanity. There is mystical contemplation with insensibility sometimes, or vague answers, or incoherent raving, or sudden wild shrieking. There are intermediate forms between this and chorea, and the attacks sometimes alternate with true epileptic seizures. The same author says epileptic insanity in children resembles the same disease in adults (see Chap. III.).

Melancholia may occur in childhood. The child is constantly wailing and whining; every impression causes a painful feeling. There may also be delusions and suicidal tendency.

Mania (delirium of young children) may occur in childhood in connection with convulsions (the most usual pathogenetic factor), blows on the head, intestinal worms, and onanism. There may be extreme violence and even homicidal tendencies. "The question of hereditary taint is in reality the important question in an examination of the insanity of early life" (Maudsley).

With regard to the cortical cells of the brain their sparseness, or extensive degeneration, gives rise to mental weakness as shown by diminished or originally slight power of perception, apprehension and attention, defective memory, weak judgment, absence or weakness of volition or of inhibitory power. These conditions, pathological and clinical, are found in idiocy, terminal dementia, general paralysis, chronic alcoholic insanity, senile dementia, chronic epileptic insanity. In these forms there are frequently motor and sensory disturbances. The pathological conditions in the acquired forms at least are led up to by a series of changes (*vide supra*) commencing apparently with vascular disturbances, though, according to Maudsley, these are themselves originated by alterations (inappreciable, macroscopically or microscopically at present) in the contents of the cerebral cortical cells. Meynert ("Lectures" at Gen. Hosp., Vienna, Session 1876–77) believes that although in a general way cerebral hyperæmia may be looked upon as the basis of mania and cerebral anæmia of melancholia, yet cerebral hyperæmia is not equivalent to maniacal excitement, nor cerebral anæmia to melancholic depression. There must be an alteration in the state of the nervous elements.

Some authorities hold the view that general epilepsy is cortical, but Mendel thinks it is primarily bulbar. According to Hughlings-Jackson unilateral epilepsy is caused by an irritative lesion in or near the motor cortex. The hyperæmia arising from such a lesion induces over-nutrition of the large pyramidal cells resulting in periodical explosions of nervous energy. B. Lewis thinks the small pyramids (sensory cells) of the cortex inhibit the large motor cells of the third layer. He found the former degenerated in general epilepsy. Meynert attaches much importance to the atrophy of the pyramids of the stratum convolutum. The cell degenerations described by these two authors may be secondary. Brown-Séquard found that lesions of the right hemisphere were much more frequently accompanied by convulsions and conjugated deviation than those of the left. It is of some interest in this connection that hysterical symptoms (tremor, anæsthesia, paralysis) are exceedingly frequent on the left side of the body as compared with the right (Brown-Séquard). Transitory mania resembles epileptic insanity in some respects, but occurs only once in a lifetime.

Many cases of mental disease develop out of physiological states. In some cases hallucinations and illusions (hence called hypnagogic, Morselli) develop out of dreams, or arise during the half waking, half sleeping state, and constitute in suitable soil, the germs of many delusions. Sadness of disposition, extreme scrupulousness, religious zeal, suspicion, pride, ambition, etc., may, under the influence of the environment in a structurally predisposed individual, easily overstep the physiological limits.

Meynert's theory of the production of hallucinations is that there is irritation of the ganglia, the impressions so produced being received by the cells of a cerebral cortex, of which the activity is diminished. Luys ("Le Cerveau") believes they are caused by a diseased condition of the optic thalami. Tamburini's view is that they are caused by derangement or irritation of the cerebral cortical sensory areas (Ireland). The last theory is most in accord with what is at present known of the anatomy, physiology, and pathology of the brain.

Hallucinations are certainly caused in some cases by irritation of the ganglia, or of some part of the sensory tract between the sense organ and the cerebral cortex. J. L. Koch ("Allg. Zeit.") would call such hallucinations illusions, but this would only create confusion. According to Despine ("L'Encéphale") whether hallucinations originate in the cortex cerebri or in the centripetal tracts, they take a centrifugal course from the former to the organs of sense. This would explain the fact that prisms cause diplopia with reference to hallucinations just as if they were real objects. Arterial spasm sometimes causes hallucinations (Meynert). In addition to the methods already mentioned, hallucinations may also arise out of the delusions. The form taken by hallucinations is influenced to a great extent by the disposition or humour of the patient.

The suicidal tendency of melancholia may originate from the patient having sustained some shock to his secondary individuality (objects connected with ambition, love of wealth, family affection, patriotism, etc.). All men try to escape from the pain which is the most intense; the secondary individuality has become more valuable to the individual than the primary; he sacrifices the latter to escape from the anguish caused by the injury to the former (Meynert).

Angry excitement arises from hyperæsthesia of the nervous elements, as shown by the increased acuteness of the senses (Meynert). Meynert thinks it probable that exaltation arises from an excess of oxygen in the encephalon.

Certain symptoms are seldom absent in insanity, viz. :—

(1,) In acquired insanity in the early stages: (a,) Insomnia (not due to pain), of which something has already been said ; (b,) Alteration in disposition and conduct (perversion, or, less commonly, exaggeration) due to over-action of part or parts of the brain ; (c,) Prolonged sense reaction-time, the result of defective nutrition of the fibres and cells, or caused by the pressure exerted on the fibres by the distended blood-vessels ; (d,) Physiognomical alterations, arising from emotions, circulation, innervation, etc. As showing the intimate relation which exists between almost all the emotions and their outward manifestations, it is an interesting fact that even the simulation of an emotion tends to arouse that emotion in the mind (Darwin, "Expression of the Emotions in Man and Animals").

(2,) In chronic and congenital insanity : (a,) Mental or moral weakness, owing to degeneration or congenital defect of brain structure ; (b,) Muscular weakness due to the same cause ; (c,) Prolonged sense reaction-time ; (d,) Physiognomical deterioration or alteration in consequence of vaso-motor defects or irregularities ; mal-innervation or irregular innervation of the facial muscles gives rise to incompatible expressions; the long continuance of some facial expression corresponding to disagreeable emotions; changes in the skin itself. The facial deterioration is more marked in females.

I mention the following cases as constituting an example of the complex nature of the pathogenesis of many, if not all, of the acquired primary insanities:—Eleven patients, recently treated by me in private practice, were suffering from climacteric neurosis. They were not insane, but on the borderland—such cases as one never meets in asylums. They were all females, married, and aged between 43 and 50. Hereditary taint was ascertained in two or three, and it probably existed in more. All suffered more or less from menstrual irregularities. Five of them had had no children after the age of 30, two none after the age of 40, and *four were childless.* There would thus seem to be some causative relationship between sterility and climacteric neurosis. Several had been in the habit of drinking strong tea three or four times a day. Most of them evinced shyness, avoiding alike society, fresh air, and sunshine; true this was a symptom, but one likely to prove very damaging to the mental equilibrium, and to become an etiological factor in the production of real insanity, if not resisted. All suffered from more or less severe dyspeptic symptoms. One was so nervous and apprehensive that she ran out of the house the moment she saw me take up a stethoscope. In three or four there was a craving for alcoholic stimulants: this was acknowledged and deplored by two. It is an alarming symptom, as the alcoholic habit can be very readily established at this period of life. Alcoholic indulgence, still further, hastens the patient's progress to insanity, and must be prevented at all hazards. Several suffered from insomnia, which is easily remedied at this stage, but greatly aggravates matters if neglected. The therapeutical indications were manifestly to remove the removable causes and symptoms, and so assist the patients to pass through the critical physiological period. Six of the cases have recovered, two have gone to distant towns, and three have been already relieved and will probably recover, attaining the post-climacteric age without the necessity of any but home treatment.

CHAPTER VIII.

THERAPEUTICS AND HYGIENE.

Prophylaxis.

This should be commenced in infancy, where there is hereditary predisposition to mental disease. These neuropathic children ought to be brought up by a healthy wet nurse. Rooms, clothing, and baths must not be too warm. Especial care should be taken during the first dentition to avoid cerebral hyperæmia and convulsions. The children ought to be much in the fresh air, and have a nutritious unstimulating diet, free from tea, coffee, or spirits. They should be early taught obedience, the character strengthened, passionate ebullitions and sentimentality discouraged, and quietude and self-control inculcated. Where the intellect is precocious it should be restrained, where backward, treated with patience; there must be no forcing. If the parents are perverse, hypochondriacal, or hysterical, the children ought, if possible, to be removed from them and educated in the private house of a medical man, teacher, or clergyman in the country. Everything likely to promote sexual development should be avoided. About the time of puberty much care is required, and any disease (chlorosis, etc.) that may arise must be energetically treated. Novel reading, as well as religious fanaticism, is to be avoided. Males ought to marry early, but females should be previously fully-developed in order to be able to combat the dangers of gestation and the puerperium. Suckling must not be allowed, or only for three months at most. During the puerperium the treatment should be strengthening. The occupation of these persons should not be exciting; they ought to live a quiet, regular life, attend to the functions of the digestive organs, and avoid the abuse of pleasures of any kind (Krafft-Ebing, "Lehrb. der Psych.," pp. 281–283).

A well-regulated system of gymnastic exercises will be found to diminish or even to check the morbid propensities from which children of neurotic diathesis suffer. In youth riding, driving, walking, cycling, rowing, and swimming are useful.

Remedial Treatment.

I.—THE IMMEDIATE RELIEF OF URGENT SYMPTOMS.
II.—ULTIMATE CARE AND TREATMENT.

I.—*Relief of symptoms.*
A.—Insomnia. An actual attack of insanity may often be warded off by relieving the premonitory insomnia. In some forms of insanity (acute delirium, acute mania), insomnia may assist in the induction of fatal exhaustion. Bucknill and Tuke consider that as a general rule it should not be allowed to continue more than three consecutive nights without the administration of a hypnotic.

Chloral Hydrate.—Clouston recommends 10 to 25 grains with from ʒss to ʒj of pot. bromid. at night in states of mental exaltation. Spitzka gives equal parts of chloral and bromide of potassium in mania. I have generally found from 20 to 30 grains of chloral act well. In some cases a dose of gr. xv will do, and if it fails, it can be repeated in three-quarters of an hour. In apyretic delirium tremens an otherwise healthy man may have 40 grains at bedtime, or 30 grains may be given and repeated in an hour if required; these doses have induced the critical sleep in eight cases under my care in private practice. Contraindicated in acute delirium, states of mental depression, visceral disease, especially cardiac and pulmonary, and febrile delirium tremens.

Occasionally it causes excitement instead of sleep. Occasionally it does not act hypnotically till the night following that on which it is administered.

The patient may be induced to take it in a little beer or spirit and water. When patients, suffering from neurasthenia or mild melancholia, wake very early and cannot go to sleep again, give 40 grains two or three times a week when they awake, until the habit is broken (Stretch Dowse, "Brain and Nerve Exhaustion," p. 56).

Paraldehyde.—This may be given when chloral is contraindicated. Dose ʒss to ʒij. Medium dose ʒj given at bedtime with a little syrup of orange, tragacanth mucilage and water.

Sulphonal.—When chloral is contraindicated. Dose, as a hypnotic, gr. xv to ʒss, or even ʒj given early in the evening in a cupful of hot fluid, preferably milk. Having no smell and only a very slightly bitter taste, it may be given with the food when the patient refuses to take anything he knows to be medicated. During the past two years I have given, in private practice, sulphonal as a hypnotic in more than forty cases of insomnia from various causes, influenza, alcoholism, climacteric neurosis, puerperal insanity, acute delirium, adolescent insanity, hysteria, grief, etc., and have found it act well in all but a few cases, the only ill-effect being vomiting in one case. In a case of alcoholism in a female it failed to induce sleep, and here a dose of chloral (gr. xl) had the desired effect. Sulphonal was given in doses of gr. xv to gr. xxx for females, and gr. xx to gr. xl, or occasionally ʒj for males. The addition of hyd. c. cret. (gr. i., ii. or iii.) seemed to increase the certainty of its action.

Krafft-Ebing speaks well of *Hypnone* in doses of 15 to 20, or even 30 drops. It procures several hours' deep refreshing sleep without disagreeable after effects. Conolly Norman ("Jour. Ment. Sci.," Jan., 1887) recommends its subcutaneous use. Surzycki found it act very unsatisfactorily in doses of ♏ $1\frac{1}{2}$ to ♏ $7\frac{1}{2}$ ("Prov. Med. Jour.," July, 1891).

Henbane.—Recommended by Clouston in doses of ʒj to ʒss of the tincture as a temporary expedient in the very agitated cases of melancholia.

Cannabis Indica.—Clouston has found a mixture of the tincture (from x min.), and bromide of potassium (from xx grs.) do more good than any other narcotic, and act less injuriously on the appetite in cases of melancholia.

Hyoscyamine.—Hypodermically; when the patient will not or cannot be got to take medicine. Acts in half-an-hour to an hour, and sleep lasts from two to eight hours, according to the dose. Acts more certainly and quickly, and in smaller doses than when given by the mouth. Dose (of Merck's crystalline) as a hypnotic for strong adult, gr. $\frac{1}{50}$ (hypodermically), increased rapidly or slowly, according to the effect (Brown, "Braithwaite's Retrospect," vol. lxxxvii). I usually commenced with gr. $\frac{1}{20}$ (hypodermically) for a man, and gr. $\frac{1}{50}$ for a woman, and in men have increased the dose to gr. $\frac{1}{4}$ without ill effects. It should not be given to weak or exhausted patients.

Hyoscine Hydrobromate.—Fischer ("Braithwaite's Retrospect") gave hypodermically $\frac{1}{120}$ to $\frac{1}{60}$ in several cases with success. Fischer also used hyoscinum hydrochloricum in doses of $\frac{1}{132}$ to $\frac{1}{66}$, and even $\frac{1}{44}$ of a grain satisfactorily, both as a calmative and hypnotic ("Medical Annual," 1889).

Da Costa ("Jour. Nerv. and Ment. Dis.") says hyoscine is of great value for occasional use, but must not be given continuously for a long time. It produces disagreeable symptoms, and if long given, hastens dementia. It may be given in cases of epileptic mania, transitory mania, maniacal furor, excitement of general paralysis, ordinary mania with marked destructive tendencies, melancholic frenzy, uncontrollable impulses, as a calmative.

Piscidia Erythrina (Jamaica Dogwood) and the *Bromides* diminish sensory hyperæsthesia, and by removing somatic and psychical irritation often procure sleep indirectly.

Methylal.—In alcoholic cases Krafft-Ebing gives ♏ 1½, hypodermically, in a syringeful of distilled water from one to three times in the twenty-four hours. It takes effect in an hour or two. Sleep is deep and refreshing without ill after effects. (See treatment of "Delirium Tremens.")

Amylene Hydrate.—Found by Scharschmidt, Girtler, Von Mering, and others, to act satisfactorily. Use not contraindicated in cardiac or digestive affections. Dose, ♏ xx to xlv or l, or even xc., given with wine, peppermint, or extract of liquorice ("Med. Annual"). Surzycki prefers it to sulphonal in cases of insomnia with general excitement or delirium. It never causes any unpleasant effects even in large doses (4 to 8 grammes). *Urethan* is much inferior to it, and to sulphonal, according to Surzycki ("Prov. Med. Jour.," July, 1891).

Chloralamide.—Gordon and others have found it very satisfactory in the treatment of the insomnia of old age, hysteria, and pulmonary diseases. It is most useful where there is little or no pain. Dose 30 to 45 grains. ("Brit. Med. Jour." and "Jour. Ment. Sci.").

Auxiliary measures (which may also be tried first, especially in mild cases); keep head high and cool and feet warm; wet packing; warm foot baths; warm drinks at bed-time; hot milk; warm beef tea; warm beef tea and spirit; warm spirit and water; stout; raw onions; dark and quiet room; galvanic current through head; head to north; in states of inanition, head low. (See "Pulse," chap. II. of this work.) *Change of scene.*

In mild cases of "agrypnia," Krafft-Ebing recommends quin. valer., in doses of gr. jss. Lactic acid, in large doses, acts as a mild hypnotic. In cases of mental overwork and worry, strychnine, in small repeated doses, will sometimes induce sleep.

Hypnotism, especially the Nancy method (by "suggestion"), has been found to act well in some hysterical, toxic, neurasthenic, convulsive and paralytic cases.

B.—Motor Excitement, Noisiness, Destructiveness.

Spitzka considers *Conium* "the best, most reliable, and safest drug" to control motor excitement in mania. He uses Squibb's fluid extract and gives 20 minims as an initial dose, and then from 10 to 15 minims every hour or half-hour until the excitement is subdued. He says larger doses may be given to patients whose tolerance has been tested.

In the restlessness and excitement of melancholia Clouston has found a mixture of tinct. cannab. indic. (from x min.) and pot.

bromide (from xx grs.) do the most good, whilst it at the same time does the least harm to the appetite for food.

In strong, violent, maniacal patients, small doses ($\frac{1}{100}$th to $\frac{1}{20}$th of a grain) of Merck's crystalline hyoscyamine may be given hypodermically and repeated several times a day according to the effect. This will be also found useful when removing to an asylum, or elsewhere, a violent patient who refuses to take medicine.

Krafft-Ebing (pp. 294-297) considers *Opium*, administered subcutaneously or per rectum, one of the most important sedatives: (1,) It diminishes psychical hyperæsthesia and præcordial anxiety. In this way it very often acts hypnotically; (2,) It stimulates the vaso-motor nerves, thereby contracting the vessels; (3,) It has trophic actions on the central nervous system and promotes nutrition. Blandford, Schüle, Ziehen, Tellegen, strongly advocate the use of opium in melancholia. It induces sleep and relieves mental anguish. It acts best in anæmic conditions and in elderly persons.

The vaso-motor and sedative actions of *Morphia* are more powerful than those of opium, but the trophic actions are absent. Morphia is best given hypodermically in the psychoses. It is indicated in: (1,) Melancholy conditions with neuralgic or vaso-paretic symptoms; (2,) Paranoia with physical persecutory delusions, auditory hallucinations, hallucinatory paranoia; (3,) The irritable excitement of declining mania, and the angry, passionate outbursts of imbeciles; (4,) The episodical excitement of general paralytics; (5,) The episodical excitement of chronic forms; (6,) Periodical mania and circular excitement with vaso-motor prodromata (small tense *pulsus celer*); here it should be given in large doses.

Codeia is not constipating and stupefying like opium, and does not induce a crave like morphia. Its strength as a narcotic is about one-third of that of morphia; the muriate may be given internally, and the phosphate subcutaneously. It was tried successfully eighteen years ago in excited cases in Norfolk County Asylum. Tinct. of *Veratrum Viride* was used about the same time in the same asylum in the episodical excitement of chronic mania, and was found to act beneficially. Since then it has been recommended by Mickle in general paralysis. In 1877, Meynert recommended the use of veratrin in the episodical excitement of general paralysis.

As auxiliaries, abundance of fluid nourishment, milk, beef tea, raw eggs; out-door exercise, where not contraindicated by pyrexia or physical weakness or the immediate surroundings of the patient. - Where there is constipation or diarrhœa, caused by constipation, ʒj ol. ricini or gr. v. calomel may be given, or from gr. ½ to gr. 2 aloin may be administered hypodermically, at first, the action of the bowels being afterwards maintained by diet, exercise, and, if necessary, cascara, or occasionally tamar indien.

Willoughby ("Lancet," 1889, vol. i., p. 1030) reports the good effects of *Pilocarpine* in a case of threatening mania; and Lyon ("Jour. Nerv. and Ment. Dis.," N.Y., 1889, vol. xvi., p. 254) in hystero-epilepsy and maniacal excitement. In 1888 I injected gr. $\frac{1}{20}$ pilocarpine nitrate subcutaneously in a man of insane diathesis and heredity, suffering at the time from maniacal motor excitement. He went to sleep half-an-hour afterwards, and slept all night and well on into the next day. He made a good recovery, and was at work in a few days. He thought the substance injected was morphia. He was quite conscious during the excitement, which somewhat resembled some of the stages of Charcot's hystero-epilepsy. About the end of 1889 I injected gr. $\frac{1}{30}$ into the arm of a young man suffering from confusional excitement of a somewhat hysterical kind, with motor restlessness, caused by fright at night. He was at business next day. However, in his case I gave gr. xxx sulphonal about half-an-hour after the hypodermic injection.

Weatherly ("Jour. Ment. Sci.," July, 1891) gives the dose of hyoscine as gr. $\frac{1}{300}-\frac{1}{100}$, increased very cautiously to $\frac{1}{30}$. He says it should be given in repeated small doses, and its effects carefully watched. He considers it to be safer than hyoscyamine. According to him its uses are to act as a *mental alterative* in troublesome cases, and as a calmative in tremor, and in motor excitement *when not hysterical*. Pilocarpine is an antidote. (See "Pilocarpine.")

C.—**The Refusal of Food and inability of the patient to feed himself.**

The refusal of food may arise from: (*a*,) Mere anorexia (loss of appetite); (*b*,) Subjective depressive ideas, as in melancholia, the patient believing himself unfit to live, or labouring under the delusion that his bowels are stopped up, etc.; (*c*,) From delusions that he has been commanded by the Supreme Being not to take food or drink, or that people are attempting to poison him; (*d*,) From mere ill-temper, whim, or caprice.

In cases of delusional insanity (*c*,) there may be gastric disorder indicating gastric rest; other symptoms will have already given rise to preparations for the patient's permanent treatment; eating is often resumed voluntarily, and these patients are generally physically stronger than the melancholic ones (*b*). The latter, as well as those suffering from anergic stupor, and therefore *unable to feed themselves*, should not be allowed to go more than two, or at most three whole days without feeding.

Class (*a*,) will require general dietetic and therapeutical measures. Class (*d*,) will, as a rule, resume eating after two or three days' abstinence. Maniacs and general paralytics generally resume eating voluntarily. Some recommend that melancholic and stuporous cases, when they refuse food, should be fed forthwith. The anergic patients may be fed by simply pouring fluid food into the back of the mouth by means of a wooden spoon or a metal feeding cup. The melancholiacs, especially the stuporous ones, may be easily fed through the nose by means of an ordinary breast exhauster (Saunders' form), about three inches of the tube being left attached to the glass reservoir and passed along the floor of the nasal cavity. Fluid food, milk, beef tea, switched eggs, etc., may be poured into the aperture in the reservoir intended to be applied to the nipple; the reservoir may be then tilted until a tablespoonful or so has gone down the pharynx, and this can be repeated frequently, allowing the patient time for rest and breathing; half a pint of fluid may be given in this way without alarming the friends of the patient. Feeding may be done in this manner three or four times a day, and after the first time the quantity may be increased and medicines, port wine, brandy, etc., may be added. As a rule, this method of feeding can only be utilised as a temporary expedient, as the patients, or many of them, learn after a time to eject the fluid by the mouth. For more effective, though much more formidable-looking methods, viz., feeding by means of long nasal tube and stomach tube, see "Ultimate Care and Treatment." Sometimes a patient can be induced to take food after being forcibly fed once, either by the nose or mouth.

Tellegen and others advocate the use of *opium* in melancholic sitophobia. Raspail and Voisin recommend washing out the stomach.

D.—Suicide or Homicide.

When there is a tendency to either or both of the above acts, the patients should be watched night and day by trustworthy persons, one or more being constantly on duty according to the physical strength of the patient, or the acuteness of the case. Mechanical restraint is not advisable, and is not even always effective, Griesinger, mentioning a case where the patient committed suicide whilst in a straight waistcoat. If the patient has actually wounded himself, restraint may be necessary for surgical reasons. All weapons, razors, knives, fire-irons, cords, poisonous substances, etc., should be kept out of the patient's way; and he should be removed as far as possible from deep water, steep staircases, railway trains, carriages, etc.

Tigges and Tellegen believe that *opium* diminishes the suicidal inclination in melancholia. The latter says it is especially useful and reliable in private practice. He advises to commence with small doses of *morphia* subcutaneously ("Jour. Ment. Sci.," Oct., 1891).

II.—Ultimate Care and Treatment.

This falls into—

A.—HOME TREATMENT.

B.—PRIVATE CARE (SINGLE PATIENTS).

C.—ASYLUM TREATMENT AND CARE.

A.—Home Treatment.

This term is applied here to the care and treatment, with medical aid, of semi-insane or uncertified insane persons by the near relatives, husband, wife, father, mother. It is not intended to imply, as will be seen below, that the patient is to remain in his usual abode.

Mental diseases are much more amenable to treatment during their prodromal and initial stages than when fully developed, therefore, in the chapter on Symptomatology, the symptoms of these stages have been relatively fully described. Change of character, irritability, insomnia, are the most prominent of these early symptoms. As in all other diseases, the *first indication* in the treatment of insanity, incipient or developed, is, if possible, to *remove the cause*. In attempting to remove the immediate cause of the symptoms the pathogenesis must be taken into account.

(1,) It may be necessary to diminish the encephalic blood supply: (*a*,) By reducing the heart's activity, using for this purpose digitalis, aided by morphia, hydrocyanic acid, cold baths, cold compresses over the cardiac region. Acute gastric catarrh and strong sexual excitement contraindicate the prolonged administration of digitalis; its cumulative action should not be forgotten; (*b*,) By dilating the peripheral blood-vessels. This plan especially recommends itself in cases of prolonged venous hyperæmia of the brain. Warm baths, wet-packing, wet-rubbing, dry cupping, purgatives (saline mineral waters, aloes, rhubarb, cascara, etc.), fulfil this indication; (*c*,) By causing the encephalic blood-vessels to contract. To this end apply cold compresses or the ice-cap to the head, or cold compresses or ice along the course of the great cervical vessels. General mustard baths, mustard foot-baths, large mustard poultices to skin of body, act in a reflex manner on the vessels of the pia, and cause contraction after some preliminary dilatation (Heidenbain, Schüller, Krafft-Ebing); this method is useful for the removal of venous hyperæmia. Vascular constringents are nicotine, hyoscyamus, nux vomica, belladonna, quinine, lead, caffeine, the bromides, small doses of opium and morphia, and, most important of all, ergot, either in the form of liquid extract or hypodermically, as Bonjean's or Wernich's ergotin. The latter drug is indicated in half gramme to gramme doses subcutaneously, once or twice daily (Krafft-Ebing) in excitement due to congestion, as in simple mania, acute mania, paralytic frenzy, transitory mania, and certain stages of acute delirium.

(2,) It may be advisable to augment the blood-supply to the brain: (a,) By increasing cardiac activity. In conditions of debility and exhaustion connected with the functional psychoses, spirits, in the form of good old wine, beer, and punch, are given with benefit, as they improve brain nutrition, encourage sleep, and diminish or retard tissue metamorphosis. When the heart's action is very weak, prescribe tea, coffee, brandy and egg mixture, ether, ethylic alcohol. In collapse and threatened syncope, subcutaneous injections of sulphuric ether or of oil of camphor (1 to 10) are excellent; (b,) By widening the bloodvessels. Warm poultices or warm water caps to the head, short cold rubbings, short shower-baths, short hip-baths. Ether, chloroform, opium and morphia in large doses, and above all, amyl nitrite, dilate the vessels. The last drug is also a cardiac stimulant, dose gtt. 4 to 6 by inhalation (Krafft-Ebing). It is beneficial in hemicrania, angina pectoris, bronchial asthma, and some cases of epilepsia vaso-motoria. Krafft-Ebing thinks it useful in passive and stuporous melancholia, with extremely defective circulation, as it relaxes the initial vascular cramp and strengthens the heart, thus causing the anæmic brain to receive more blood; (c,) By facilitating the flow of blood to the brain. Rest in bed with head low is excellent in states arising from inanition, and will often procure rest more quickly than narcotics (Krafft-Ebing, pp. 290-293).

Certain causes of mental disease, e.g., mental or physical overexertion, alcoholic or sexual excess, certain bodily disorders (anæmia, neuralgia, menstrual disorders, etc.), and privation, are, or should be, removable. The forms of insanity capable of being cured or ameliorated by the treatment of the underlying bodily disorders or functional over-action, are insanity from abdominal disorders, anæmic insanity, insanity of Bright's disease, consecutive insanity, diabetic insanity, epileptic insanity, insanity with exophthalmic goître, hysterical insanity, lactational insanity, masturbational insanity, hypochondriacal melancholia, insanity of myxœdema, metastatic insanity, insanity of oxaluria and phosphaturia, pellagrous insanity, phthisical insanity, podagrous insanity, insanity of paralysis agitans, post-connubial insanity, primary mental deterioration, puerperal insanity (septicæmia, inflammation, diminished excretion and secretion, loss of blood, etc.), rheumatic insanity, anergic stupor, syphilitic insanity, toxic insanity, traumatic insanity, uterine or amenorrhœal insanity. Apart from any direct causal relationship with the mental disease, bodily derangements, general or local, should be carefully watched for, and when noticed, promptly treated. Other causes, such as disappointment of various kinds, adversity, fright, shock, etc., do not admit of this indication being fulfilled, but the patient may be removed from the disagreeable surroundings more or less intimately associated with the cause. *Change of scene*, then, is the *second indication* in the treatment of mental diseases, and a very important one it is in almost all forms, frequently (combined with the first indication) preventing an attack passing beyond the prodromal or initial stage. Even change from one asylum to another nearly always acts beneficially and is often curative; in at least two cases of many months' duration I have known recovery take place shortly after transfer. One was a case of paranoia (persecutional) with auditory hallucinations, the other of acute mania tending towards chronicity with othæmatoma.

(3,) The patient should have, and if possible, in a fresh, bracing atmosphere by the seaside or on the hill-tops, much *out-door exercise*, short of actual fatigue. In cases accompanied by decided pyrexia or great physical weakness, this would be of course contraindicated.

(4,) The *diet* should be generous and consist in great proportion of nutritious and easily assimilable fluids, such as milk, switched eggs, beef tea, soups, etc. From two to six quarts of milk may be given in the twenty-four hours, constipation being avoided by giving fruit in the morning, and, if necessary, drachm doses of the ext. cascar. sagrad. liq. at night.

(5,) The *treatment of symptoms.*—*Insomnia* may be combated by the methods already mentioned (pp. 243-245), but in prodromal, incipient, or borderland cases, sulphonal seems very useful, as it produces a soothing and quieting effect after the sleep induced by it has passed off, and it does not seem to act so deleteriously as chloral on melancholiacs. For irritability, restlessness, erotic tendency, epileptic seizures, give pot. bromid. (from xx grs.) with tinct. cannab. indica (from x min.) three or four times a day. (See "Treatment of Epilepsy, Neurasthenia," etc., below).

In a case of neurasthenia with nocturnal emissions, I gave ʒj of the liquid extract of salix nigra every night at bedtime with good effect. In three cases of severe dysmenorrhœa, aggravated by influenza, one ½-drachm dose relieved, and three or four doses at intervals of four hours completely arrested the excruciating pain.

Præcordial anxiety may, when mild, be relieved by warm baths, sinapisms over the epigastrium, ac. hydrocyan., ext. belladon., and, in an ill-nourished, anæmic patient, rest in bed. In severe cases, opiates subcutaneously. If the pulse is small and not frequent, give at the same time acetic ether; if it is frequent with excited cardiac action, tinct. digital. In masturbatory cases, especially neurasthenic ones, chloral hydrate may cut the attacks short (Krafft-Ebing, *op. cit.*, p. 309).

Meynert says baths of 22° C. or a little higher are excellent, soothing, especially, anxious melancholiacs.

In auditory hallucinations from hyperæsthesia of the acoustic centres, Krafft-Ebing recommends a trial of the galvanic current, and says he has obtained favourable results from the administration of morphia.

For depression, nervous exhaustion, anorexia, neuralgia, give syrup of the hypophosphites, ʒj three times a day in a wine-glassful of water before food. Or give iron, arsenic, strychnine, phosphorus, ol. jecoris, ergot, opium, zinc, cocaine, damiana, or quinine, singly or in various combinations, according to the case. Valerianate of zinc in the form of pills, beginning with one-grain doses, and $\frac{1}{30}$ of a grain of the phosphide three times a day after food may be given. I have found this pill with a mixture containing arsenic, very beneficial in neurasthenia. For the relief of severe neuralgia or headache, 4 grains of antifebrin will be found useful, and this dose may be given every six hours, or antipyrin in 8-grain doses every hour until pain is relieved or until three doses have been given (Kingsbury). In a case of mine of severe facial neuralgia, where the above drugs and others failed, phenacetin in 20 grain doses twice a day succeeded. Where there is pain arising from organic disease (tumour, inflammation, etc.) nothing equals morphia, given hypodermically. For other symptoms see *ante* ("Immediate relief of urgent symptoms").

Electricity.—De Watteville ("Medical Electricity," 2nd ed., p. 174) says it may be used as a remedial agent in the following diseases and conditions, viz., chorea, hysteria, epilepsy, neurasthenia, insomnia, cephalalgia; cerebral anæmia and hyperæmia; cerebral hæmorrhage and ischæmic softening; cerebral sclerosis or degeneration. The galvanic current should be used as a rule. It should be very weak or weak, the electrodes being applied to the head or head and neck for a period (to be cautiously increased if the current is well borne) limited at first to one to three minutes. Wiglesworth has found it useful in stupor.

Tigges ("Allg. Zeit.") treated 201 cases of insanity with electricity; many were relieved, and 24 (of whom seven were stuporous) of the melancholic cases cured.

Massage—Savage says it is rarely, if ever, useful in ordinary cases of insanity; its chief use is in mental depression with physical weakness, loss of flesh, and deficient action of the gastro-intestinal tract. Hack Tuke found it beneficial in one case, and curative in one of suicidal religious melancholia, with eroticism. It is much used by Weir-Mitchell, Stretch Dowse and others in neurasthenia, though its application in this complaint is considered useless or even injurious by some practitioners. Jacoby has found it valuable in neuralgia.

Hypnotism.—Aug. Voisin has by means of hypnotic suggestion cured persons suffering from the following conditions and symptoms, viz., hallucinations and delusions, disturbances of special and general sensation, suicidal ideas, acute and furious mania, mania and agitation during catamenia, dipsomania and morphinomania, onanism, infantile depravity, amenorrhœa in the insane. Relapses were not more than a tenth of the cures ("Jour. Ment. Sci.," 1889). Castelli and Sumbroso cured by this means a case of hysterical melancholia ("Jour. Nerv. and Ment. Dis.," 1886). Hack Tuke ("The Influence of the Mind upon the Body") speaks of the cure of hysterical conditions and even of organic paralysis by Braidism. By the latter method the patient looks at a bright object held above the eyes until the hypnotic state is induced; this is the method used by Charcot and his school. Passes, loud noises, etc., will also bring about the condition. In the Nancy or "suggestion" method, of which Bernheim is the chief advocate, the patient sits in a chair and is told by the physician in a confident tone that he will soon go to sleep, that his sleep will be natural, that he is feeling drowsy, etc. This plan was known to, and sometimes practised by, Braid and Faria. Many cases of neuralgia have been cured by both methods.

Krafft-Ebing believes that hypnotic suggestion is a valuable addition to the therapeutics of functional nervous diseases. Wetterstrand tried it in 718 cases of various diseases and conditions. Of these only 19 were not susceptible. It acted satisfactorily in *petit mal* and in the incipient stages of the less unfavourable psychoses. Brilliant results were obtained in anæmia arising from leucorrhœa and dyspepsia, and Wetterstrand advises beginners to commence with cases of this kind. He and others have found hypnotic suggestion unfailing in the treatment of incontinence of urine in children ("Jour. Ment. Sci."). R. P. Smith and A. T. Myers report unfavourably as to its therapeutical effects in 21 cases of insanity ("Jour. Ment. Sci.," April, 1890).

Where an attack of insanity has supervened suddenly, or where the prodromal or initial symptoms have been overlooked, or have not been amenable to treatment, the possibility of home treatment will depend mostly on the patient's means, but also partly on the character of the mental affection, and on the physical condition of the sufferer. A patient of strong physique, with marked suicidal or homicidal tendencies, or who obstinately refuses food, will require a number of attendants, and in the latter case three, or at least two, medical visits daily. On the other hand, cases of simple melancholia, and of primary deterioration, of simple mania, of ordinary acute mania in non-muscular persons, of puerperal mania, if food is not refused, of alcoholic insanity, acute or chronic, where deprivation of alcohol can be ensured, of primary confusional insanity, of anergic stupor, etc., etc., can be treated outside an asylum at comparatively little expense, yet not so economically as in one of these institutions. Nevertheless, the stigma still attaching to individuals who have been patients in asylums, and to a greater or less extent to their relatives, still acts as a strong deterrent to asylum treatment. Uncertified insane patients must not be kept for profit, even by relatives. Any uncertified alleged lunatic can now be visited by the Commissioners (see chap. ix.) Home treatment having been decided upon, a detached house in a healthy, bracing neighbourhood, hilly or seaside should be selected. The patient should have two rooms on the ground floor, one as a sitting room, the other as a bedroom. The latter should be well warmed by means of covered hot water coils, the bed should be on the floor, the window well shuttered on the inside, if necessary the floor may be covered with straw mattresses, of which there should be a sufficient number to change; all chairs, fire-irons, gas-pipes, brackets or ornaments should be removed. In the sitting room, all articles that might be used as weapons or missiles should be removed, yet at the same time the room ought not to be made too bare and gloomy.

When *forcible feeding* is required for any length of time, it will be necessary to use the stomach tube, either nasal or oral. The oral preferably, as by the nasal method an alarming amount of dyspnœa, discomfort and cyanosis is produced even when the tube has not passed into the larynx, having been guided past the fauces and the laryngeal aperture by means of stilettes, as in the French method. The nasal tube must also, to pass the nostril, be smaller than the oral. The oral method having been selected, it will be requisite to have at least four attendants if mechanical restraint is to be avoided. If the patient is quiet, as in some cases of melancholia, he may be fed sitting in a chair, with the help of one or two attendants. In other cases the patient having been laid on a bed on the floor with his head and shoulders raised by pillows, one attendant at the top should hold the patient's head firmly between his hands, another should hold the arms, and a third the legs. When the patient is very obstructive and violent, it is safer to have a restraining or feeding chair. (In certified cases the use of such a chair must be recorded and reported.) The physician standing on the left side of the patient should then introduce between the teeth the closed point of a steel screw mouth dilator, by screwing apart the points of which the mouth may be widely opened, and so retained, the mouth dilator being now used as a gag and held by a fourth attendant. In unresisting toothless cases no gag is required. The physician standing on the patient's right side, can then readily introduce a well-oiled soft rubber stomach tube (occasionally a stiff one may be required), or a siphon stomach pump, guiding the end of it down the back of the pharynx with the left forefinger. If the stomach tube is chosen, the fluid food can be poured into the funnel-shaped end, if the siphon, the funnel-shaped end can be slightly weighted and put in the vessel containing the food, placed above the level of the patient's head, and the fluid introduced by the usual method of using the siphon stomach pump. At first half a pint of fluid will be sufficient, but this can be rapidly increased to a pint, a pint and a half, or even a quart, given twice or three times a day. The fluid may consist of milk, eggs, beef tea, minced and pounded meat, thin arrowroot, broth, soups, combined in various proportions according to the state of the patient's bowels, etc. Orange juice, pulp of grapes, medicine, spirits, wine, etc., may be added. It has been recommended to add salt in order to cause appetite for food.

Minor details.—It is advisable to run some tepid water through the feeding tube before using it. The ends of the mouth dilator should be covered with clean pieces of lint each time it is used, and no permanent india-rubber covering should be employed, as it is liable to cause inflammation, abscess, and even necrosis. In introducing it, it is well to take advantage of gaps in the teeth. The patient's clothes should be protected by a sheet or serviette.

When the tube is being introduced for the first time it generally causes some cyanosis and appearance of choking, but this effect soon ceases, and the tube being in the œsophagus can be passed until many inches of its length have disappeared, but if by mistake it has entered the larynx only a few inches can be passed. Further, if the tube is in the larynx, air will be inspired through it; this is impossible if it is in the stomach, though gaseous matters may pass out by the tube and expiration be thus simulated.

Kräpelin recommends auscultation of the stomach whilst air is blown into the feeding-tube in order to be quite certain that the latter is not in the trachea.

On the flexible tubes now there is nearly always a ring about fourteen inches from the end, and when this ring is at the patient's lips, the end of the tube will be near the cardiac orifice of the stomach. With soft tubes care should be taken that the end does not catch at the constriction of the pharynx or at the epiglottis or larynx, and the tube get coiled up in the pharynx and mouth instead of passing down the œsophagus. It is advisable to have a stiff tube always in readiness. In withdrawing the tube pull it slowly for five or six inches, agitating it at the same time so as to empty it; then draw the end of it quickly through the pharynx and mouth. The patient should be watched for some time to prevent him trying to induce vomiting. The tube and gag should be well washed and disinfected.

I have generally found twice a day sufficiently often for forcible feeding. An emaciated melancholiac was fed only twice a day for many months. He had out-door exercise regularly. He gained over 30 lbs. in weight and came round to take food voluntarily.

Out-door exercise, in-door amusement, entertainments (unless in acute cases), occupation, literature of various kinds, regularity of meals and sleeping hours, attention to gastro-intestinal functions, suitable clothing, kind and gentle, yet firm treatment, are absolutely necessary. Some useful employment on a farm or garden, or in workshops, kitchen, laundry, or bedrooms, is very beneficial, and has proved curative in many cases. Most asylums, private as well as public, have now adopted this system to a greater or less extent.

The patient's morbid ideas and fixed delusions should never be agreed with, and it is almost useless to contradict them; therefore, the best plan is to turn the subject as soon as possible, or divert the patient's attention to some amusement or occupation.

The non-medicinal or mental or moral treatment is of extreme value. The condition of the psychical atmosphere (the impressions on all the senses including the muscular) is as important to the patient whose cerebral structure is diseased as the physical atmosphere is to the sufferer from pulmonary mischief. The highest mental faculties are formed out of the lowest, and these are easily accessible. It is therefore only reasonable to expect that in derangement of the former the latter will be capable of being acted upon so as to influence the former beneficially. In addition to the methods already mentioned of acting on the brain through the senses, the influence of music and of coloured light (blue depressing, red exciting) should not be forgotten. Cerebral localisation might also be brought to bear. W. Carter reports ("Liverpool Med. Chi. Jour.") a case of aphasia rapidly cured by verbal re-education, combined with constant use and vigorous exercise of the left arm and hand. A system of education might be useful in the psychoses as well as in aphasia and idiocy. Sollier ("Progrès Médical") says the recovery rate in the French provincial asylums has fallen in a few years from 27 to 20. He thinks this low recovery rate (about half the English) is owing principally to the neglect of non-medicinal treatment. In one English institution, Barnwood House Hospital for Mental Diseases, the recovery rate as shown by the last report is 68 per cent. Recovery rates are calculated on the yearly admissions.

Hypnotics should only be used occasionally, and the same may be said of calmative medicines. When the patient is "wet and dirty," he should be frequently taken to the night chair or closet, and wet or soiled clothes ought to be changed immediately. Retention of urine should be watched for, and, in fact, the medical man should investigate the state of the internal organs, including the abdominal and pelvic viscera, every alternate day, for the patient may be "wet" or even "dirty" whilst urine is accumulating in the bladder, and the bowels are becoming loaded with scybala, and there may be serious thoracic mischief without the usual subjective signs such as pain, cough, dyspnœa.

B.—*Private Care (Single Patients).*

Under this head are included cases kept in unlicensed houses, preferably the residences of medical men, for a consideration. Formerly, only one case could be taken care of in this way, but by the Lunacy Act of 1890 the commissioners can give permission for more than one insane patient to be kept in an unlicensed house. Every person of unsound mind in such paid, private care must be duly certified in accordance with the provisions of the new Lunacy Acts (see next chapter).

Weatherly ("The Care and Treatment of the Insane in Private Dwellings") considers the following *cases suitable* for this method of treatment. Many cases of simple melancholia, some cases of acute primary dementia (anergic stupor), cases of acute delirious mania (acute delirium), semi-insane persons with exaggerated eccentricities, but free from dangerous habits, the majority of cases of recurrent mania, harmless chronic cases of insanity, convalescing patients, and "almost all cases of insanity in the early stage, except those coming under the head of acute mania, acute melancholia, melancholia with stupor, erotomania. and cases having distinct homicidal and suicidal tendencies . . . . provided, of course, that adequate means of properly caring for them are present. With regard to chronic cases of mental disease, the cases I consider unsuitable for domestic treatment—or rather care—are those of homicidal and suicidal mania, melancholia with stupor, general paralysis of the insane, epileptic insanity and erotomania, and also some of the many cases of idiocy and imbecility."

But where the patient's means are adequate, any form of insanity may be treated in this way, as in that case the requisite number of attendants may be retained, padded rooms fitted up, carriage exercise and other means of recreation obtained, and unlimited changes of clothing provided. Erotic tendency may be controlled by the bromides, or where the patient is not strong physically, by the liquid extract of salix nigra. With regard to mechanical restraint, this is seldom resorted to in the best asylums, and then only for surgical reasons or to prevent self-injury. It may be least injuriously effected by means of leather gloves, fastening at the wrist, care being taken that they do not abrade the skin, each glove having a loop running on a belt loosely worn round the waist. Or by means of the camisole, suggested by Magnan ("Camisole. Recherches sur les Centres Nerveux") which consists of a kind of combination suit, the long sleeves being attached by straps to the outside of the trousers instead of being tied round the chest, as in the ordinary straight waistcoat. By Magnan's method respiration is not interfered with. It should, however, be remembered that, according to Section 40 of the Lunacy Act 1890, "mechanical means" of restraint are to be "such instruments and appliances as the Commissioners may, by regulations to be made from time to time, determine."

Seclusion, meaning fastening a patient in a room by himself between 7 a.m. and 7 p.m. ("compulsory isolation in the day-time") can very rarely have the excuse in domestic care put forward for its use in asylums, viz., the removal of the patient from the propinquity of exciting and excitable fellow-patients, but when resorted to must, like restraint, be entered in the medical visitation book ("Medical Journal") by the medical attendant, who must there state the time, duration, means, and reasons (see chap. IX). Shower baths and plunge baths, which should never be used as instruments of torture, are frequently asked for by patients, and either the latter or sponge baths for the purposes of cleanliness, are almost indispensable. In cases of masturbational insanity, primary mental deterioration, simple melancholia, and acute mania, a short shower bath will often do good. Cold baths are useful in general paralysis (*vide infra*).

In acute delirium and senile insanity a warm bath of half-an-hour to an hour's duration will often procure sleep without medicine or other means, or with the help only of some warm beef tea and spirits.

A patient not actually residing in the house of a medical man should have constant, and if possible, resident medical supervision. The medical attendance should not be limited to the statutory fortnightly visit. Before undertaking the treatment of such a case the practitioner ought to have a distinct understanding that the attendants are to be under his sole direction and obedient to him. This is absolutely necessary, as however amenable to discipline attendants may be when employed in an asylum, they are liable to become too independent when nursing single patients.

C.—ASYLUM TREATMENT.

Asylums are divided into—

1.—PAUPER ASYLUMS.

2.—LUNATIC HOSPITALS.

3.—LICENSED HOUSES OR PRIVATE ASYLUMS.

1.—*Pauper asylums.*

Under this head may for practical purposes be included county and borough asylums, criminal asylums, and the lunatic wards of workhouses. In Ireland many of the public asylums are, or rather, were designated "District Hospitals for the Insane." In the United States some of them are more happily called "State Hospitals." (The name of a public institution might seem to be of little importance, but as a matter of fact, respectable working men complain bitterly that letters from or about their relatives in some of the English county asylums have their envelopes endorsed *pro bono publico*, with the words "County Asylum"). This is the only form of treatment available for the poorer classes, and even for many persons belonging to the professional class. In these large public establishments patients have the advantage of discipline, the example of patients once troublesome, now settled down, extensive farm and gardens, varied occupation and amusements. On the other hand, some of the institutions contain too many patients under one medical superintendent, and the medical staff is numerically inadequate to perform the amount of work, clerical and other, expected of it, and at the same time to allow its members to study their cases clinically day by day, as in other special hospitals. Some of the county asylums now take private patients at moderate weekly payments. Even when the patient has been made a pauper, his friends are permitted or required to contribute to his maintenance if they are in a position to do so.

It would be a great advantage to the clinical study of mental diseases if all recent cases of insanity drawn from the poorer classes were first received in hospitals or asylums like St. Anne's, in Paris (but situated just outside the large towns and with adequate farms, etc.), whence they could be forwarded to the larger asylums, only those being retained whose illness would probably be of short duration, or who were required for special study and treatment. St. Anne's, where considerable numbers of newly admitted patients are seen and carefully examined by the physicians every morning in the presence of the pupils, forms an excellent training school, especially when taken in conjunction with the opportunities for pathological study afforded by the chronic cases of the Salpêtrière and Bicêtre. In Vienna, a portion of the very large general hospital is devoted to the reception and treatment of insane patients. This is convenient for clinical instruction, but does not seem to me to be the most beneficial arrangement, so far as the patients are concerned. There was at one time an insane ward in Guy's Hospital, but it was not a success. Most of the insane need ample space for recreation and work.

"Every public asylum should be available for scientific research and clinical teaching of insanity to students of medicine, and to qualified practitioners" (Care and Treatment Committee, Med. Psych. Assoc.). It is also recommended by the same committee that provision should be made for the treatment of *out-patients;* that *Recent Cases* should, unless obviously incurable, be received in a special ward or block, or building; that there should be a special *Infirmary ward* or block.

2.—*Lunatic Hospitals.*

The inmates of these institutions are derived from the upper and middle classes, and the terms vary from £500 or even £1000 a year to ten shillings a week, although there are very few paying the latter sum or anything near it. The ordinary payments are from $1\frac{1}{2}$ to 6 guineas. Originally intended for middle class patients whose friends dislike the idea of sending them to pauper asylums, they have now come to compete with the highest class of private asylums. As instruments of treatment for the curable, and as homes for the incurable, they are excellent. Some of them have several villa residences in their grounds where the patient, if necessary, can live apart from other sufferers and yet remain under the supervision of the medical superintendent and his staff. The grounds, farm and gardens are generally extensive. The number of inmates conduces to the frequency of associated entertainments and amusements such as concerts, dances, theatrical performances, cricket, tennis, bowls, golf, boating parties, etc., etc., as well as cards, billiards, chess, draughts, dominoes. These asylums are managed by boards of governors, and the receipts are all expended on the patients or the buildings in which they live. The medical superintendent is a salaried official, and the governors derive no pecuniary profit whatever from the institution. These establishments are visited and carefully inspected twice a year by the Commissioners in Lunacy.

Voluntary boarders who have never been in an asylum, can be taken, so that facilities may be given for the treatment of incipient cases of insanity where the patient feels his own weakness. Although very advantageous for the treatment of those who can afford the terms, there is something else wanted for poor professional men, clerks, and tradesmen who, recoiling from making their relatives paupers, yet cannot afford to pay the (to them) large sums required by the lunatic hospitals and private asylums of the present day.

A few asylums placed near the large centres of population and built to accommodate about 200 patients, the payments varying from a minimum of five shillings a week to a maximum of fifteen, inclusive of everything, would meet, to use a hackneyed expression, " a long-felt want."

That in addition, there should be separate asylums for dipsomaniacs and drunkards who have not yet deteriorated into chronic alcoholic lunatics, I think anyone will admit who has had experience of inebriate cases in private asylums and in private outside practice ; and that further, the patient should under certain circumstances be compelled to enter such an asylum for a given time ; the time to be extended afterwards, if necessary, by the medical and legal authorities ; and lastly, that these asylums for inebriates should not be kept for gain but be established on the original lines of the present lunatic hospitals.

3.—*Private Asylums (Licensed Houses).*

Bucknill and Tuke (p. 650) are of opinion that whilst small asylums may do for chronic lunatics, for the curative treatment of recent cases "an asylum (including under this term lunatic hospitals and licensed houses) containing at least thirty or forty patients should be chosen, and one containing four or five times that number should be preferred. A certain minimum number of fellow-patients is needful to establish that system of method and discipline which forms a great part of the curative influence of asylum treatment." Where the patient's means are only moderate, he will, if very troublesome, be better treated in an asylum than under domestic or private care. Under the head of troublesome would come cases of acute mania, acute melancholia, persecutory delusional insanity (persecutional paranoia), most cases of general paralysis, cases with strong suicidal, homicidal or erotic tendencies, and cases that are persistently wet, dirty or destructive.

It is best to select an asylum in which the medical superintendent exercises undivided control. The asylum should not be so near the patient's home as to interfere with his walks, drives, etc.; nor yet so far from his relatives as to make visiting inconvenient for them.

That many patients are very comfortable in these establishments is proved by the facts that those who recover frequently re-visit them when well; that some patients return of their own accord when they begin to feel ill, getting themselves medically certified, or returning as voluntary boarders; that many old chronic patients elect to remain inmates of these asylums rather than be removed to their homes or to other institutions.

Wherever treated the patient should be allowed as much liberty as is compatible with safety (in Scotland the open-door system is being successfully carried out in several asylums); when convalescent, he ought to be permitted to go about on parole, and before being legally discharged, he should be allowed to go out on probation: he should not be sent immediately to the locality where he became ill, and should not be allowed at once to resume his avocation.

The formation of After-Care Associations is a step in the right direction. These are doing good work on the Continent and in London.

The Gheel, or insane village system, if it could be established in this country, would be a great improvement on asylum treatment so far as many insane patients are concerned.

TREATMENT OF SPECIAL FORMS OF INSANITY.

Choreic Insanity.—Dresch, Pianese and others, think chorea is microbic, and use sodii salicyl. as a bactericide.

Climacteric, Adolescent, Pubescent, Masturbational and **Senile Insanity** are benefited by the bromides, piscidia erythrina, and salix nigra (see chap. VII.).

Coarse Brain Disease (Insanity from) may in some cases (arising from abscesses, tumours, spicula, etc.) be benefited by cranial operations based on cerebral localisation. Since 1876 Macewen has performed successfully many such operations for the relief of various conditions and symptoms. Bergmann has also operated successfully. In cases of brain tumour, Horsley urges that after pot. iodid. has been given in large doses for six weeks without very notable and real improvement, exploratory operation should be resorted to. It affords relief even if cure by removal is impossible ("Brit. Med. Jour."). Hack Tuke ("The Influence of the Mind upon the Body," p. 413) reports some of Braid's cases in which organic paralysis was diminished or cured by hypnotism. In cerebral hæmorrhage Horsley advocates pressure on or ligature of the common carotid, De Watteville recommends electricity, and W. Carter ("Liverpool Med. Chi. Jour.") small doses of morphia. Hayes Agnew, speaking from the experience of the Philadelphia surgeons, advocates trephining in cerebral abscess and in intra-cranial traumatic hæmorrhage ("Prov. Med. Jour.").

Delirium (Acute).—Most authors advise that hypnotics and sedatives should be used carefully and sparingly if at all. The patient should have warm baths of half-an-hour to an hour's duration at bedtime, cold being applied to the head and a careful watch kept by the medical attendant during the time the patient is immersed. The patient must have abundant fluid food, milk, eggs, beef tea, etc., given forcibly if not taken voluntarily. He should be kept in bed in a darkened room, the windows being well guarded, and everything that might cause irritation or injury removed.

When the baths cannot be given, packing in a warm, wet sheet, with blankets over all may be substituted, cold as before being applied to the head.

Acute maniacal cases tending towards acute delirium should be treated in much the same way. The object is to relieve cerebral congestion, whilst at the same time maintaining the physical strength.

Epilepsy and Epileptic Insanity.—The bromides are still the sheet anchor in the treatment of the neurosis, and frequently beneficial in the excitement and irritability occurring in the psychosis. Their good effects are increased and their deleterious ones diminished or avoided by intercombination, and by the simultaneous administration of such therapeutic agents as arsenic, zinc, the carbonates, ammonia, the iodides, cannabis indica, cascara sagrada, bark, bitter limes.

Borax, in initial doses of 10 grains three times a day, occasionally succeeds when the bromides fail. Mairet ("Prog. Méd.") recommends borax where there is coarse brain disease, and the bromides in idiopathic epilepsy. In children I have found equal parts of the three bromides with liq. cascar., suavis and syr. senn., act well. Potts and Wood give (to adults) gr. 6 of antipyrin and gr. 20 of amm. bromid. three times a day with good results ("Braithwaite's Retrospect").

In the treatment of the neurosis Alexander recommends percussion of the spine; electricity to the spine; the removal of causes of irritation, gastric, intestinal or other; trephining over the seat of injury in traumatic cases; shortening the round ligaments (Alexander's operation) in suitable cases of uterine flexion: no stimulants or tobacco, or as little as possible; a mild nutritious diet. Removal of the superior cervical ganglia of the sympathetic: of twenty-four cases operated on by Alexander during a period of four to six years, six were cured, ten improved, especially mentally; five unimproved; two died; and one was lost sight of ("The Treatment of Epilepsy," 1889).

Jacksonian epilepsy, arising from gross lesions, has been frequently cured by means of cerebral surgical operations.

Leonte and Bardesco consider that trephining is indicated where monoplegia and convulsions co-exist, and that it should be done early ("Brit. Med. Jour.," Nov., 1891).

The results of the excision of the epileptogenous focus in cases of focal epilepsy without gross lesion, an operation first performed by Horsley, are distinctly encouraging, both as to the fits and the mental condition, the improvement in the latter being immediate and progressive ("Brit. Med. Jour.")

Keen says the results of cortical excision have been, on the whole, very encouraging in traumatic epilepsy ("Braithwaite's Retrospect," July, 1891).

In the *status epilepticus*, cathartic drugs or enemata; inhalation of chloroform, or administration of chloral and bromides by mouth or rectum.

Exophthalmic Goître (Insanity of).—For exophthalmic goître; galvanism, place positive pole in sub-aural space and apply mild labile current from negative over goitre and closed eyelids. Three to ten milliampères, three to five minutes daily ("Medical Annual," 1889, p. 124). Cardew ("Lancet," April 7th, 1891) uses a current of a still lower maximum strength for six minutes three times a day. He applies the anode to nape of neck and moves the kathode up and down side of neck.

General Paralysis of the Insane.—To give any reasonable chance of success the *prodromata* must be attacked; the cause, if possible, removed, change of scene, rest, etc., prescribed.

Medicines.—Crichton-Brown has found benefit in the developed disease from the administration of physostigma faba. Julius Mickle formerly recommended tinct. fer. perchlor., but latterly has given full doses of pot. iodid., with or without small doses of bromide, treatment which seems to ameliorate somewhat the condition of the patient, diminishing excitement and pain. He still gives iron when the patients are feeble, or quiet, or in the later stages of the disease. Boubila ("Mercredi Médical") strongly recommends the chloride of gold and sodium, $\frac{1}{500}$ to $\frac{1}{100}$ of a gramme.

Other Measures.—Magnan treated a number of patients at St. Anne's thirteen years ago by applying tartar emetic ointment to the scalp at short intervals of time until a deep sore was produced, but the results were not encouraging. Lately Claye Shaw has treated two cases with apparent benefit at Banstead, by trephining; and he quotes another case from the Brookwood Asylum in which this operation was followed by "marked relief to several prominent symptoms" ("Brit. Med. Jour.," Sept., 1891). Rey ("Progrès Médical," 15 Août, 1891) trephined in a case; the patient was able to return home in a month and a half, and remained calm and quiet.

Voision recommends cold baths. So does Meynert, in the early stages.

As general paralysis is primarily a vaso-motor disease and secondarily one of connective tissue proliferation, it would seem *a priori* that vascular constringents, such as ergot, strychnine, pot. bromid., quinine, antipyrin, etc., ought to be beneficial in the incipient stage of the disease, whilst hydrarg. perchlor. (as a resolvent of connective tissue) combined perhaps with pot. iodid. should do good in the later stages.

Hysterical Insanity.—Cases have been reported ("Jour. Nerv. and Ment. Dis.," etc.) in which hypnotism has effected a cure. De Watteville says electricity may be used as a remedial agent in the neurosis ("Medical Electricity," p. 174).

Idiocy and Imbecility.—Many of these patients are much improved by the system of education now adopted in the idiot asylums, but there is still a lack of facility for the treatment of children suffering from epileptic idiocy. In two cases of microcephalus operated on, one by Lannelongue, the other by Horsley, there was marked improvement after craniectomy. Other cases have not done so well ("Brit. Med. Jour.," Sept., 1891). McClintock has recently operated successfully.

Melancholia.—Opium (including its preparations and derivatives) so strongly recommended in the treatment of this affection has been discarded by Clouston, Julius Mickle and other authors, and many other alienists. In active melancholia Krafft-Ebing strongly recommends morphia and the aqueous extract of opium hypodermically, the latter to be preferred for its trophic action, where nutrition has failed. Wilks also recommends the use of opium. *Vide supra* as to opinions of Blandford, Tellegen, etc. Kiernan has found quebracho act very beneficially in cases accompanied by præcordial pain. Clouston extols the use of quinine in melancholia. Meynert advocates the exhibition of iron, quinine, amyl nitrite, and (when there is exudation) saline diuretics.

Myxœdema (Insanity of).—The "British Medical Journal," Nov. 29th, 1890, reports a case of grafting the thyroid of an animal for myxœdema in Paris by Walther. The grafting was performed under the right breast of a woman of 40. The operation was followed by considerable improvement, speech became clearer, and the gait much better. Murray has found hypodermic injections of an extract of the thyroid gland of a sheep to be followed by improvement ("Brit. Med. Jour.," Oct., 1891).

Neurasthenia.—Krafft-Ebing recommends a nutritious diet, rich in protein and fat; the use of tonics in the widest sense of the word; aërotherapy (residence on mountains); hydro-therapy (rubbing, fresh and salt water baths, etc.); electricity (general faradisation, electric baths, etc.): iron, arsenic, strychnine, phosphorus, ergot, opium, zinc, cocaine, damiana, or quinine, according to the case; as sedatives, piscidia erythrina (fluid extract) and the bromides; as hypnotics, paraldehyde in the first place, then amylene hydrate, and sulphonal; chloral hydrate should very seldom be used; in neurasthenia gastrica, forcible feeding may be required. In *neurasthenic masturbatory melancholia*, tonics with opium; hydro-therapeutic treatment; careful oversight to guard against masturbation.

In *neurasthenic insanity from fixed ideas* (folie du doute, partial emotional aberration, etc.); baths, cold and sea; climatic treatment; general faradisation; tonic medicines, iron, quinine, ergotin, arsenic. For the mental suffering, company, travelling, diversions, agreeable and pleasant, yet not arduous, occupation. The cerebral impressionability is further diminished by pot. bromid., ʒj–ʒjss daily. During the attacks, pot. brom., ʒjss–ʒijss, morphia injections, chloral hydrate, alcohol, consoling assurances by some trusted friend.

In *neurasthenic masturbatory paranoia*, morphia and pot. bromid. act beneficially on the hyperæsthesiæ, paralgiæ, and hallucinations.

The five essential points in Weir-Mitchell's treatment of neurasthenia are: (1,) Seclusion; (2,) Rest; (3,) Massage; (4,) Electricity; (5,) Dietetics and Therapeutics. Rest in bed for six weeks or two months. Massage for six weeks at least. Diet at first skimmed milk and beef tea; eggs, oysters, brandy, meat, bread, butter, etc., to be added in a few days (Stretch Dowse, "Massage").

Paranoia.—Krafft-Ebing recommends the use of morphia in certain cases (see "Relief of Symptoms"). In six cases (paranoia, dementia, and paranoia passing into dementia) Burckhardt opened the skull and removed a portion of the cortical gray substance with a sharp spoon. In the majority of the cases the most troublesome symptoms (abusiveness, aggressiveness, hallucinations) were relieved. One patient died of convulsions six days after the operation (ablation of portion of temporo-sphenoidal cortex). In another, ablation of portions of the right superior parietal and supra marginal gyri (see "Morbid Anatomy of Symptoms") caused paralysis of the left arm, lasting several weeks, and transitory paresis of the left leg ("Jour. Ment. Sci., Oct., 1891).

Pellagra and Pellagrous Insanity.—Arsenic has been found useful.

Puerperal Insanity.—Maintain strength by giving, if necessary forcibly, abundance of milk, raw eggs, beef tea, port wine, etc. The bromides, chloral and other depressants should be avoided. Two cases of mine have done well in private practice with 10 grains of salol and 5 of quinine, given every six hours, ʒss of sulphonal in hot milk every evening, and ♏x tinct. cannab. ind. every four hours; the vagina being at the same time syringed twice daily with a warm, weak solution of permanganate of potash. In both these cases the temperature was supra-normal, and the lochia were offensive.

Syphilitic Insanity.—Give from 10 to 30 grains of pot. iodid. three times a day, preferably the latter dose in a tumblerful of milk two hours after food, the dose being preceded by a little brandy if there is a tendency to coryza. The dose may be increased to 100 grains or more, and mercury may at the same time be given either hypodermically or epidermically.

Toxic Insanity.—(1,) Remove the cause; (2,) Eliminate the poison by aperients, diaphoretics, diuretics, and in the case of lead, pot. iodid.; (3,) Procure sleep by chloral; (4,) Maintain the strength by diet and tonics; (5,) Allay excitement by means of baths, amm. bromid., cannab. indica, digitalis; (6,) In alcoholic and opium insanity endeavour to diminish the crave; in the case of the former, capsicum and gentian are of some benefit. Kola nut and red cinchona obtained a reputation which has not been maintained. Strychnine has been used with success subcutaneously in full doses for the alcoholic crave, and in delirium tremens.

Delirium Tremens.—Krafft-Ebing cautions against debilitating measures. The indications are to maintain the strength and procure sleep. To fulfil the former give milk diet, and in complicating or febrile cases, add wine or brandy. To procure sleep, Krafft-Ebing recommends in young strong people, chloral hydrate, gr. 15–30, with or without gr. ¼ of morphia, every three or four hours for two or three doses; if this plan fails give ext. opii gr. ½ every three or four hours till the desired effect is obtained; he recommends the drug to be used hypodermically (1 in 20), but if this cannot be managed, it may be administered in a clyster or suppository; it should be continued in small doses ($\frac{1}{6}$–$\frac{1}{3}$ gr.) for a few days after the critical sleep. In complicating and pyretic cases give opium with abundance of wine or spirit; or methylal (which is not a cardiac depressant) ♏ 1½ every two or three hours subcutaneously (1 in 10) ad effectum (after fourth or fifth injection as a rule); or paraldehyde ℨiij daily, or amylene hydrate, ℨiss daily, until sleep is induced. When there is extreme exhaustion give ℨj to ℨjss of zinc acetate daily, well diluted, in a mucilaginous vehicle. Krafft-Ebing treats the alcoholic psychoses on the same lines, with the addition of ergotin or digitalis if indicated. In all forms quinine should be given during convalescence.

Morphinismus.—The daily dose should be diminished very gradually, except that the first reduction may amount to a half or a third of the daily dose. The strength should be maintained by giving milk and brandy freely. Several physicians have treated cases of morphinism and of alcoholism successfully with hypnotic suggestion. If the patient continues to take morphine clandestinely after his supposed cure, an examination of the urine will reveal the fact.

Cocainism (Chronic).—Clouston recommends amm. bromid., brandy or wine, tea and coffee; possibly paraldehyde or sulphonal for two or three nights.

For a list of the mental diseases capable of being cured or ameliorated by the treatment of the *underlying bodily disorders* or *functional over-action*, see above (Section A., *Home Treatment*).

CHAPTER IX.

LEGAL REGULATIONS AND FORENSIC PSYCHIATRY.

CERTIFICATION OF INSANE PRIVATE PATIENTS IN ENGLAND AND WALES—VOLUNTARY BOARDERS—LAWS AS TO KEEPING SINGLE PATIENTS IN ENGLAND AND WALES—CHANCERY PATIENTS—UNCERTIFIED LUNATICS—PAUPER LUNATICS—LUNATICS (NOT PAUPERS) NOT UNDER PROPER CARE AND CONTROL OR CRUELLY TREATED OR NEGLECTED—WANDERING LUNATICS—CRIMINAL LUNATICS—CERTIFICATION OF THE INSANE IN SCOTLAND—CERTIFICATION OF THE INSANE IN IRELAND—CERTIFICATION OF THE INSANE IN THE STATE OF NEW YORK—CERTIFICATION OF THE INSANE IN THE STATES OF CONNECTICUT, PENNSYLVANIA, MASSACHUSETTS, AND ILLINOIS—TESTAMENTARY CAPACITY OF THE INSANE—EVIDENCE (TESTIMONY) OF THE INSANE—LEGAL TESTS OF INSANITY AND LEGAL RESPONSIBILITY OF THE INSANE.

CERTIFICATION OF INSANE PRIVATE PATIENTS IN ENGLAND AND WALES.

It having been decided that the patient is to enter a lunatic hospital, a licensed or unlicensed house, or a county asylum (as a private patient) and one having been selected, the first step is to write to the proprietor or medical superintendent for the requisite statutory forms. These consist of a form of "Urgency Order," one of "Petition for an Order for reception of a Private Patient," two of "Statement of Particulars," one annexed to the Petition, and one to the Urgency Order, three of "Certificate of Medical Practitioner," one of which should have a space for the Urgency Certificate below the space for "Facts communicated by others." The Medical Certificates must now be on separate sheets of paper, and, though essentially the same as formerly, are somewhat differently worded since the passing of the Lunacy Act, 1890.

When the patient is violent, homicidal, or suicidal, refuses food, is restless, sleepless, and noisy at night, or is greatly excited and very restless and noisy in the daytime, he can be removed, if fit physically for removal, on the "Urgency Order," accompanied by *one* Medical Certificate containing a statement or Certificate of Urgency added below the "Facts communicated by others."

53 Vict. c. 5, Sched. 2, Form 4.

FORM OF URGENCY ORDER FOR THE RECEPTION OF A PRIVATE PATIENT.

(a) Or hospital, asylum, or as a single patient.

(b) Name of patient.

(c) Lunatic, *or* an idiot *or* a person of unsound mind.

(d) Some day within two days before the date of the order.

(e) Husband, wife, father, father-in-law, mother mother-in-law, son, son-in-law, daughter, daughter-in-law, brother, brother-in-law, sister, sister-in-law, partner, *or* assistant. *[If not the husband or wife, or a relative of the patient, the person signing to state as briefly as possible:—1. Why the order is not signed by the husband or wife, *or* a relative of the patient. 2. His or her connection with the patient, and the circumstances under which he or she signs.]

(f) Superintendent of——the ——asylum,—— hospital, *or* resident licensee of the——house [describing the asylum, hospital, *or* house by situation and name.]

I, the undersigned, being a person twenty-one years of age, hereby authorise you to receive as a Patient into your house *(a)*_____

*(b)*_____ _____

as a *(c)*_____

whom I last saw at _____ _____

on the *(d)*_____ day of_____ 189 .

I am not related to or connected with the Person signing the Certificate which accompanies this Order in any of the ways mentioned in the margin.*(e)* Subjoined [*or* annexed] hereto is a Statement of Particulars relating to the said _____ _____

(Signed)

Name and Christian }
Name at length }

Rank, Profession, or } _____
Occupation (*if any*) }

Full Postal Address - _____

*How related to or }
connected with the } ___
Patient - - - }

Dated this__ _____ day of_____ 189 .

To *(f)*_____

_____

Form 2.

STATEMENT OF PARTICULARS REFERRED TO IN THE ABOVE (OR ANNEXED) ORDER.

If any Particulars are not known, the Fact is to be so stated.

[Where the patient is in the petition or order described as an idiot, omit the particulars marked*.]

The following is a Statement of Particulars relating to the said _____

Name of Patient, with Christian Name at length _____
Sex and Age _____
*Married, Single, or Widowed _____
*Rank, Profession, or previous Occupation (if any) _____
*Religious Persuasion _____
Residence at or immediately previous to the date hereof _____
*Whether First Attack _____
Age on First Attack _____
When and where previously under Care and Treatment as a Lunatic, Idiot, or Person of Unsound Mind _____
*Duration of existing Attack _____
Supposed Cause _____
Whether subject to Epilepsy _____
Whether Suicidal _____
Whether Dangerous to Others, and in what way _____
Whether any near Relative has been afflicted with Insanity _____
Names, Christian Names, and full Postal Addresses of one or more Relatives of the Patient _____
Name of the Person to whom Notice of Death to be sent, and full Postal Address, if not already given _____
Name and Postal Address of the usual Medical Attendant of the Patient _____

(Signed) (g)

(g) When the petitioner or person signing an urgency order is not the person who signs the statement, add the following particulars concerning the person who signs the statement.

Name, with Christian Name at length _____
Rank, Profession, or Occupation (if any) _____
How Related to or otherwise Connected with the Patient _____

53 Vict. c. 5.—Sched. 2, Form 8.

CERTIFICATE OF MEDICAL PRACTITIONER.

(a) Insert residence of patient.
(b) County City, or Borough, as the case may be
(c) Insert profession or occupation, if any.
(d) Insert the place of examination, giving the name of the street, with number or name of house, or should there be no number, the christian and surname of occupier.
(e) County, city, or borough, as the case may be.
(f) Omit this where only one certificate is required.
(g) A lunatic, an idiot, or a person of unsound mind.
(h) If the same or other facts were observed previous to the time of the examination, the certifier is at liberty to subjoin them in a separate paragraph.
(i) The names & christian names (if known) of informants to be given, with their addresses & descriptions.
* Or, not to be.
(k) Strike out this clause in case of a patient whose removal is not proposed.
(l) Insert full postal address.

In the matter of _____ of (a) _____ in the (b) _____ of _____ (c) _____ an alleged lunatic.

I, the undersigned_ ____ do hereby certify as follows:

1. I am a person registered under the Medical Act, 1858, and I am in the actual practice of the medical profession.

2. On the ____ day of 189 at (d) ___ in the (e) _____ of _____ (separately from any other practitioner) (f) I personally examined the said _____ and came to the conclusion that he is (g) ____ ____ and a proper person to be taken charge of and detained under care and treatment.

3. I formed this conclusion on the following grounds, viz.:—

(a.) Facts indicating Insanity observed by myself at the time of examination (h), viz. _____

(b.) Facts communicated by others (i), viz. _____

If an Urgency Certificate is required it must be added here (See Form 9).

4. The said _____ appeared to me to be* _____ in a fit condition of bodily health to be removed to an asylum, hospital, or licensed house (k).

5. I give this certificate having first read the section of the Act of Parliament printed below.

(Signed) _____
of (l) ____ ____ _____

Dated this _____ day of ____ _____ 189

LUNACY 8.

(53 Vict. c. 5, ss. 4, 11, 16, 28, 29.)

Extract from section 317 *of the Lunacy Act*, 1890.

Any person who makes a wilful misstatement of any material fact in any medical or other certificate, or in any statement or report of bodily or mental condition under this Act, shall be guilty of a misdemeanour.

53 Vict. c. 5.—Form 9.

STATEMENT ACCOMPANYING URGENCY ORDER.

I certify that it is expedient for the welfare of the said _____ [*or* for the public safety, *as the case may be*] that the said _____ should be forthwith placed under care and treatment.

My reasons for this conclusion are as follows: _____

The medical practitioner who signs the certificate accompanying an Urgency Order should have examined the patient not more than two clear days prior to the patient's admission to an asylum, lunatic hospital, licensed house, or as a single patient,— for further instructions read the marginal notes of the certificate. As to persons capable legally of signing "Urgency Order," see marginal notes of form. It does not matter whether the "Urgency Order," or the medical certificate accompanying it, is signed first, but both must be signed within two days of the patient's admission to the asylum, etc. The Urgency Order remains in force seven days, or until the pending petition is finally disposed of.

In any case, whether the patient is removed forthwith on an urgency order, or whether it is not considered necessary to do so, a "Petition for an Order for reception of a Private Patient," must be sent to a County Court Judge, a Stipendiary Magistrate, or one of the Justices appointed annually at quarter sessions to hear such petitions. There is a list, compiled by Sutherland, published at a small price, of the specially appointed J.P.'s who are entitled to make orders for the reception of private patients. But the recently passed Lunacy Act, 1891 (54 and 55 Vict. c. 65) is calculated to diminish the inconveniences hitherto experienced in having orders signed. By Sect. 24 of this new Act, the judicial authority is empowered to act when he has not jurisdiction in the place where the alleged lunatic is; he can also transfer a petition to another judicial authority who is willing to receive it. If a Justice has not been specially appointed, an order signed by him will be valid if approved by a judicial authority within fourteen days. It is also practically provided that the justices of a county or borough may specially appoint all the justices thereof to sign orders for the reception of private patients. The jurisdiction of any judicial authority is to continue until a fresh appointment is made. The Lord Chancellor may empower the chairman of a board of guardians to sign orders for the reception of pauper lunatics. The petition should, if possible, be presented by the husband or wife of the alleged lunatic, and if not, the reasons why it is not must be stated, as well as the circumstances under which the petition is presented by the petitioner. The judicial authority must make an order forthwith or appoint a time, within seven days, for the consideration of the petition. He may then either make an order, dismiss the petition, or adjourn its consideration for any period not exceeding fourteen days for further evidence or information. For further particulars see form and read marginal notes.

53 Vict. c. 5—Sched. 2, Form 1.

PETITION FOR AN ORDER FOR RECEPTION OF A PRIVATE PATIENT.

(a) — a justice of the peace for —, or his honour the judge of the county court of —, or stipendiary magistrate for —.

(b) Full postal address, and rank, profession, *or* occupation.

(c) At least twenty-one.

(d) A lunatic, *or* an idiot, *or* person of unsound mind

(e) Asylum, *or* hospital, *or* house, *as the case may be.*

(f) Insert a full description of the name and locality of the asylum, hospital, or licensed house, or the full name, address, and description of the person who is to take charge of the patient as a single patient.

(g) Some day within 14 days before the date of the presentation of the petition.

h) Here state the connection *or* relationship with the patient.

IN THE MATTER OF _____ a person alleged to be of unsound mind.

To (*a*) _____ The Petition of _____ of (*b*) _____ in the County of _____

1. I am _____ (*c*) years of age.

2. I desire to obtain an Order for the Reception of _____ as (*d*) _____ in the (*e*) _____ of _____ situate at (*f*) _____

3. I last saw the said __ _____ at _____ on the (*g*) _____ day of _____ 189 .

4. I am the (*h*) _____ of the said _____ [*or if the petitioner is not connected with or related to the patient, state as follows*:] I am not related to or connected with the said _____

The reasons why this Petition is not presented by a relation or connection are as follows: _____

The circumstances under which this Petition is presented by me are as follows : _____

5. I am not related to or connected with either of the persons signing the certificates which accompany this petition as (*where the petitioner is a man*) husband, father, father-in-law, son, son-in-law, brother, brother-in-law, partner, or assistant (*or where the petitioner is a woman*), wife, mother, mother-in-law, daughter, daughter-in-law, sister, sister-in-law, partner, or assistant.

6. I undertake to visit the said_____personally or by some one specially appointed by me at least once in every six months while under care and treatment under the Order to be made on this Petition.

7. A Statement of Particulars relating to the said _____accompanies this Petition. *If it is the fact, add:*

8. The said_____has been received in the_____ Asylum, *or* Hospital, *or* House, *as the case may be,* under an Urgency Order dated the_____

The petitioner therefore prays that an Order may be made in accordance with the foregoing Statement.

(*i*) Full christian and surname.

Signed (*i*)_____

Dated this_____day of_____189 .

When neither certificate is signed by the usual medical attendant the reason must be stated by the petitioner (see Form below).

WHEN NEITHER CERTIFICATE IS SIGNED BY THE USUAL MEDICAL ATTENDANT.

53 Vict., c. 5, s. 81.

(*a*) Name of patient.

(*b*) To be signed by the petitioner.

I, the undersigned, hereby state that it is not practicable to obtain a Certificate from the usual Medical Attendant of (*a*)_____for the following reason, viz. :_____

Signed (*b*)_____

_____189 ,

If a previous petition has at any time been dismissed, the facts relating to its dismissal are to be stated in the fresh petition (see Form below).

WHEN A PREVIOUS PETITION HAS BEEN DISMISSED.

53 Vict. c. 5, s. 7 (4).

(*a*) Name of patient.

(*b*) Name of asylum, hospital, licensed house, or single charge.

I, the undersigned, hereby state that a former Petition for the reception of (*a*)_____into (*b*)_____was presented to_____(*c*)_____in the month of_____189 and dismissed.

(c) Justice of the peace for ——, or judge of county court of ——, or stipendiary magistrate for ——.

Herewith is a copy (furnished by the Commissioners in Lunacy) of the Statement sent to them of the reasons for its dismissal.

(*Signed*) _____

——————————————————— 189 .

NOTE.—This Copy is to be obtained from the Commissioners in Lunacy by the Petitioner at his own expense.

[An Order for Reception of a Private Patient is to be obtained upon a private application by Petition to a Judge of County Courts, or Stipendiary Magistrate, or Metropolitan Police Magistrate, or specially appointed Justice of the Peace. (*Vide supra* as to operation of Lunacy Act, 1891). The Petition is to be presented, if possible, by the husband or wife, or by a relative (*i.e.*, a lineal ancestor or a lineal descendant, or lineal descendant of an ancestor not more remote than great grandfather or great grandmother) of the Lunatic, and is to be accompanied by a Statement of Particulars and two Medical Certificates on separate sheets of paper. One of the Medical Certificates accompanying the Petition must, if practicable, be by the usual Medical Attendant of the Lunatic; if not by him the reason must be stated (see Form above). If a previous Petition has at any time been dismissed, the facts relating to its dismissal are to be stated in the fresh Petition (see Form above); and the Petitioner must obtain from the Commissioners in Lunacy a copy of the statement sent to them of the reasons for its dismissal, and present this copy with his Petition. The Reception Order (which will not remain in force for more than seven days after its date), the Petition, the Statement of Particulars, and the Medical Certificates must be sent to the Superintendent or Proprietor of the Asylum, Hospital, or House where the patient is to be received.]

53 Vict. c. 5—Sched. 2, Form 8.

ORDER FOR RECEPTION OF A PRIVATE PATIENT TO BE MADE BY A JUSTICE APPOINTED UNDER THE LUNACY ACT, 1890 (*see above as to effect of Sect. 24 of Lunacy Act, 1891*), JUDGE OF COUNTY COURT OR STIPENDIARY MAGISTRATE.

(a) A justice for —— specially appointed under the Lunacy Act 1890, or the judge of the county court of ——, or the stipendiary magistrate for ——.

I, the undersigned, _____ being (a) _____ upon the petition of _____ of (b) _____ in the Matter of _____ a Lunatic (c) _____ accompanied by the Medical Certificates of _____ and _____ hereto annexed, and upon

(b) Address and occupation.	the undertaking of the said (d)_____to visit the said
(c) Or an idiot or person of unsound mind.	_____personally or by some one specially appointed by the said (d)_____once at least in every six months
(d) Name of petitioner.	while under care and treatment under this Order, hereby
(e) Asylum or hospital, or house or as a single patien	authorise you to receive the said_____as a Patient into your (e)_____
	And I declare that I have [or have not] personally seen the said _____ before making this Order.
(f) To be addressed to the medical superintendent of the asylum or hospital, or to the resident licensee of the house in which the patient is to be placed.	Dated this_____day of_____189 . (Signed) (a)_____ A Justice for——————appointed under the above-mentioned Act [or the Judge of the County Court of——————or a Stipendiary Magistrate]. To (f)_____

A member of the managing committee of a lunatic hospital must not sign reception orders or certificates.

An order on petition must be acted upon within seven clear days after its date, but "summary reception orders" may be suspended. (See "Lunatics not under proper Care," etc., and "Wandering Lunatics," *postea*).

The petition should be accompanied by a "Statement of Particulars" similar to that accompanying the urgency order, and by two medical certificates (on separate sheets of paper) one of which should, if possible, be under the hand of the patient's regular medical attendant. The statement of particulars may or may not be made by the person presenting the petition, or in the case of the urgency order by the person making the order, but it (the statement of particulars) must be signed by the person making it. When this person is not he who signed the petition or urgency order, certain particulars must be added, viz., the address, etc. of the person signing (see Form *ante*).

The medical men must have personally examined the patient separately not more than seven days before the petition is presented, and the "Facts indicating insanity" must be those observed at the time of examination, though after sufficient facts have been stated to prove insanity, other facts observed at preceding or subsequent dates may be added (see Form above). The facts should be clear and explicit; delusions should be sought for, given, and stated to be delusions; hallucinations (of which sense or senses and in what form); morbidly defective memory (facts illustrating this defect); abnormal loquacity; incoherence (examples of such incoherence); noisiness; destructiveness; objectionable habits; violence (in what way violent); attempts to commit suicide (by what means); homicidal attempts (in what way and upon whom); emotional exaltation; emotional depression; sleeplessness; restlessness and motor excitement; refusal of food, fluid or solid. The practitioner who makes and signs the urgency certificate can make and sign *one* of the other two certificates and found both his documents on the same examination of the patient, and it is not necessary that the wording should differ.

In the medical certificates, names, dates, and addresses, should be carefully and fully entered, and attention should be given to the marginal notes indicated by the small letters or numbers in the body of the certificate. If a certificate cannot be obtained from the usual medical attendant, the fact must be stated in writing by the petitioner to the judge or magistrate, such statement to form part of the petition (see Form above).

The following persons are disqualified from signing medical certificates, either urgency or other:—The petitioner or person signing the urgency order, the husband or wife, father or father-in-law, mother or mother-in-law, son or son-in-law, daughter or daughter-in-law, brother or brother-in-law, partner or assistant of the petitioner or person (signing the urgency order).

Neither of the persons signing the medical certificates for the reception of a patient shall be father or father-in-law, mother or mother-in-law, son or son-in-law, daughter or daughter-in-law, brother or brother-in-law, sister or sister-in-law, or the partner or assistant of the other of them (see under heading "Single Patients," *postea*). Medical visitors are not to sign for reception into a licensed house or hospital unless directed to visit the patient by a judicial authority, etc.

According to Section 330 of the Lunacy Act, 1890, a person who has before the passing of this Act, signed or carried out an order or a medical certificate that a person is of unsound mind, or who presents a petition after the passing of the Act, or does anything in pursuance of this Act shall not be liable to any civil or criminal proceedings if such person has acted in good faith, and with reasonable care. If any proceedings should be taken, they can be stayed by a summary application to the High Court or a Judge thereof, upon such terms as to costs, etc., as the Court or Judge may think fit, "if the Court or Judge is satisfied that there is no reasonable ground for alleging want of good faith or reasonable care."

No prosecution for a misdemeanour under Section 317, Lunacy Act, 1890 (see extract below Medical Certificate form) can take place except by order of the Commissioners or by the direction of the Attorney-General, or the Director of Public Prosecutions.

Lunacy Act, 1890 (53 Vict. Ch. 5), Sect. 331.—(1,) "An

action brought by any person who has been detained as a lunatic against any person for anything done under this Act shall be commenced within twelve months next after the release of the party bringing the action, and shall be laid or brought in the county or borough where the cause of action arose, and not elsewhere; (2,) If the action is brought in any other county or borough or is not commenced within the time limited for bringing the same, judgment shall be given for the defendant."

Examination of Patient in Asylum or of Single Patient.

According to Section 49 of the Lunacy Act, 1890, any person, whether a relative or friend or not of a patient who is detained in any asylum, etc., may apply to the Commissioners to have such patient examined by two medical practitioners, and if the Commissioners are satisfied that it is proper to grant such order, they may do so. If, after two separate examinations with an interval of at least seven days between such examinations, the two medical practitioners certify that the patient may, without risk to himself or injury to the public, be discharged, the Commissioners may order the patient to be discharged at the expiration of ten days from the date of the order.

VOLUNTARY BOARDERS.

The superintendent or proprietor of a licensed house may now, with the previous consent in writing of two of the Commissioners, or where the house is licensed by the justices, of two of the justices, receive and lodge as a boarder for the time specified in the consent *any person* who is desirous of voluntarily submitting to treatment. After which time (unless it is extended by further consent) such boarder must be discharged. Any relative or friend of a patient may be received under the same conditions.

The intending boarder must himself apply to the commissioners or justices for their consent.

A boarder may leave a licensed house by giving twenty-four hours' notice to the superintendent or proprietor of his intention to do so, and if prevented from so doing, is entitled to recover £10 from the superintendent or proprietor for each day, or part of a day, during which he is detained.

LAWS AS TO KEEPING "SINGLE PATIENTS" IN ENGLAND AND WALES.

A "Single Patient" is a person received for profit into an unlicensed house as a lunatic under Certificates.

Under special circumstances the Commissioners may allow more than one patient to be received as Single Patients into the same unlicensed house (Section 46, Lunacy Act 1891). Otherwise it is a misdemeanour to detain two or more lunatics in an unlicensed house.

"Any person who for payment takes charge of or receives to board or lodge any person as a lunatic shall be deemed to be a person deriving profit from the charge of a lunatic within the meaning of Lunacy Act, 1845" (Lunacy Acts Amendment Act, 1889, Section 35, "Jour. Ment. Sci.").

Duties of Person having care or charge of Single Patient.

(The paragraphs marked with an asterisk apply also to the proprietors of private asylums in so far as private patients are concerned.)

*(1,) To receive with the patient the medical certificate or certificates, along with the urgency order or petition and judge's order, and the statement of particulars, for the description of which see the preceding pages. To receive a patient without these documents is a misdemeanour except where no profit is derived from the charge, or in the case of chancery patients or of the transfer of a patient. Blank forms, etc., may be purchased through a bookseller from Eyre and Spottiswoode, East Harding Street, Fleet Street, E.C., or from Shaw and Sons, Fetter Lane.

The following persons are disqualified from signing medical certificates in addition to those already mentioned under the heading "Certification," etc., viz.:—(a,) The person taking care or charge of the patient; (b,) Any person interested in the payments on account of the patient; (c,) The person who is going to act as "Medical Attendant" of the patient; (d,) The husband or wife, father or father-in-law, mother or mother-in-law, son or son-in-law, brother or brother-in-law, daughter or daughter-in-law, or the partner or assistant of either of the above persons.

It is a misdemeanour to receive a patient under a certificate signed by any of the foregoing disqualified persons.

*(2,) To transmit to the Commissioners within one clear day of the patient's reception, a notice of admission together with a true and perfect copy of the documents on which the patient has been received—neglecting to do this is a misdemeanour. If the Commissioners find any of the certificates, etc., incorrect or defective, they may send the copies back so that the originals may be amended (with the consent of the judicial authority) by the persons signing them. The certificates, etc., must be duly amended (as marked on the copies by the Commissioners) within fourteen days of the patient's reception, otherwise his discharge may be ordered by any two of the Commissioners. Amendments should be initialed by the person making them.

*(3,) When the patient has not been seen by the judge who signed the reception order, such patient should receive within

twenty-four hours of his admission a notice in writing of his right to see a magistrate or judge (other than the judicial authority who made the order) within seven days, according to the following Form :—

*NOTICE OF RIGHT TO PERSONAL INTERVIEW.

Take notice that you have the right if you desire it, to be taken before or visited by a judge of county courts, magistrate or justice. If you desire to exercise such right, you must give me notice thereof by signing the enclosed form on or before the   day of   .

Dated   (Signed),   C.D.,
                                Proprietor of   House.

*NOTICE OF DESIRE TO HAVE A PERSONAL INTERVIEW.

Address   Dated

I desire to be taken before or visited by a judge, magistrate or justice having jurisdiction in the district within which I am detained.

(Signed),

* Unless within twenty-four hours after admission the "Medical Attendant" signs and sends to the Commissioners the following :—

CERTIFICATE AS TO PERSONAL INTERVIEW AFTER RECEPTION.

I certify that it would be prejudicial to A.B. to be taken before or visited by a judge of county courts, magistrate or justice.

(Signed),   C.D.,
Medical Attendant of the said

Reception orders now remain in force for one year, then for two years, then for three years, then for successive periods of five years under certain conditions (see duties of "Medical Attendant" *postea*).

(4,) To cause the patient to be visited at least fortnightly by a medical practitioner who has not made either of the certificates, and who derives no profit from the charge, and who is styled the "Medical Attendant" of the patient. Failure to comply with this regulation is a misdemeanour unless the Commissioners have given permission for the visits to be paid less frequently. A medical attendant must be appointed even if the person in charge of the patient is a medical practitioner.

(5,) To make (if a medical practitioner) entries in the Medical Visitation Book or Medical Journal at least once in two weeks, when the fortnightly visits of the medical attendant have been permitted by the Commissioners to be made less frequently.

* (In private asylums these entries must be made weekly.)

FORM OF MEDICAL VISITATION BOOK OR MEDICAL JOURNAL FOR SINGLE PATIENTS. AUTHORISED BY THE COMMISSIONERS IN LUNACY, DECEMBER 1st, 1879.

Date.	Mental Condition. What evidence of Insanity. * "The first entry after admission to be a sketch of previous history of case, and full particulars of mental and bodily condition, and not to be entered here, but on blank pages to be left for the purpose at the beginning of book.	Bodily health and condition.	Restraint or seclusion since last visit. When and how long? By what means and wherefore?	Visits of Friends. Date of Visit. Name of Friend.	State of House and Furniture, Bed and Bedding. Supply and condition of Wearing Apparel.	Dietary proper. If not, state the reason.	Employment, Exercise, and Amusements.

Archbold (Glen's Edition, 1877, p. lx.) suggests that a common copy book having a few pages ruled neatly to the required pattern, and with the proper head lines, is quite sufficient.

*(6,) The medical visitation book, petition, orders and certificates must be so kept that they may be accessible to the commissioners whenever they may visit the patient.

* The making of an untrue entry in the Medical Journal is a misdemeanour. Upon every visit of a Commissioner or Visitor the medical journal must be produced.

*(7,) To send to the Commissioners at the expiration of a month after the reception of the patient a report as to the mental and bodily condition of the patient in such form as they may direct. He must also report to the Commissioners at any time, when required.

*(8,) To send notice to the petitioner, or person who made the last payment, as soon as the patient recovers. The notice must state that unless the patient is removed within seven days from the date of the notice, he will be discharged. If the patient is not so removed he must be forthwith discharged.

*(8a,) To send notice of the discharge, removal or death of the patient to the Commissioners within two clear days of such discharge, removal or death. Failure to send these notices or making a false statement therein is a misdemeanour. "Failure to send" entails a penalty not exceeding £50 in the case of a single patient.

* FORM OF NOTICE OF DISCHARGE.

I hereby give you notice that   , a private patient, received into this house on the   day of   189 , was discharged therefrom (a)   by the Authority of on the   day of   189 .
    (Signed),
        Proprietor (or Superintendent) of   House.
    Dated this   day of   189 .
To the Commissioners in Lunacy.
    (a) Recovered, or relieved, or not improved.

*(9,) To transmit within forty-eight hours of the death of the patient a duly certified copy of the medical attendant's statement of the cause of death, etc., to the Commissioners, to the person who signed or obtained the reception order, to the registrar of deaths for the district, and (within two days) to the county or borough coroner. Failure to do so entails a penalty not exceeding £50 in the case of a single patient, and is a misdemeanour in all cases.

*(10,) To give effect to the Commissioners' order within seven days of the visit of two of them (one legal, the other medical) for the discharge of the patient.

(11,) To give effect to any direction given by any two of the Commissioners that the medical attendant of a single patient shall cease to act in that capacity, and that some other person be employed in his place. Failure to give effect to this direction is a misdemeanour.

*(12,) To show any Commissioner at his request any part of the house or grounds. Refusal to do so is a misdemeanour.

One or more of the Commissioners once at least in every year must visit every unlicensed house in which a single patient is detained, unless such patient be so received by a person deriving no profit from the charge, or by a committee appointed by the Lord Chancellor, and report to the Commissioners on the treatment and state of bodily and mental health of the patient.

A Commissioner or one or more Visitors of the county or borough may, at all reasonable times, on the request of the Commissioners, visit a single patient and report to the Lord Chancellor or the Commissioners on the patient's treatment, health, and payments. The maximum penalty for obstructing a Commissioner or Chancery visitor is £50.

*(13,) To forward unopened all letters written by the patient and addressed to the Lord Chancellor, or to any Judge in Lunacy, or to a Secretary of State, or to the Commissioners or any Commissioner, or to the person who signed the order for the reception of the patient, or on whose application or petition such order was made. Letters addressed to other persons than those mentioned may be forwarded or detained at the discretion of the person who has charge of the patient, but if detained, must be endorsed by such person with the reasons for detention and laid before the Commissioner or Commissioners at his or their next visit. Neglect entails a penalty not exceeding £20.

*(14,) If the patient escape, to transmit to the Commissioners within two clear days a written notice of the escape, and on his recapture, a written notice of the same also within two clear days. Penalty for omission, a sum not exceeding £10.

If the patient be not retaken within fourteen days after his escape he can only be brought back after being re-certified and a fresh order, etc., obtained. A patient escaping from one of the three kingdoms (England, Scotland, Ireland) to another of them, can now be retaken on a warranty granted by the Lunacy authorities of the country from which he has escaped after the warrant has been countersigned by any Justice or Sheriff of the country to which he has resorted.

(15,) If a medical practitioner, he must on the 10th of January, or within seven days from that time in every year, report in writing to the Commissioners the state of health, bodily and mental, of such patient, with such other circumstances as he may deem necessary to be communicated to the Commissioners. He must also furnish a report when required at any time or from time to time by the Commissioners, specifying such particulars as the Commissioners may direct ("Archbold," p. 464).

The whole Act (16 and 17 Vict. c. 96) under which the January report had to be made, is now repealed (53 Vict. c. 5, Sched. 5). (See "Duties of Medical Attendant" *post.* for reports now to be made). It should, however, be stated that the last section of the last-mentioned Act says that the repeal is not to affect a practice established. If the reception order has expired through failure to send in the special reports now required, it is a misdemeanour, having knowledge of that fact, to detain the patient.

(16,) To comply with the regulations as to entries in the Medical Visitation Book. Liable to a penalty not exceeding £5 for failure to do so.

(17,) If not a medical practitioner, he or she must keep a note of the days on which seclusion is resorted to, and of the length of time on each occasion, and must produce such note to the medical attendant on his or her next visit, to be entered in the Journal (Medical Visitation Book).

*(18,) According to Section 40 of the Lunacy Act, 1890, mechanical means of restraint must not be applied except for surgical or medical treatment, and to prevent the lunatic from injuring himself or others.

In every case a medical certificate must be signed describing the means used and the reasons for it.

CERTIFICATE AS TO MECHANICAL MEANS OF RESTRAINT.

* I, the undersigned C.D. (the Medical Superintendent, or a medical officer of the   Asylum, or the   Hospital, or the Medical Proprietor or Attendant of the   House, or the Medical Officer of the   Workhouse, or the Medical Attendant of A.B., a lunatic under care or treatment at   , *as the case may be*) certify that I have examined A.B., a lunatic in the said (asylum, hospital, house, or workhouse *or* the said A.B., *as the case may be*), and that in my opinion mechanical means of bodily restraint were (or are) necessary in his case for purposes of surgical or (medical) treatment (or to prevent him from injuring himself or others). The necessary means are *(state them)*.

I found my opinion on the following grounds *(state them)*.

(Signed)

* This certificate must be signed by the medical attendant or by the medical practitioner having care and charge of the patient. A full daily record of every case [of mechanical restraint] must be kept and a copy, together with copies of the restraint certificates, sent to the Commissioners quarterly.

Wilful contravention of this Section is a misdemeanour.

(19,) When he proposes to change his residence and to remove the patient with him, he must give to the Commissioners and to the person who signed the order or petition for the order for the reception of the patient seven clear days' notice of the proposed change, with the exact address and designation of the new residence.

(20,) In order to send the patient to any specified place or places for the benefit of his health for any definite time under proper control, it is necessary to obtain the consent of one of the Commissioners, and before such consent is given it is required that the approval in writing of the person who signed the reception order or petition, or by whom the last payment was made, be produced to such Commissioner, unless he on cause being shown dispenses with the same.

Removal to a place not specified in the consent is equivalent to a discharge, and necessitates readmission by fresh order and certificates. If the patient remains away longer than the specified time, or any permitted extension of that time, he is also discharged.

* (20,) Applies also to Metropolitan private asylums, for an absence exceeding 48 hours, but provincial ones obtain consent for an absence exceeding that time from the Visiting Justices.

* (21,) By obtaining from one of the Commissioners his consent to an order of transfer [made by the person who signed the petition or order or made the last payment] the patient can be transferred to the charge of another person without the necessity of fresh certificates.

*(22,) It is a misdemeanour for any person having charge of a single patient to abuse, ill-treat or wilfully neglect such patient in any way; such persons are liable on conviction on indictment to fine or imprisonment or both, or to forfeit for every such offence on a summary conviction thereof before two justices, any sum not exceeding £20 nor less than £2.

According to Archbold (Glen's Ed., 1877, page 460), a husband having the care and charge of his wife, is not a person having the care and charge of a lunatic within the statute (16 and 17 Vict. c. 96, *now repealed*), the provisions of which do not apply to persons whose care or charge of a lunatic is of a purely domestic nature, or arises from natural duty only. But a person was indicted and convicted for wilfully neglecting his lunatic brother. By sect. 206 of the Lunacy Act, 1890, uncertified alleged lunatics may now be visited by the Commissioners. (See " Uncertified Patients " *post.*)

"If . . . . . any person having the care or charge of any single patient, or any attendant of any single patient, carnally knows or attempts to have carnal knowledge of any female under care or treatment . . . . . as a single patient, he shall be guilty of a misdemeanour, and, on conviction on indictment, shall be liable to be imprisoned with or without hard labour for any term not exceeding two years; and no consent or alleged consent of such female thereto shall be any defence to an indictment or prosecution for such offence" (53 Vict., c. 5, s. 324).

(23,) Should the person in charge become dangerously ill, the friends of the patient should at once be communicated with, in order that arrangements for a transfer may be made (Weatherly, p. 102). But by Sect. 59, Lunacy Act, 1890, the Commissioners may, upon the death of a person having charge of a single patient, order the transfer of the patient to some other person. They may do so either upon the application of the person having authority to discharge the patient, or if he does not apply within seven days after the death, upon their own motion.

* (24,) The person who has charge of the patient should not allow him to transact any business. A person under certificates cannot legally transact business.

Upon the written request of the Commissioners, or any two of them, single patients may be visited by visitors appointed for the county or borough, and any such visitor being a medical practitioner is entitled to remuneration. The Commissioners can also under their common seal specially appoint persons to visit urgent cases and report thereon; every such person has all the powers of a Commissioner; this may be done shortly after the patient's admission, when the Commissioners cannot immediately visit such patient. The Commissioners or others may be required by the Lord Chancellor or a Secretary of State to visit a lunatic or alleged lunatic, examine, inspect, and report. Penalty for obstructing any person so authorised other than a commissioner or regularly constituted visitor, £20.

Any two of the Commissioners may order the removal of a patient from the care of one person to that of another, or to an asylum.

Duties of the Medical Attendant.

(The paragraphs marked with asterisks apply also to the Medical Superintendents or Medical Attendants of private asylums.)

The Lunacy Act 1845, defines a "medical attendant" as a duly qualified and registered physician, surgeon, or apothecary, who keeps any licensed house or in his medical capacity attends any licensed house or any asylum, hospital, or other place where any lunatic is confined.

* The medical attendant of a single patient must be a registered medical practitioner, not deriving and not having a partner, father, son, or brother, who derives any profit from the care or charge of such patient.

* The person who is to be the medical attendant, must not sign either of the certificates on which the patient is received.

* (1,) The medical attendant must after two clear days and within seven, from the date of reception of the patient, transmit to the Commissioners a report or statement of the patient's mental and bodily condition.

FORM OF STATEMENT.

I have this day (some day not less than two, or more than seven clear days after the admission of the patient) seen and examined   received into this house on the   day of   189 , and hereby certify that with respect to mental state, he (or she)   and that with respect to bodily health and condition he (or she)
              Signed
                  Medical attendant of
              Dated

(2,) The medical attendant must visit the patient once, at least, in every two weeks, unless by an order of the Commissioners the visits are permitted to be made less frequently. This permission is not, as a rule, accorded until the patient has been visited once by a Commissioner.

* (3,) He must now also sign a certificate according to the form already given, when mechanical restraint is used.

* (4,) When the person in charge is not a medical practitioner, the medical attendant must keep a full daily record of cases where mechanical restraint is used, and transmit it to the Commissioners quarterly.

* (5,) Mechanical restraint must now only be used for medical or surgical reasons, or to prevent the lunatic injuring himself or others. Wilful contravention of these regulations as to restraint is a misdemeanour.

* (6,) The medical attendant of every single patient must at the expiration of one month after the reception of the patient, send to the Commissioners a report as to the mental and bodily condition of the patient in such form as they may direct.

(7,) The medical attendant must at each visit enter in the medical visitation book (for the form of which, see under the "Duties of person having care, etc.," *ante*), the date of each of his visits, and a statement of the several particulars required as to the condition and circumstances of the patient and of the house.

* Making an untrue entry in the medical visitation book is a misdemeanour, and failure to comply with the regulations as to the entries therein, entails a penalty not exceeding £5.

(8,) He must on the 10th January, or within seven days thereof in every year, report in writing to the Commissioners, the state of health, mental and bodily, of the patient, and such other circumstances as may be deemed necessary to be communicated. Such annual report should give all these particulars fully, even although no change may have occurred since the previous report.

* (9,) By sect. 7 of the Lunacy Act, 1891, an order for the reception of any patient shall remain in force for one year, after that for two years, and after that for three years, then for successive periods of five years, provided the medical attendant report specially to the Commissioners as to the bodily and mental state of the patient, and certify that the patient remains of unsound mind, and is a proper person to be detained under care and treatment. This special report must be sent not more than a month, and not less than seven days before the end of each period. If the lunatic has been so found by inquisition, the above periods must date from May 1st, 1890, and the reports are to be sent to the Masters.

* (10,) The person making the special report, must give the Commissioners any further information concerning the patient they may require.

* Knowingly to detain a patient after the reception order has expired, is a misdemeanour. This applies to the person who has charge of the patient.

* An order of transfer is not deemed a reception order under this section, and the original reception order, unless continued, will expire at one of the above-mentioned periods.

* (11,) The Commissioners may at any time require from the medical attendant of a single patient a report in writing as to the patient, in such form and specifying such particulars as the Commissioners direct, and such report shall be in addition to any periodical reports required to be sent to the Commissioners.

* (12,) In the event of the death of the patient, the medical practitioner who attended him in his last illness, must prepare and sign a statement of the cause of death and the duration of the disease of which the patient died.

* (13,) This statement must be entered in the medical visitation book and a copy of the statement, certified by the person in charge of the patient, must be transmitted by him to the coroner for the county or borough, within two days after the death.

19

* FORM OF NOTICE AND CERTIFICATE TO BE SENT TO THE COMMISSIONERS WITHIN FORTY-EIGHT HOURS OF THE DEATH OF THE PATIENT.

I hereby give you notice that _____ _____ _____ a Private Patient received into this house on the _____ day of _____ _____ _____ 189 died therein on the _____ day of _____ _____ 189

(Signed) _____

Proprietor (or Superintendent of) _____ House _____

Dated this _____ day _____ 189 _____

* (14,) And I further certify that _____ _____ was present at the death of the said _____ _____ _____ and that the apparent cause of death of the said _____ _____ _____ [ascertained by post mortem examination (if so)] _____ _____ was _____

(Signed) _____ _____ _____ _____

_____ _____ Medical Attendant of the said _____ _____

To the Commissioners in Lunacy.

* In Medical Statement (p. 288) for the words "received into this house on" substitute the words "the patient mentioned in the Notice of Admission dated."

The offices of the Commissioners in Lunacy are at 19, Whitehall Place, London, S.W.

Statutory forms, books, etc., may be obtained from Messrs. Knight & Co., 90, Fleet Street, London, as well as from the publishers already mentioned.

PERSONS FOUND LUNATIC BY INQUISITION (CHANCERY PATIENTS).

The Judge in Lunacy may upon application direct an inquisition whether a person is of unsound mind and incapable of managing himself and his affairs. For this, medical affidavits and medical evidence are required. The medical affidavits must be divided into distinct paragraphs, but this is generally attended to by the lawyer managing the case. Under certain circumstances the inquiry is held before a jury.

When the property does not exceed £2000 in value or the income thereof £100 per annum, the Judge in Lunacy may (if the person's insanity and incapacity are proved to his satisfaction) allow it to be dealt with for the maintenance of the patient without an inquisition. When the property is under £200 in value, any county court judge of the patient's district may authorise the clerk of the guardians, a relieving officer, or other person to deal with it.

A chancery patient can be admitted into any asylum, or as a single patient into any unlicensed house, without medical certificates; an order signed by the committee of the person, with an office copy of the order for the appointment of such committee annexed thereto is sufficient authority for the reception of such patient; in cases where no committee of the person has been appointed, the patient may be sent to an asylum, etc., by order of one of the Masters in lunacy.

The individual (unless he is the committee of the person) who undertakes the care of a chancery patient in his house for profit, is liable to the same responsibilities and duties as are enforced in the case of a single private patient.

The fortnightly visitation of the medical attendant is not required in the case of single chancery patients. The periodical reports (see *ante*) should be sent to the Masters in lunacy, instead of the Comissioners in lunacy.

Single chancery patients are visited at least, four times every year for the two years following the inquisition by one of the Lord Chancellor's visitors, and afterwards at least twice each year (the intervals between the visits never to exceed eight months), and at least once every year by one of the Commissioners in lunacy.

UNCERTIFIED LUNATICS.

Any person who keeps without profit an uncertified insane patient, may be required by the Commissioners to send reports by a medical practitioner of the mental and bodily condition of the patient and also to send other particulars. The Commissioners may visit such patient and exercise all their powers except those of discharge. They may report to the Lord Chancellor who may discharge the patient, or remove him to an asylum.

PAUPER LUNATICS.

A pauper patient is defined as one who is "maintained wholly or in part by any parish, union, county, or borough."

When a patient is unable to meet the expense of certification, and of maintenance in a private asylum, or as a private patient in a county asylum (12/- a week and upwards), the best method is to refer the friends to the relieving-officer who will see to the patient's certification, visitation by the union medical officer, or, in case of urgency, immediate and temporary admission into the workhouse. For such admission to be legal there must be proper accommodation in the workhouse, and the patient (according to Sect. 20, Lunacy Act, 1890) must not be detained more than three days, proceedings to be taken in the meantime to have him legally certified. When the union medical officer becomes aware of the presence of a pauper lunatic in his district, he must give written notice of the fact to the relieving officer within three days, or, in the absence of a relieving officer, to the overseer. The relieving officer, or when there is no relieving officer, the overseer who comes to such knowledge by this, or other means, must within three days give notice to a justice of the county or borough in which the patient is situated. A justice must not now sign an order for the reception of a pauper into an asylum, etc., or a workhouse, until he is satisfied that the alleged pauper is either in receipt of relief, or in such circumstances as to require relief for his proper care. A person visited by a medical officer of the union at the expense of the union, is deemed to be in receipt of relief. An order for the reception of a pauper or wandering lunatic into an asylum, etc., cannot now be signed by an officiating clergyman (as such), and a relieving officer or overseer, but must ordinarily be signed by a magistrate, who must be satisfied that there is no room in the public asylums of the patient's county before the said patient can be legally sent to another county or to a private asylum as a pauper patient. Two or more Commissioners may call in a medical practitioner, and by order direct a pauper lunatic to be received in an institution for the insane. See p. 273 as to chairman of board of guardians signing orders. There is only one medical certificate required for an insane pauper, and it is essentially the same as those required for a private patient. There are no urgency orders or certificates and no "petition." The statement only differs in one or two particulars, *e.g.*, the insertion of the union to which the pauper patient is chargeable, and the omission of the name, etc., of the usual medical attendant; it is signed by the relieving officer or overseer.

53 Vict. c. 5, Sched. 2.—Form 12.

ORDER FOR RECEPTION OF A PAUPER LUNATIC, OR LUNATIC WANDERING AT LARGE.

(*a*) **Name of Patient.**
(*a*1) **Residence or Occupation.**
* If not "in receipt of relief" strike out these words.
† If in receipt of relief strike out the words in brackets.
(*b*) Lunatic *or* an idiot *or* a person of unsound mind.
(*c*) Where the order directs the

I, _____ having called to my assistance _____ of _____ a duly qualified medical practitioner, and being satisfied that (*a*) _____ (*a*¹) _____ _____ is a pauper in receipt of relief* [or in such circumstances as to require relief for h proper care and maintenance †] and that the said _____

is a (b) _____ and a proper person to be taken

charge of and detained under care and treatment, *or* that

(a) _____

—————————————————————————

is a lunatic, and was wandering at large, and is a proper person to be taken charge of and detained under care and treatment (c) _____

hereby direct you to receive the said _____

—————————————————————————

—————————————————————————

as a patient into your (d)_____. Subjoined is a statement of particulars respecting the said_____

    (*Signed*)_____

      A Justice of the Peace for_____

  Dated the_____day of_____189 .

To (e)_____

Side notes:
lunatic to be received into any asylum, other than an asylum of the county or borough in which the parish or place from which the lunatic is sent is situate, or into a registered hospital or licensed house, it shall state that the justice making the order is satisfied that there is no asylum of such county or borough or that there is a deficiency of room in such asylum; or (as the case may be) the special circumstances, by reason whereof the lunatic cannot conveniently be taken to any asylum for such first-mentioned county or borough.
(*d*) Asylum *or* hospital *or* house.
(*e*) The superintendent of the asylum for the county or borough of——: *or* the lunatic hospital of——; *or* proprietor of the licensed house of ——; describing the asylum, hospital, or house.

Under certain circumstances (accommodation, suitable condition of patient, etc.) a pauper lunatic may be detained in a workhouse for more than fourteen days against his will, on the order of a magistrate, such order being made upon the application of the relieving officer, supported by a medical certificate signed by a practitioner not being an officer of the workhouse, and by the certificate of the workhouse medical officer. The latter certificate enables the workhouse authorities to detain the patient for fourteen days from its date, against his will, as also does the temporary removal order of a justice.

Pauper lunatics not recovered, and chronic harmless pauper lunatics may, under certain circumstances, be removed to and kept in a workhouse, the chronic cases remaining on the books of the asylum. A pauper lunatic may be boarded out with a relative or friend, by the authorities of an asylum, his maintenance being paid for by the guardians. The allowance to the friend must not exceed the cost of maintenance in the asylum.

An order for the reception of a patient as a pauper authorises his detention if he should be afterwards classified as a private patient, and *vice versa*.

LUNATICS (NOT PAUPERS) NOT UNDER PROPER CARE AND CONTROL, OR CRUELLY TREATED AND NEGLECTED.

Any person knowing of such a lunatic should acquaint a constable, overseer, or relieving officer of the district or parish who must give information on oath of the presence of such lunatic to a justice within three days. When a particular relieving officer is directed to do the duties in respect of lunatics, every other relieving officer must report to him. When a specially appointed justice (see p. 273) is informed by any person on oath that there is such a lunatic in his jurisdiction he may visit him, or authorise two medical practitioners to do so, examine, and certify. The justice after calling to his aid two medical practitioners, may make an order for the reception of the lunatic into an asylum. The justice may suspend the execution of his order for 14 days. The medical practitioners, or either of them, may also require the suspension of the execution of the order, if the patient is unfit physically for removal. In the latter case he may be received in the asylum within three days after the date of a medical certificate stating that he is fit to be removed. A relative or friend may take charge of him if the justice is satisfied that he (the patient) will be well treated.

WANDERING LUNATICS (PERSONS DEEMED LUNATICS, WANDERING AT LARGE).

The constable, etc., must, immediately he knows of such a person (whether he be a pauper or not) being in his district, apprehend him and take him before a justice having jurisdiction in the county or borough. The justice on receiving information on oath, may compel the constable, etc., to bring such person before him or some other justice having jurisdiction in the district. One medical certificate is required. Regulations as to suspension of order, as in case of lunatic not under proper care and control.

It is now provided by Lunacy Act, 1891, that this and the preceding class of patients are to be classified as paupers until it is ascertained that they are entitled to be classified as private patients.

CRIMINAL LUNATICS.

1. *Dangerous Lunatics.*

A person mentally deranged discovered under circumstances that denote a purpose of committing some crime, may be examined by two justices with medical assistance, and sent to the county asylum, or, if in a fit condition, handed over to a friend who agrees to become responsible for his safe care.

2. *Prisoners found to have been insane at the time of committing their offences* (whether treason, murder, felony, or misdemeanour).

If the jury acquit the prisoner on the ground of insanity at the time the offence was committed, he is ordered by the Court to be kept in custody until an order is made by the Crown for his safe custody during Her Majesty's pleasure, when he will be removed to a specified place (now generally Broadmoor Asylum). Lunatics guilty of the less grave of the above offences may be sent to a county asylum or given into the custody of a responsible friend.

Recently a lunatic pleaded guilty to felony and was sentenced by the judge to imprisonment and penal servitude, "*for the prisoner's own protection*"!

3. *Prisoners found to be insane during custody before trial.*

The justices or directors must make inquiry as to the sanity of the criminal with the aid of two duly qualified medical practitioners. If they certify him insane, the Secretary of State may direct him to be removed to an asylum.

4. *Prisoners found to be insane at the time of trial.*

If a prisoner appears to be insane on *arraignment*, and a jury impanelled to try the question of his sanity find him insane the trial will not proceed, but the prisoner will be kept in custody till Her Majesty's pleasure be known. The Crown may then make an order for his safe custody at a specified place, during Her Majesty's pleasure.

If a prisoner is found insane at the *trial* the proceedings are as above. If a prisoner, brought before a court to be discharged for want of prosecution, appears to be insane, a jury is impanelled and the proceedings are as above.

5. *Prisoners found to be insane during Custody after Conviction.*

Where there is reason to suppose that a prisoner under sentence of death is insane, the Secretary of State may order an inquiry by medical men as to his insanity, and direct his removal to an asylum for insane prisoners (Broadmoor Criminal Lunatic Asylum). If the prisoner is confined for any other criminal offence, the inquiry may be ordered by two of the visiting justices of the prison in which he is confined, or two justices for the county, etc., in which he is, if not in a prison to which visiting justices are appointed.

STATEMENT RESPECTING CRIMINAL LUNATICS TO BE FILLED UP AND TRANSMITTED TO THE MEDICAL SUPERINTENDENT WITH EVERY CRIMINAL LUNATIC.

Name _____
Age _____
Date of Admission _____
Former occupation _____
From whence brought _____
Married, single, or widowed _____
How many children _____
Age of youngest _____
Whether first attack _____
When previous attacks occurred _____
Duration of existing attack _____
State of bodily health _____
Whether suicidal or dangerous to others _____
Supposed cause _____
Chief delusions or indications of insanity _____
Whether subject to epilepsy _____
Whether of temperate habits _____
Degree of education _____
Religious persuasion _____
Crime _____
When and where tried _____
Verdict of jury _____
Sentence _____

CERTIFICATION OF THE INSANE IN SCOTLAND.

An urgent case may be admitted into and detained in an asylum for a period not exceeding three days on one medical "Certificate of Emergency," accompanied by the "Request" of a relative or friend. Ordinarily there are required two medical certificates the forms for which are, with the exception of one or two remarkable verbal variations, not unlike those used in England prior to the passing of the recent Lunacy Bill, a "Statement" also somewhat similar to that formerly in use in England, a "Petition" to the Sheriff or Steward, and an "Order" by that authority.

Private and pauper patients are certified and placed in asylums in the same way. The certifying medical men must see the patient on the day of certification. The date of the petition must be within fourteen clear days following the dates of the medical certificates, and the order must be dated within fourteen days prior to the reception of the lunatic. Where there is property to be administered two medical certificates as to the patient's insanity and incapacity, and a petition from a near relative stating the amount of the property, are presented to a judge of the Court of Session, and after they have been intimated in the Court for eight days, if there is no opposition a *Curator Bonis* is appointed to manage the lunatic's property and act for him.

"A patient can be treated, with a view to cure, anywhere out of an asylum for twelve months without formal certificates, if a medical opinion to that effect and intimation is sent to the Commissioners in Lunacy" (Clouston, "Ment. Dis." 1st Ed. p. 610).

If a lunatic (*single patient*) is detained in a private house for more than a year, or for profit, an order from the Sheriff or the sanction of the Board must be obtained. If the lunatic is a pauper the inspector of the poor applies for the order, and the Sheriff may grant it on one medical certificate. On January 1st, 1891, 23·7 per cent. of all pauper lunatics were "*boarded out*" in private dwellings. *Voluntary boarders* can be received into asylums. The written consent of a Commissioner must be obtained previously. For this the boarder must apply in writing.

N.B.—This Form shall remain in the keeping of the Superintendent of the Asylum, after the Order of the Sheriff is obtained. [FORM A.]

25 and 26 Vict. Cap. 64, Sect. 14.

PETITION TO THE SHERIFF TO GRANT ORDER FOR THE RECEPTION OF A PATIENT INTO AN ASYLUM.

(a) Sheriff or Steward.
(b) Shire or Stewartry.

Unto the Honourable the (a) _____ of the (b) of _____ and his Substitutes,—

The petition of ___ humbly showeth that it appears from the subjoined Statement and accompanying Medical Certificates that ___

(c) State degree of Relationship or other capacity in which Petitioner stands to Lunatic.

your Petitioner's (c) _____ is at present in a state of Mental Derangement, and a proper person for treatment in an Asylum for the Insane. May it therefore please your Lordship to authorise the transmission of the said _____ to the _____ _____and to sanction __ __ __admission into the said Asylum.

To be signed by the party applying _____ ___ _

(d) The date of the Petition must be within fourteen clear days following the dates of the Medical Certificates.

Dated this _ ___ (d) day of __ _ One thousand eight hundred and __

STATEMENT.

If any of the Particulars in this Statement be not known, the fact to be so stated.

1.—Christian Name and Surname of Patient at length - - - - - - _____
2.—Sex and Age - - - - - _____
3.—Married, Single, or Widowed - - - _____
4.—Condition of Life, and previous Occupation (if any) - - - - _____
5.—Religious Persuasion, so far as known - - _____
6.—Previous Place of Abode - - - _____
7.—Place where found and examined - - _____
8.—Length of time Insane - - - _____
9.—Whether First Attack - - - - _____
10.—Age (if known) on First Attack - - _____
11.—When and Where previously under Examination and Treatment [*name of Establishment into which last received, and Year of Reception or approximation thereto*] - - _____
12.—Duration of Existing Attack - - - _____
13.—Supposed Cause - - - - _____
14.—Whether subject to Epilepsy - - _____
15.—Whether Suicidal - - - - _____
16.—Whether Dangerous to others - - _____
17.—Parish or Union to which the Lunatic (if a Pauper) is Chargeable - - - _____
18.—Date of becoming Chargeable - - - _____
19.—Christian Name and Surname, and Place of Abode of nearest known Relative of the Patient, and degree of Relationship (if known), and whether any Member of h Family known to be or to have been Insane - - - - - _____

20.—Special circumstances (if any) preventing the insertion of any of the above particulars - _____

I certify, that, to the best of my knowledge, the above particulars are correctly stated.

Dated this _____ day of _____ One thousand eight hundred and _____

(To be Signed by the party applying) _____

MEDICAL CERTIFICATE.

(a) Set forth the qualification entitling the person certifying to grant the certificate; e.g. Member of the Royal College of Physicians, Edinburgh.
(b) Physician or Surgeon, or otherwise, as the case may be.
(c) Insert the street, and number of the house (if any), or other like particulars.
(d) Insert Designation and Residence, and if a Pauper state so.
(e) Lunatic, or an insane person, or an idiot, or a person of unsound mind.
(f) State the facts.

(g) State the information, and from whom received.

I, the undersigned, _____
being a (a) _____
and being in actual practice as a (b) _____
do hereby certify, on soul and conscience, that I have this day, at (c) _____
in the County of _____
separately from any other Medical Practitioner, visited and personally examined (d) _____
and that the said _____
a (e) _____
and a proper person to be detained under Care and Treatment, and that I have formed this opinion upon the following grounds, viz.:—

1. Facts indicating Insanity observed by myself: (f)

2. Other Facts (if any) indicating Insanity communicated to me by others: (g) _____

Name and Medical Designation _____

Place of Abode _____

Dated this _____ day of _____
One thousand eight hundred and _____.

The certifying medical men must have no pecuniary interest in the asylum in which the lunatic is to be placed; but one of them may be a medical officer of such asylum, provided it is not a private asylum.

CERTIFICATE OF EMERGENCY.

(This Certificate authorises the detention of a Patient in an Asylum for a period not exceeding three days without any Order by the Sheriff.)

(a) State medical qualification.
(b) State place of examination.

I, the undersigned, _____ being (a) _____ hereby certify, on soul and conscience, that I have this day, at (b)_____ in the County of_____ , seen and personally examined_____ , and that the said person is of unsound mind, and a proper Patient to be placed in an Asylum.

And I further certify that the case of the said person is one of Emergency.

(Signed)_____

Dated this_____ day of _____ One thousand eight hundred and_____ .

(The following should be filled up in every case in which a Certificate of Emergency is acted on.)

I hereby request the Superintendent of the_____ to receive therein _____ to whom the foregoing Certificate of Emergency refers.

Relationship or other capacity ⎱
in which Applicant stands ⎰ _____
to Patient

Signature and Address _____

Date_____

ORDER TO BE GRANTED BY THE SHERIFF FOR THE TRANSMISSION AND RECEPTION OF THE LUNATIC.

(a) State whether Sheriff, Sheriff-Substitute, Steward, or Steward-Substitute.
(b) State whether a County or Stewartry.

I,_____(a) _____ of the (b)_____ of_____ having had produced to me, with a Petition at the instance of (c)_____

(c) Insert Name and Designation.

(d) Describe him, and if a Pauper, state so.

(e Lunatic, or an Insane Person, or an Idiot, or a Person of unsound mind.

(f) Public, District, Parochial, or Private.

Certificates under the hands of _____
and _____, being two Medical Persons duly qualified in terms of an Act, intituled "An Act for the Regulation of the Care and Treatment of Lunatics and for the Provision, Maintenance, and Regulation of Lunatic Asylums in Scotland," setting forth that they had separately visited and examined (d) _____
and that the said _____ is a (c) _____
and a proper Person to be detained and taken care of, DO HEREBY AUTHORISE you to receive the said _____
as a patient into the (f) _____ Asylum of _____
and I authorise _____ Transmission to the said Asylum accordingly; and I transmit you herewith the said Medical Certificates, and a Statement regarding the said _____
which accompanied the said Petition.

Dated this _____ day of _____ One thousand eight hundred and _____.

(g) Public, District, Parochial, or Private.

To the Superintendent of the (g) Asylum of _____

CERTIFICATION OF THE INSANE IN IRELAND.

I.—Admission into Private Asylums.

There are required two medical certificates, a "statement" and an "order"; the latter is signed by a private individual who may or may not be a relative of the patient. The patient must be admitted within seven days after the date of the medical certificates. (See Form below.)

The medical certificates must be signed and dated on the day the patient is examined. The disqualifications for signing are practically the same as in England (*vide supra*). If a patient has been admitted on one certificate, another must be obtained within fourteen days of the date of the first.

Lunacy inquisitions, ordered by the Court of Chancery where the property of the insane, or alleged insane, person is valued at more than £1,000, are carried out as in England. An inquisition taken, or a writ of supersedeas issued, in the one country may be acted upon in the other under an order of the Irish Lord Chancellor or the English Judge in Lunacy.

STATEMENT AND ORDER TO BE ANNEXED TO THE MEDICAL CERTIFICATES AUTHORISING RECEPTION OF AN INSANE PERSON.

The Patient's true Christian and Surname at full length	_____
The Patient's Age	_____
Married or Single	_____
The Patient's previous occupation (if any)	_____
The Patient's previous place of abode	_____
The licensed House or other Place (if any) in which the Patient was before confined	_____
Whether found Lunatic by Inquisition, and Date of Commission	_____
Special Circumstance which shall prevent the Patient being separately examined by two Medical Practitioners	_____
Special circumstance which exists to prevent the insertion of any of the above particulars	_____

Sir,—Upon the Authority of the above Statement, and the annexed Medical Certificates, I request you will receive the said _____ as a Patient into your House.

I am,

Name	_____
Occupation (if any)	_____
Place of Abode	_____
Degree of Relationship (if any) to the Insane Person	_____

To _____ Proprietor of the Private Lunatic Asylum at _____

FORM OF MEDICAL CERTIFICATE.

I, the undersigned, hereby certify, That I separately visited and personally examined _____ the Person named in the annexed Statement and Order on the _____ day of _____ One Thousand Eight Hundred and _____ and that the said _____ is of unsound Mind, and a proper Person to be confined.

(*Signed*) Name_____

Physician, Surgeon, *or* Apothecary_____

Place of Abode_____

☞ The admission of the Patient must be within seven days after the date of the Medical Certificates, agreeably to Act 5 and 6 of Victoria.

"*Single Patients*" may be kept in unlicensed houses, but, unless they are chancery lunatics, they must not be received without the above documents if any profit is derived from the charge.

II.—*Admission into District Lunatic Asylums* (*District Hospitals for the Insane*). (See Form E below.)

Privy Council Regulations.

Rule XI.—Persons labouring under mental disease, for whom papers of application are filled up in the prescribed forms, to the satisfaction of the Board, and who shall be duly certified as insane by a registered physician or surgeon, who shall state the grounds on which he forms his opinion, shall be admissible into District Asylums, after having been examined by the Resident Medical Superintendent, or, in his absence, by the visiting physician or surgeon.

Rule XVI.—No patient, other than a "dangerous lunatic" shall be admitted without the sanction of the Board, except by order of the Lord Lieutenant, or the Inspectors of Lunatics or one of them, or in case of urgency, when any three Governors or the Resident Medical Superintendent, or in his absence the visiting Physician of the asylum may admit upon their or his own authority, stating on the face of the order, the ground thereof, provided always that when a patient has been admitted under this rule, the Resident Medical Superintendent, or in his absence the Visiting Physician, shall submit the case to the *special* consideration of the Board at its next meeting for the decision of the Governors thereon.

Regulations of April 24th, 1885, as to military patients.—All soldiers serving in Ireland who shall be duly certified to be of unsound mind, may be temporarily admitted into any district lunatic asylum.

Form E.

DISTRICT LUNATIC ASYLUM.—FORM OF APPLICATION FOR ADMISSION.

No Patient can be admitted without a previous application made to the Asylum, and a form of admission obtained which must be accurately complied with.

The following Declarations, Certificates, Forms and Engagement, are to be filled up and transmitted to the Resident Medical Superintendent previously to the Lunatic being sent to the Institution, and no Lunatic will be received until it shall be notified to some of the friends of the Lunatic that there is a vacancy for his or her reception.

No application will be attended to which does not state the Name, Residence and Occupation, and degree of relationship of the two next male relatives, and the next female relatives of the Patient (when such exist), according to the annexed form, when it is possible to give those particulars.

It is requested that a person will accompany the Patient to the Asylum, who is able to give the best information respecting his or her disease, former mode of life, etc.: and it is expected that the Lunatic will be properly *clad*.

Patients are not admitted on Sundays.

DECLARATION.

County of _____) I, _____ of _____ in the County of _____
To wit   ) by Occupation a _____ do solemnly and sincerely
declare that _____ residing at _____ in the Parish of _____ and
County or City of _____ is Insane, and has been so for _____ and
that the said _____ is destitute, and has no friend who is willing or able
to support _____ in a Private or other establishment for insane, and
that _____ has been resident in the County or City of _____ for
_____ and that the information in the annexed form is correct.

And I make this solemn Declaration, conscientiously believing the same to be true, and by virtue of the provisions of an Act made and passed in the Sixth Year of the Reign of his late Majesty King William the Fourth (5 & 6 Wm. IV., c. 62), intituled an "Act to Repeal an Act of the present Session of Parliament, intituled an 'Act for the more effectual Abolition of Oaths and Affirmations, taken and made in various departments of the State, and to substitute Declarations in lieu thereof, and for the more entire suppression of voluntary and extra-judicial Oaths and Affidavits, and to make other provisions for the abolition of unnecessary Oaths.'"

Declared to by me _____

| Stamp to the value of One Shilling to be affixed here. | Made and Subscribed at _____ in said County, or City of _____ before Me, a Justice of the Peace for said County, or City, this _____ day of _____ 18 . |

_____ Justice.

The following Form must be filled up by the Friends of the Lunatic.

NAMES OF THE TWO NEXT AKIN TO THE LUNATIC.

Relative's Names   -   -   -   _____
  Residence   -   -   -   _____
  Occupation   -   -   -   _____
  Degree of Relationship -   -   _____
Lunatic's Age   -   -   -   _____
Religion   -   -   -   _____
Original disposition and habits of life _____
Place of Birth   -   -   -   _____
Place of Abode   -   -   -   _____
Occupation or Trade   -   _____

Whether Single or Married, and if a Female whether she has had Children - - - - -

Whether any near Relative has been Insane - - - -

How long ill, and if violent - -

What Education - - -

CERTIFICATE OF A MAGISTRATE AND CLERGYMAN, OR POOR LAW GUARDIAN.

We certify hereby that we have personally inquired into the case of the above-named _____ who is represented as residing in the Parish of _____, and County or City of _____ and we believe _____ to be a Lunatic in destitute circumstances, and that the statement of _____ is correct.

    Name of Magistrate _____

    Name of Clergyman or Poor Law Guardian _____

Date _____

MEDICAL CERTIFICATE.

I certify that _____, whom I visited on _____ day of _____ and into whose case I specially and personally inquired, is now insane, and I am of opinion, from the nature of _____ malady, that _____ is a fit subject for speedy admission into the _____ District Lunatic Asylum, and that I have filled up the annexed form to the best of my knowledge and belief.

    Signature of Physician _____

    Residence _____

Date _____

Species of Insanity - - -

Probable Cause of Derangement -

Prominent Symptoms - - -

Whether affected by Bodily Disease -

Whether Idiotic or Epileptic - -

Facts indicating Insanity as observed by me - - - -

    Signature _____

To the Board of Governors of
_____ _____ District Lunatic Asylum.

GENTLEMEN,—In consideration of your receiving into and maintaining in the above Asylum_____ as a Patient, I hereby undertake that within one week from my receiving notification from the Inspectors of Lunatic Asylums or the Board of Governors, that the said_____ is no longer a fit person to be accommodated therein, I shall remove the said_____ from the Asylum, and in the event of my failing so to do, I hereby agree to be responsible to the Board for any expense they may incur in having the said_____ removed therefrom, and also for all the costs of his/her maintenance in the Asylum after the expiration of said week, and until he/she shall have been finally removed therefrom.

Signature _____

6d. Stamp.

COMMITTAL WARRANT OF A DANGEROUS LUNATIC OR A DANGEROUS IDIOT.

To be Signed by Two Magistrates sitting together.

In pursuance of Act 30 & 31 Vict., c. 118.

(*a b c*) Here state Name and Address of each Informant, and Date of each Information.

(*d*) Here state Name of Lunatic or Idiot.

(*e*) Here state Place of Abode of Lunatic or Idiot.

(*f*) Here state Position in Life of Lunatic or Idiot.

(*g*) Here state Name of Place and County, County of a City, County of a Town, City, or Town as case may be, in which discovery and apprehension took place.

(*h*) Here state the facts from which it appears that the person was discovered and apprehended under circumstances denoting a derangement of mind, etc.

(*i*) Here state Name of Medical Officer.

(*j*) Here state Address of Medical Officer.

County of_____ } By Two or more Justices of the Peace in and for said County.
to wit
_____

To the Resident Medical Superintendent of the Asylum at_____

Whereas, by Information sworn before us by (*a*)___
of (*b*)____
on the (*c*)_____ day of_____ 18_
it has been proved to our satisfaction that (*d*)
_____
of (*e*)_____
by occupation a (*f*)_____

LEGAL REGULATIONS AND FORENSIC PSYCHIATRY. 307

(k) If the Medical Officer whom the Justices call to their assistance is the only Medical Officer of the Dispensary District in which the Justices shall be at the time, then fill the blank left at *k* as follows in italics, and insert at *l* the name of such Dispensary District, and at *m* the County, County of a City, County of a Town, City, or Town in which such Dispensary District is situate, namely, "*The Medical Officer of the* l——*Dispensary District, situate in* m——, *and being the Dispensary District in which we now are.*"

If there is more than one Medical Officer of the Dispensary District in which the Justices shall be at the time, the nearest available Medical Officer of such District is to be called by the Justices to their assistance; and in that event the blank left at *k* is to be filled up as follows in italics—inserting at *l* the name of such Dispensary District, and at *m* the County, County of a City, County of a Town, City, or Town in which such Dispensary District is situate, namely, "*The nearest available Medical Officer of the* l——*Dispensary District situate in* m——, *and being the Dispensary District in which we now are.*"

If there is no Medical Officer or no available Medical Officer of the Dispensary District in which the Justices shall be at the time, the nearest available Medical Officer of any neighbouring Dispensary District is to be called by the Justices to their assistance; and in that event the blank left at *k* is to be filled up as follows in italics—inserting at *l* the name of the Dispensary District of such Medical Officer, and at *m* the County, County of a City, County of a Town, City, or Town in which such Medical Officer's Dispensary District is situate, and at *n* the name of the Dispensary District in which the Justices shall be at the time, and at *o* the County, County of a City, County of a Town, City, or Town in which the Dispensary District in which the Justices shall be at the time is situate, namely, "*The nearest available Medical Officer of the* l —— *Dispensary District situate in* m ——, *being a neighbouring Dispensary District to*

has been discovered and apprehended at (*g*)_____ ___

_____

under circumstances denoting a derangement of mind, and a purpose of committing an indictable crime, that is to say (*h*) ____ _____

_____

And whereas we have called to our assistance (*i*)_____

of (*j*) _____

who is (*k*)_____

_____

And whereas the said (*i*)_____ ___

has duly examined the said (*d*)_____

and has duly certified by the Medical Certificate annexed hereto that the said ___ _____

is now a dangerous (*p*)___ _____

And whereas we have seen and examined the said (*d*)___ _____ _____

and upon the evidence aforesaid and our view and examination aforesaid are satisfied that the said (*d*) .   ___   _   - _

is now a dangerous (*p*)___ ___   __ __ _

We therefore direct that the said (*d*)_____

shall forthwith be taken to the said District Lunatic Asylum at (*q*)_____ which is the Lunatic Asylum for the said County (*r*)_____

in which County (*r*) _____

the said (*d*)__   __ __ _____

was discovered and apprehended as aforesaid.

And we hereby, in Her Majesty's name, charge and command you, the aforesaid Resident Medical Superintendent of the said Asylum to receive and detain in the said Asylum the body of the said (*d*) _____ and there safely to keep until removed therefrom, or otherwise discharged by due course of law, and for your so doing, this shall be your sufficient Warrant and Authority.

308 LEGAL REGULATIONS AND FORENSIC PSYCHIATRY.

the n —— *Dispensary District, situate in* o ——, *and in which last-mentioned Dispensary District we now are.*"

(*p*) 'Lunatic' or 'Idiot,' as case may be.

(*q*) Here insert name of Asylum.

(*r*) Or 'County of the City,' or 'County of the Town,' as case may be.

Given under our Hands and Seals, at ____
this ____ day of _____ 18__ .

_____ J.P.   Seal.

_____ J.P.   Seal.

The attention of the Magistrates is particularly requested to the proceedings required under the provisions of the 10th clause of the Act 30 & 31 Vict., c. 118.

The following Forms must be filled up by the Medical Officer who has personally examined the Lunatic or Idiot:—

I.—MEDICAL CERTIFICATE.

(*a*) Here state Lunatic or Idiot, as the case may be.

I certify that _____, whom I visited on _____ day of _____, and into whose case I specially and personally inquired, is now a dangerous (*a*) _____ and I am of opinion, from the nature of h__ malady that __he is a fit subject for speedy admission into _____ Lunatic Asylum, under the provisions of the Act 30 and 31 Vict. c. 118, s. 10.

Date _____ 18 __

Signature of Medical Officer ____

Residence _____

Dispensary District _____

II.—STATEMENT OF PARTICULARS OF CASE.

(*b*) Here state Lunatic or Idiot, as the case may be.

Species of Insanity -   -   -
Probable Cause of Derangement -   ____
Prominent Symptoms -   -   -
Whether affected by Bodily Disease -
Whether Idiotic or Epileptic -   -
Facts indicating that the Patient is a } ____
Dangerous (*b*) _____   }

I hereby certify that this Form is filled up correctly, to the best of my opinion and belief.

Date _____

Signature of Medical Officer _____

The following Forms must be filled up by the Friends of the Lunatic or Idiot:—

If no Friends of the Lunatic or Idiot are known this Form may be filled up by the police so far as their information will enable them to do so.

NAMES OF THE TWO NEXT AKIN TO THE LUNATIC OR IDIOT.

Relatives' Names - - -
   Residence and Post Town - -
   Occupation - - - -
   Degree of Relationship - -
Age of Lunatic or Idiot -
Religion - - - -
Place of Birth - - - -
Place of Abode - - - -
Occupation or Trade, and whether means of his own - -
Whether Single or Married, and if a Female, whether she has had Children - - - - -
Whether any near Relative has been Insane - - - -
How long ill, and if violent - -
Habits of Life, Temperate or Intemperate, etc. - - -
Education - - - -

Date _____

Signature _____

Privy Council Regulations of January 26th, 1876, as to the admission of Paying Patients into District Lunatic Asylums.

Rule XXVI. Amended.—No such patient shall be admitted into any district lunatic asylum, so long as there shall be unsatisfied and legitimate claims for the admission of lunatic poor who have no available means of their own, and whose friends are unable or unwilling to contribute towards their care and maintenance in the asylum. The Resident Medical Superintendent shall submit the application for the admission of every such patient to the Board of Governors, which application shall be in Form H hereunto annexed, and shall state the amount stipulated to be paid by the friends of such patient towards his support, and shall be accompanied by a certificate signed by a magistrate and a clergyman, stating that they are personally acquainted with the circumstances of such patient and his friends, and that to the best of their knowledge and belief, such patient is unable and his friends are unable or unwilling to pay for his care and maintenance in a private lunatic asylum.

If the Board of Governors shall approve of the admission of such patient upon the terms stipulated, the Resident Medical Superintendent shall forthwith submit the application to the Inspectors of Lunatics; and no such patient shall be admitted into any district asylum without the previous sanction of such Inspectors, or one of them, *provided always that in case of patients admitted to the Asylum, otherwise than under the foregoing provisions of this Rule, the Governors may, at any time subsequently, with the sanction of the Inspectors, or one of them, and on receipt of such a certificate as is in the last paragraph mentioned, approve of terms of payment to be made by the friends of such patient for his support.*

Every agreement for payment shall be made with the Resident Medical Superintendent, and shall not exceed the average of the general cost, nor be less than one-half of the average cost, for the care and maintenance of patients in the district lunatic asylum to which such patient shall be admitted; and the amount so stipulated shall be payable in advance by half-yearly instalments, provided always that in special cases the Board of Governors may authorise such an alteration in the charge as they think proper, not less in any case than one-fourth the average cost.

Paying patients shall be subject to the same rules and regulations as other patients in regard to their treatment, care, and maintenance. (General Rules and Regulations for the Management of District Lunatic Asylums in Ireland.)

Form H.

FORM OF APPLICATION FOR THE ADMISSION OF A PAYING PATIENT INTO A DISTRICT LUNATIC ASYLUM.

DECLARATION.

County of _____ } I, _____ residing at _____ in _____ do
    To wit.   } solemnly and sincerely declare that _____
of _____ in the County of _____ has, for some time past, been in a state of insanity and mental derangement; and that the said _____ is unable to pay for _____ care and maintenance, and has no friend who will support the said _____ in a Private Lunatic Establishment, and that _____ has been a resident of the said County of _____ for the last _____ years; and that the annexed certificates and forms are

correctly filled up. And I hereby undertake to pay, at the rate of £____ per annum, at such times and in such way as the Board of Governors may desire, for the support and maintenance of the said _____ in the _____ District Lunatic Asylum until _____ is discharged therefrom.

And I make this solemn declaration, conscientiously believing the same to be true, and by virtue of the provisions of an Act made and passed in the Sixth Year of the Reign of his late Majesty King William the Fourth (5 & 6 Wm. IV., c. 62), intituled 'An Act to repeal an Act of the present session of Parliament, intituled "An Act for the more effectual abolition of Oaths and Affirmations, taken and made in various departments of the State, and to substitute declarations in lieu thereof, and for the more entire suppression of voluntary and extra-judicial Oaths and Affidavits, and to make other provisions for the abolition of unnecessary Oaths."'

Declared to by me _____ _____

(An Impressed 2/6 Stamp here.)

Made and subscribed at __ __ __ __ in said County of _____ __before me, a Justice of the Peace for the said County, this _____ day of _____ 18 .

_____ __ ___ Justice.

MEDICAL CERTIFICATE TO BE SIGNED BY TWO PHYSICIANS.

I certify that _____ whom I visited on _____ day of _____ and into whose case I specially and personally inquired, is now insane, and I am of opinion, from the nature of _____ __malady, that _____ is a fit subject for speedy admission into _____ District Lunatic Asylum, and that I have filled up the annexed form to the best of my knowledge and belief.

Signature of Physician _____
Residence _____ _____
Signature of Physician _____
Residence ___ __ ____ __ __ _____

Species of Insanity - - - _____
Probable Cause of Derangement - _____
Prominent Symptoms - - - _____
Whether affected with Bodily Disease _____
Whether Idiotic or Epileptic - - _____
Facts indicating Insanity as observed } _____
   by me - - - - }

               (Signature)

MAGISTRATE AND PARISH CLERGYMAN'S CERTIFICATE.

We certify that we have specially inquired into the case of_____ who has resided in the Parish of_____ County of_____ for the last _____years; that we do believe_____ to be a lunatic, and to the best of our knowledge and belief such patient is unable, and that the friends of the said Lunatic are unable or unwilling, to pay for_____ care and maintenance in a Private Lunatic Asylum, and we recommend _____as a fit subject for admission as a paying patient into the _____District Lunatic Asylum.

      Given under our hands, this_____ day of _____18

                                         Magistrate.

                                      Parish Clergyman of_____

Engagement to Remove to be entered into by the Applicant for the Lunatic's Admission, if required by the Board of Governors, pursuant to Rule 15.

At any time, on receiving notification from the Inspectors or Board of Governors of the _____District Lunatic Asylum that the above-named _____is no longer a fit person to be accommodated therein, I hereby undertake to remove from the Asylum the said_____ within one week from the date of receiving such notification, as aforesaid ; and if not so removed within a fortnight, I hold myself responsible to the Board for any expense incurred by the removal of said .

                     Signature __

THE FOLLOWING FORM MUST BE ACCURATELY FILLED UP ON BEHALF OF THE PATIENT BY APPLICANT.

Lunatic's Age - - - - _____
Religion - - - _____
Place of Birth - - - _____
Place of Abode - - - _____
Occupation or Trade - - - _____
Whether Single or Married, and if a Female, whether she has had Children - - - - - _____
Whether any near Relative has been Insane - - - - _____
How long ill, and if violent - - _____

  Dated _____ day of _____ _____ 18

         Signed_____

It is requested that a person will accompany the Patient to the Asylum, who will be able to give the best information respecting the disease, former mode of life, habits, propensities, etc.—and that such Patient be furnished with a good strong suit of clothes, a change of linen, and other articles.

CERTIFICATE REQUIRED UNDER AMENDED XXVI. PRIVY COUNCIL RULE FOR CHANGING A NON-PAYING PATIENT INTO A PAYING ONE.

  We hereby certify that we are personally acquainted with the circumstances of _____ and _____ friends, and that to the best of our knowledge and belief _____ is unable, and _____ friends are unable or unwilling, to pay for _____ care and maintenance in a private Lunatic Asylum.

       Magistrate _____

       Clergyman _____

Dated_____

  To the Governors, _____ _____ District Asylum.

  The speediest method of admission into the district asylums in Ireland is to have the patient certified as a "dangerous" lunatic or idiot, that is, that he is a lunatic or idiot who apparently has the purpose of committing an indictable crime (*vide supra*

"Committal Warrant, etc."). The information is sworn by a constable or constables before two magistrates (sitting together) who see and examine the patient and call to their assistance a dispensary medical officer.

Attention has been called ("Med. Press and Circ.," Sept. 9th, 1891) to the fact that of 3,095 persons certified insane in Ireland during the last official year, 2,165, or 70 per cent., were certified as "dangerous lunatics," whereas in Scotland only five persons, or 0·2 per cent. of those admitted during the year belonged to the same category. The friends of "dangerous lunatics" are exempt from payment in Ireland.

It is to be inferred from the last (40th) Report of the Inspectors of Lunatics that there are now no "single patients" in Ireland, and that insane paupers are not "boarded out" as in Scotland and England. National institutions (similar to the Royal Albert and Earlswood in England, and Larbert and Baldovan in Scotland) for the training and education of idiots and imbeciles, are much required.

CERTIFICATION OF THE INSANE IN THE STATE OF NEW YORK.

(Chapter 126, Laws of 1890.)

Sect. 1.—In accordance with the provisions of this section the board for the establishment of State insane asylum districts and other purposes, has divided the State into State insane asylum districts.

Sect. 5.—Each of the State asylums for the insane shall receive patients, whether in an acute or chronic condition of insanity, from the district in which the asylum is situated.

Sect. 9.—In case any insane person, his relatives, guardians or friends may desire that he become an inmate of any State asylum situated beyond the limits of the district where he resides, and there be sufficient accommodation there to receive him, he may be received there in the discretion of the chairman of the State Commission in Lunacy and the superintendent of such asylum. Any expenses of removal, in such case, must be borne by said insane person's guardians, relatives, or friends, as the case may be.

COMMITMENT OF PATIENTS.

(Chapter 446, Laws of 1874.)

Sect. 1.—No person shall be committed to or confined as a patient in any asylum, public or private, or in any institution, home or retreat for the care and treatment of the insane, except upon the certificate of two physicians, under oaths, setting forth the insanity of such person. But no person shall be held in confinement in any such asylum for more than five days, unless within that time such certificate be approved by a judge or justice of court of record of the county or district in which the alleged lunatic resides, and said judge or justice may institute inquiry and take proofs as to any alleged lunacy before approving or disapproving of such certificate, and said judge or justice may, in his discretion, call a jury in each case to determine the question of lunacy.

Sect. 2.—It shall not be lawful for any physician to certify to the insanity of any person for the purpose of securing his commitment to an asylum, unless said physician be of reputable character, a graduate of some incorporated medical college, a permanent resident of the State, and shall have been in the actual practice of his profession for at least three years. And such qualifications shall be certified to by a judge of any court of record. No certificate of insanity shall be made except after a personal examination of the party alleged to be insane, and according to forms prescribed by the State Commission in Lunacy (these blank forms may be obtained free upon application to the State Commission in Lunacy, county clerks, superintendents of the poor, and the superintendents of asylums or hospitals for the insane), and every such certificate shall bear date of not more than ten days prior to such commitment.

Sect. 3.—It shall not be lawful for any physician to certify to the insanity of any person for the purpose of committing him to an asylum of which the said physician is either the superintendent, proprietor, an officer, or a regular professional attendant therein.

(*Chapter 283, Laws of 1889, as amended by Chapter 273, Laws of 1890.*)

Sect. 7.— * * * * One year after the date of the passage of this act (May 14, 1889), it shall not be lawful for any medical examiner in lunacy to make a certificate of insanity for the purpose of committing any person to custody unless a certified copy of his certificate has been so filed and its receipt in the office of the Commission (State Commission in Lunacy), as above provided has been acknowledged.

In addition to the medical certificate of two physicians sworn to and approved by a judge in the county in which the patient resides, there must be presented at the time of admission of a patient either an order from the superintendent of the poor, or the county judge, or a bond (in case of private patient) guaranteeing the amount charged for care and treatment.

Admission of Private Patients to the State Hospital for the Insane.

On and after October 1, 1890, there shall be no distinction allowed between private and public patients in respect to the scale or care and accommodations furnished them.

On and after October 1, 1890, no private patient will be admitted to any State hospital except in strict accordance with the statute: Whenever there are vacancies in the asylum (State Hospital) there may be received such recent cases as may seek admission under peculiarly afflictive circumstances, or which in the superintendent's opinion promise speedy recovery and upon an order granted by the State Commission in Lunacy upon an application in writing, addressed to the Commission, of a near relative, guardian or committee of the patient. (Dr. C. W. Pilgrim, "Report of Trustees of Willard State Hospital," 1890.)

CERTIFICATION OF THE INSANE IN THE STATES OF CONNECTICUT, PENNSYLVANIA, MASSACHUSETTS, AND ILLINOIS.

In *Connecticut* there are required for the admission of a private patient into the hospital for the insane only a request signed by a guardian, near relative, or friend, and one medical certificate running thus:—" I hereby certify that I have, within one week of this date, made personal examination of A.B., of C.D., and believe him to be insane." This is subscribed and sworn to by the physician before an officer authorised to administer oaths, who certifies to the respectability of the physician and the genuineness of the signature.

In *Pennsylvania* in ordinary cases it is necessary that the medical certificate should be signed by at least two physicians actually in practice five years, who shall certify that they have separately examined the patient, and believe him to be insane and requiring the care of an asylum. They must not be related by blood or marriage to the patient, or connected medically or otherwise with the institution. This certificate must be made within one week of the examination of the patient, and within two weeks of his admission. Further, it must be sworn to or affirmed before a judge or a magistrate, who must certify the genuineness of the signature, and the standing and good repute of the signers. It is not, however, necessary that the judge or magistrate should examine the patient, or express any opinion in regard to his insanity. The order and statement are signed by the person at whose instance the patient is received.

The law of Pennsylvania allows persons to place themselves

voluntarily in an asylum for a period not exceeding seven days, on signing an agreement giving authority to detain them, which may be renewed from time to time for the same period.

In *Massachusetts* there are required the certificates of two physicians (the facts upon which their opinions are founded being specified), and an order signed by a judge certifying that he finds that the person committed is insane, and fit for treatment in an asylum. If the judge thinks it undesirable to see the patient, he may certify to that effect and still commit him. If he is in doubt he may summon a jury of six to his aid.

An urgent case may be received by the superintendent and detained for five days on an emergency certificate. This is signed by two physicians who certify that the patient is labouring under violent and dangerous insanity, and it is accompanied by an application for admission from the Mayor or one of the Aldermen of the place in which the patient resides.

In *Illinois* the alleged lunatic must be tried by a county court judge and a *jury of six*. The patient is examined by a physician employed by the Court. This official states to the Court and the jury the result of his medical examination. Evidence is also given by the relatives and friends of the person supposed to be insane. Finally the jurors retire to consider their verdict. ("The Insane in the United States," by D. Hack Tuke, "Jour. Ment. Science," April, 1885).

TESTAMENTARY CAPACITY OF THE INSANE.

In England the law is much more influenced by the will itself than by any evidence as to the mental state of its maker. If known to be in accordance with his wishes when sane, it will probably be ruled valid. If made during a period of sanity by a lunatic, it will probably be ruled valid, if consistent. Persons suffering from acute forms of insanity (mania and acute melancholia); from weak-mindedness either congenital or the result of apoplexy, epilepsy, acute mental disease, or senility; or from suspicious monomania, are very liable to have their wills upset.

In acute forms and sometimes in weak-mindedness, the failure arises from actual incapacity. In most cases of weak-mindedness, the cause of failure is the testator's susceptibility to undue influence. Persons suffering from weak-mindedness consecutive to acute insanity, may dislike their relatives or refuse to forgive them for having put them in an asylum, or they may display emotional instability, leaving their property to some institution for which they had no regard before the attack of insanity.

In senile weak-mindedness the testator will be frequently found to be querulous, exacting, exhibiting defective, often very defective memory, and so forgetting heirs-at-law, etc. The sexual passions may also be abnormally strong in senility, and, for this reason, as well as in consequence of the general mental weakness, such testators may be the subjects of undue influence.

In weak-mindedness following apoplexy, the will may be disputed because the testator is aphasic although perfect testamentary capacity may exist with aphasia; but when amnesia and aphasia are combined, the patient may be unable to make a will or may be too readily influenced. If after apoplexy the memory is retained, and there is no change in the patient which would influence his opinions of his near relatives, he should be allowed to make a will.

Before outbreaks of insanity the patient though apparently sane, may suffer from a moral perversion, causing him to suspect and dislike his relatives and make a will to their disadvantage; this is especially liable to occur when there have been previous attacks of insanity followed by confinement in an asylum.

The onset of melancholia may be associated with the revocation or alteration of a will, the testator leaving his property to charities or religious bodies to atone for his supposed past misdeeds.

The suspicious monomaniac may destroy all wills, or make a will in favour of a stranger or of charitable institutions. (Savage, "Insanity and allied Neuroses," page 473, *et seq.*). Persons suffering from any form of delusion as to the amount of property possessed by them, whether the delusion is in excess of the reality or the opposite (ambitious monomania, general paralysis, melancholia), would be likely to have their wills upset.

When a medical man is examining a patient as to his testamentary capacity Clouston (*op. cit.*) recommends that he should ascertain: (1,) Whether the patient is free from the influence of drink or drugs; (2,) Whether he understands the nature of the act he is about to perform, and the effect of the document; (3,) Whether the disposition of the property is a natural one or the result of delusions or mental weakness; (4,) Whether there is mental facility from bodily weakness or other cause, undue influence being exerted from without. Under these circumstances it is advantageous, in order to ascertain the truth, to be alone with the patient; (5,) Whether the intending testator is able to go over the particulars of the proposed disposition of his property without help, and repeat his statement correctly after an interval of, say, a quarter of an hour; (6,) Whether he knows in a general way the amount of the property he has to bequeath.

Clouston (*loc. cit.*) advises "not to let a good motive make us sanction a bad will, however natural its provisions may be, however much trouble or expense it may save."

According to Taylor ("Medical Jurisprudence") the validity of a will depends not so much upon the sanity or insanity of the testator as upon his competency or incompetency to make a will. The best test of competency is that the testator should know, at the time of signing the will, the nature and amount of his property and the just claims of his near relations. Taylor reminds the medical man that when he acts as a witness to a will, he practically testifies to the competency of the testator to make it. The same author quotes cases which show that a person may be placed under interdiction or even confined in a lunatic asylum, and yet be competent to make a will.

Maudsley remarks ("Responsibility in Mental Disease") that formerly, if an insane testator made a natural disposition of his property, a lucid interval at the time of making the will was presumed. It was at one time held that delusion voids a will. This was at first rejected in New Hampshire, U.S., and afterwards in the Court of Queen's Bench. Maudsley says it is necessary to wait for future decisions to learn whether the principle laid down in the latter case (Banks *v.* Goodfellow) is to govern the *making of contracts* by partially insane persons, or whether such contracts are to be voided in accordance with the old rule that the law voids every act of the lunatic, although the insanity may be extremely circumscribed, and although the act to be voided has been in no way influenced by the insanity.

THE EVIDENCE OF THE INSANE.

The two points to be primarily considered in estimating the value of the evidence of insane persons are :—1st, Is the patient in such a condition mentally as to know that he ought to speak the truth to the best of his ability ? and 2nd, Is he able to report the facts ? ("Journ. Ment. Sci.," January 1, 1891, page 110). In the case of Regina *v.* Hill the attendant (Hill) was found guilty of manslaughter mainly on the evidence of a lunatic who laboured under the delusion that he had spirits in his head. The case was argued before four judges, who held unanimously that this lunatic's evidence was properly admitted. In citing authorities two cases were quoted in which the evidence of persons partially or wholly insane was admitted. The witness may be examined as to his mental state, and the evidence of witnesses may be adduced as to his actual state of sanity or insanity.

Campbell ("Complaints by Insane Patients," "Journ. Ment. Sci.," October, 1881,) called attention to a case where an attendant was fined in a Scotch Sheriff's Court for assaulting a patient on the sole evidence of the patient, who was stated to be labouring under delusions. The medical evidence was to the effect that the patient's delusions had no reference to his injuries, and that his statement as to how he received the latter, could be implicitly relied on. In a recent case an attendant (Hays), who had been discharged after the inquest on a patient formerly under his care, took an action for wrongful dismissal against the Board of Governors of the Richmond Asylum. The Recorder of Dublin held the evidence of two lunatics (who testified as to the violence of Hays towards the deceased) to be practically valueless, and directed the jury to find a verdict for the plaintiff (Hays). Yet the Recorder said the witnesses had given their evidence admirably and that "he had never heard better witnesses." On appeal, this judgment was quashed by Justice Holmes on the ground that the Governors were justified in dismissing Hays or any other servant, without giving any reasons whatever. He, therefore, did not consider it necessary to call the patients as witnesses. In respect to the evidence of a person of unsound mind he said, "First I must be satisfied that the person understands the nature of an oath, and next, that he understands and appreciates the evidence he is giving." ("Journ. Ment. Science," *loc. cit.*).

LEGAL TESTS OF INSANITY AND LEGAL RESPONSIBILITY OF THE INSANE.

"A *testator* is considered to be sane if he is 'of sound mind, memory, and understanding.' In *lunacy inquisitions* the test is: 'Is the patient capable of managing himself and his affairs?' In *criminal cases* the legal test of insanity is the knowledge of right and wrong; that is to say, a criminal is considered to have been sane when he committed his crime if 'he then knew the nature and quality of the act and that it was wrong.'" (Mercier, "Sanity and Insanity," pp. 98, 99).

Vide supra "Testamentary Capacity of the Insane" *re* first test. With regard to the second test, the inquisition, so far as it finds the lunatic incapable of managing himself, may be superseded, and any order for commitment varied or rescinded.

The last test is manifestly intended to be that of the legal responsibility of insane, or alleged insane persons who have committed crimes; but the plea of uncontrollable impulse may be raised even after the accused person has been found responsible by the above test, in which event the patient's anamnesia ought to be useful

Maudsley observes (*op. cit.*) that by the judgment of the Court of Queen's Bench in the case of Banks *v.* Goodfellow, the law relating to testamentary capacity is made to agree with that relating to criminal responsibility so far as this.—That a partially insane person is competent to make a will or commit a crime ; not being declared incapable in the one case, nor exempted from punishment in the other, except when the act can be shown to be the offspring of the insanity. But they differ in these points: (1,) That while an insane delusion will invalidate a will springing from it, an insane delusion will not always invalidate a criminal offence arising from it ; (2,) That while the disordered feelings in insanity have due weight given to them in will cases, they receive no consideration in criminal trials; and (3,) That while no special test of civil capacity is enunciated as a legal principle, the whole case being decided upon its merits by the jury, a special test of responsibility is proclaimed as a legal principle in criminal cases.

The "right and wrong" test is the one applied by the majority of legal authorities, but it is not the only one that is or has been used. According to Clouston ("Ment. Dis." 1st. Ed. p. 617) Judge Tracey held that a criminal should be punished unless he was as irresponsible as a wild beast ; Lord Denman made the presence or absence of insane delusion the test ; Lord Moncrieff, "the man's habit and repute as to sanity among his fellow-men who knew him well ;" and lastly, Justice Stephen's new criminal code "proposes to make the man's power of controlling his actions the test," he being also guilty if the mental disease causing loss of self-control has been produced by his own default.

"The law now recognises only one test of lunacy, viz., Was the person whose act is in question able to understand its nature, and to pass a fairly rational judgment of its consequences to himself and others, and was he a free agent so far as that act was concerned ?" ("Journ. Ment. Sci.," July 1891, p. 422).

The above test practically contains the propositions of the new Criminal Code regarding responsibility. Taylor (*op. cit.*) points out that an act may be an offence, although the mind of the person who does it is affected by disease, or is deficient in power, if such disease or deficiency does not produce one or other of the effects mentioned in these propositions. In order to justify legal responsibility it is not sufficient to prove insanity; it must be proved that the insanity has reached a certain degree. Voluntary drunkenness is excepted from these provisions. A criminal act implies the existence of intention, will, and malice. The presence of delusion does not necessarily constitute irresponsibility, nor does the total absence of delusion entail responsibility. The surrounding circumstances must be taken into account.

Guy ("The Factors of the Unsound Mind and the Plea of

Insanity," p. 222) gives as the four principal sources of delusion leading to homicidal acts :—Religious excitement or despondency: jealousy ; domestic anxiety exaggerated into fear of starvation : and discontent transformed into an insane belief in persecution.

The forms of insanity in which crimes are committed are :— Impulsive insanity, moral insanity, periodical insanity, circular insanity, mania, epileptic insanity, alcoholic insanity, traumatic insanity, puerperal insanity, delusional melancholia, suicidal (and homicidal) melancholia, paranoia, incipient general paralysis (shop lifting, etc.), dementia, imbecility (see Chap. III). Insanity may be feigned (see "Diagnosis"), and a patient suffering from one form of mental disease may simulate another. In impulsive cases and in cases of insane temperament or diathesis without certifiable insanity, the patient's anamnesia, including both his family and personal history, should be carefully investigated.

GENERAL INDEX.

	PAGE
ALBERS, p. 84; Alexander, 264; Allbutt (Clifford), 36, 40; Arndt, 61; Archbold, 2, 286	
Abdominal disorders	49
— — insanity from (etiology)	137
Aberration, mental, grouping of forms of	131
— partial emotional	105
— — — (etiology)	143
— — — (prognosis)	184
— — — (treatment)	266
Ability to write, loss of	33
Abnormalities of speech	43
Abruptness of outbreak	11
Absence of sexual desire	216
— — two or more senses	41
Absent-mindedness	11
Absinthe, insanity from	126
Abulia	11
— congenital	157
Acquired mental weakness	132
Action and speech, monotony of	34
— and thought, want of continuity in	16
— — want of force in	23
— or speech limited	33
Actions, childishness in	15
— improvidence of	30
— motiveless	34
Active melancholia	97
Activity, unusual and useless	11
Acts and speech, inconsistencies in	31
— and words, paying attention to	36
— — — — to own	48
— extraordinary	11
— impulsive	30
— monotonous	11
Acute alcoholic insanity	120
— — (prognosis)	187
— confusional insanity	56
— delirium	60
— — (diagnosis)	165
— — (etiology)	138
— — (pathology)	194

	PAGE
Acute delirium (prognosis)	175
— — (treatment)	263
— dementia	117
— — (etiology)	144
— — (prognosis)	186
— epileptic insanity (diagnosis)	165
— mania	90
— — (diagnosis)	163
Acuteness of the senses, to investigate the	vi, 152
Admission into private asylums (Ireland)	301
— — asylums	153
— of private patients to hospital (New York State)	316
Adolescent insanity	49
— — (etiology)	137
— — (prognosis)	173
— — (treatment)	263
Affection natural, loss of	35
Affections of the will	48
Affective sensibility, perverted, etc.	11
— sentiments	79
Ageustia (or Ageusia)	11
Aggressiveness	11
Agitated melancholia	97
Agitation, excessive	11
Agoraphobia	29, 105, 158
— (prognosis)	184
Alcohol and drugs, increased reaction to	39
Alcoholic and other intoxication (diagnosis)	148
— dementia	126
— excesses, tendency to	12
— insanity	120
— — chronic	123
Alcoholic insanity (pathology)	202
— — (prognosis)	187
— paralysis	160
— pseudo-general paralysis	125
Alienists' classification	3
Alterations in the blood	14, 79
— of sentiments	41
Ambitious delusions	156

GENERAL INDEX.

Amenomania	106, 158
— (etiology)	143
— (prognosis)	184
Amenorrhœa	12
Amenorrhœal insanity	131
— — (diagnosis)	163
— — (etiology)	145
— — (prognosis)	188
Amnesia	12
Amylene hydrate in insomnia	245
Anæmia	12
Anæmic insanity	50
— — (diagnosis)	157
— — (etiology)	137
— — (prognosis)	173
Anæsthesia	12
— cutaneous	16
Anatomy, pathological	189
Anergic stupor	117, 167
— — (etiology)	144
— — (prognosis)	186
Animals, devils, etc., seen by patient	164
Annoyance at trifles	12
Anorexia	12, 160
— and sitophobia	215
Anosmia	12
Answering of questions by patient	154
Anthropophobia	105
Antipathy to friends and relations	89
Anxiety	12
Apathy	12
Aphasia	13
— (diagnosis)	149
Apoplectiform attacks	13
Apoplexy	54
Apprehensiveness	13
Arachnoid, the (macroscopical morbid anatomy)	189
Arcus senilis	13
Arteries atheromatous	13
Assertions, false and malevolent	23
Asthma, insanity of	vi, 59, 165
Astraphobia	106
Asylum, district rules for (Ireland)	309
— examination of patient in	279
— private, admission into (Ireland)	301
— removal of patient to another	287
— treatment	259
Asylums, admission into	153
— forms of insanity met with at	170

Asylums pauper	259
— private	261
Ataxia	13, 78
Atony	13
Atrophy, optic nerve changes, etc.	36
Attacks, apoplectiform	13
— epileptiform	21
Attendant, medical, duties of	287
Attention, defective	13
— — morbid anatomy of	213
Attitude, immobile	13
— insinuating	13
— listless	13
— suggestive of auditory hallucinations	13
— suggestive of delusions of grandeur	13
Auditory disturbances, morbid anatomy of	204
— hallucinations	25
Automatic ideas and words	14
Aversion to movement	35
Awkwardness in manipulations	33
BAILLARGER, p. 51; Bain, 218; Ball, 5, 24, 70, 177; Bastian, 228; Batty Tuke, 136, 191, 219; Bell, 60; Bertheir, 110; Bischoff, 193; Blandford, 100, 101, 126, 127, 142, 180, 188; Boeck, 216; Bra, 5, 15, 25, 26, 50, 51, 52, 60, 61, 62, 63, 64, 67, 68, 69, 71, 82, 87, 93, 94, 97, 105, 113, 115, 116, 118, 123, 124, 126, 129, 138, 141, 144, 177, 178, 185, 188, 194, 197, 198, 200, 216; Briand, 194, 195; Bricquet, 198; Broadbent, 38, 39; Browne (Crichton), 214; Bucknill and Tuke, 1, 3, 17, 24, 25, 26, 49, 50, 53, 54, 58, 59, 65, 83, 84, 85, 86, 89, 106, 110, 112, 135, 136, 146, 169, 172, 174, 176, 178, 179, 181, 184, 192, 216, 261	
Babbling and chattering	14
Back, weakness in	14
Bacteria in blood	79
Bang, insanity from (diagnosis)	165
Barometric and seasonal conditions	14
Bed, refusal to leave	14

GENERAL INDEX. 325

	PAGE
Bed sores	14
Being controlled, sense of, etc.	16
Biliousness	14
Birth, oddness and peculiarity from	36
Bladder, lesions of	217
Blindness	14
Blödsinn, terminaler	65
Blood, alterations in	79
— alteration of	14
— lesions of	217
Boarders voluntary	279
Bodily functions, disturbance of	14
— symmetry, want of	14
Bones, lesions of	218
Book, form of medical visitation	282
Borax in epilepsy	264
Brain, degeneration of the cortical cells of the	238
— disease, coarse	54
— hypochondriasis	95
— nutrition of the	237
— tumours	55
— weight of in insanity	192
Bra's classification	5
Bright's disease, insanity of	50
— — — (etiology)	137
— — — (prognosis)	173
— — — with (diagnosis)	163
Bromides, the in insanity	263
Bronchitis, cyanosis from	59
Brutishness	14
Bucknill & Tuke's percentages of frequency of forms of insanity	169
Bulimia	14

CALMEIL, *p.*71; Campbell, 319; Charcot, 55, 193; Charles, 221; Clifford Allbutt, 36, 40; Clouston, 5, 6, 15, 30, 40, 44, 47, 49, 50, 52, 53, 54, 56, 58, 64, 66, 80, 86, 87, 88, 89, 90, 91, 92, 93, 96, 97, 98, 99, 100, 104, 105, 108, 111, 112, 113, 116, 117, 118, 126, 127, 129, 130, 131, 137, 138, 144, 169, 173, 176, 181, 188, 217, 297, 321; Crichton-Browne, 214

	PAGE
Cachexia	14, 79
Calculation by patient	155
— defective	14
Cannabis indica in motor excitement	245
— — insomnia	244

	PAGE
Cannabis indica, insanity from	127
— — — — (diagnosis)	165
Cardiac disease	59
Carnal knowledge of female patient, penalty of	287
Cases of primary insanities	241
— for private care	257
Catalepsy	14
— causes of (etiology)	137
Cataleptic insanity	50
— — (etiology)	137
— — (prognosis)	173
Catamenia, disorders of	14
— suppression of	44
Causes, exciting, of insanity	135
— of catalepsy (etiology)	137
— — chorea (etiology)	137
— — hallucinations	239
— — insanity	137
— — — in England & Wales	134
Cell, the, in its relation to other cells of cerebrum, etc.	221
Central ganglia, cells of the	221
Cephalalgia	14
Cerebellum and cerebrum, connections of	227
— cells of the	221
— effects of removal of	229
— in animals, children, idiots, etc.	228, 230
— size and development of	233
— the (macroscopical morbid anatomy)	193
— theories as to functions of	227, 233
Cerebral hemispheres, weight of	231
— causes of insomnia	226
— localisation	vi, 202
— meningitis (diagnosis)	148
— neurasthenia	102
— rheumatism	114
— substance (macroscopical morbid anatomy)	190
Cerebrum, cells of the	221
— effects of removal of	216, 230
Certificate as to mechanical means of restraint	285
— — personal interview	281
— — for criminal lunatics	296
— — dangerous lunatic (Ireland)	308
— — non-paying patient (Ireland)	313
— — paying patients (Ireland)	310
— blank forms of, where purchased	280, 290

GENERAL INDEX.

Certificate medical (Ireland) .. 302
— — persons disqualified from signing 278
— — (Scotland) 299
— of death of patient 290
— — emergency (Scotland) .. 300
— — medical practitioner (England & Wales) 272
— — magistrates, etc. (Ireland) 305
Certification of insane patients in England & Wales 269
— — — Ireland 301
— — — State of Connecticut 316
— — — Illinois 316
— — — Massachusetts .. 316
— — — Pennsylvania .. 316
— — — New York 314
— — — Scotland 296
Chancery patients 290
Change in disposition 19
— of habits .. 24, 156, 157
— of mood 34
— of occupation 36
— of residence with patient .. 286
— of temper 45
Changes in retina 40
— in skin 42
— vascular 47
Character, change of 14
— mobility of 34
Chattering and babbling .. 14
Child, ruling instincts in the .. 238
— the, in its perceptions, etc. 221
Childhood, melancholia in .. 238
Childishness in actions .. 15
Children, delirium of young .. 131
— — — (diagnosis) .. 164, 165
— — — (prognosis) 188
— injuring 31
Chloralamide in insomnia .. 245
Chloral hydrate in insomnia .. 243
— insanity from 127
Choreic insanity 50
— — (etiology) 137
— — (pathology) 193
— — (prognosis) 173
— — (treatment) 262
— movements 15
Chronic alcoholic insanity .. 123
— — — (diagnosis) 159
— confusional insanity (diagnosis) 158
— epileptic insanity (diagnosis) 161, 165
— hysterical insanity 82

Chronic mania 92
— morphismus 155
Circular insanity 51
— — (etiology) 137
— — (prognosis) 174
Circulation, defective facial .. 22
Claustrophobia 29, 105, 158
— (prognosis) 184
Classification, Bra's 5
— by International Congress of Alienists 3
— by London College of Physicians 8
— Clouston's 5, 6
— Esquirol's .. .. v, 2
— Krafft-Ebing's .. v, 4, 9
— Morel's 2
— Savage's 8
— Skae's 4
— Spitzka's .. .. v, 6
— of mental diseases 2
Clergyman's certificate for paying patients (Ireland) 312
Climacteric insanity 53
— — (diagnosis) 164
— — (etiology) 137
— — (pathology) 193
— — (prognosis) 174
— — (treatment) 263
Clouston's classification .. 5, 6
— percentages of frequency of forms of insanity .. 169
Coarse brain disease 54
— — — insanity from (etiology) .. 137
— — — — (prognosis) .. 174
— — — — (pathology) .. 193
— — — — (treatment) .. 263
Cocainism, chronic treatment 268
Codeia in motor excitement .. 246
Colour of cerebral substance .. 190
Coma, or somnolence 15
Commissioner's classification 2
Commissioners in Lunacy, report of, 1889 134
Committal warrant (Ireland) 306
Commitment of patients (New York State) 314
Company low, disposition for .. 33
Complexion, pale and pasty .. 15
Compound hallucinations .. 26
Concentration of intellectual operations round one idea 15
— of thoughts on health, etc. 15
Conceptions 29
— imperative 104

	PAGE
Conditions, barometric and seasonal	14
Conduct etc., change in	15
— extraordinary	15
Confusion, feeling of	15
— of ideas	15
Confusional insanity	56, 57
— — chronic (prognosis)	175
— — (etiology)	138
— — (pathology)	194
— — primary or acute (diagnosis)	162
— — — (prognosis)	174
Congenital abulia	157
— imbecility	83
— mental and moral weakness	133
Congress, International, 1867	135
Conium in motor excitement	245
Connecticut, certification of the insane in the State of	316
Connections of the cerebellum with the cerebrum	227
Consciousness, abolished	15
— confused	15
— impaired	15
— temporary losses of	16
— theories concerning	227
Consecutive insanity	58
— — (diagnosis)	155, 163
— — (etiology)	138
— — (pathology)	194
— — (prognosis)	175
Consistence of cerebral substance	192
Constipation	16
Continuity, want of, in thought and action	16
Contractures, muscular	35
Conversation by patient not sustained	159, 162, 163
— sustained by patient	155
Convulsions	vii, 16
— hysterical	27
— infantile succeeding	31
Cord, vocal, paralysed	47
Countenance, expressionless	16
— expressive of distrust, etc.	16
Courage, failure of	16
Cramps	16
Cranium, the (macroscopical morbid anatomy)	189
Cretinism	83, 85
— (diagnosis)	167
— (etiology)	141
— (prognosis)	178
Crimes in connection with insanity	320

	PAGE
Criminal lunatics	294
Criminals, the brain in	237
Cruelty	16
Cursatory impulses	16
Cutaneous anæsthesia	16, 78
— sensibility, disorders of (morbid anatomy)	202
Cyanosis from bronchitis	59
— — etc., insanity of (etiology)	138
— — — (prognosis)	175
DARWIN, pp. 149, 240; Delasiauve, 198'; Down, 83	
Dangerous lunatics	294
— — (Ireland)	303
— — committal warrant (Ireland)	305
Death of patient	289
— order of frequency of forms of insanity at	170
Declaration for admission of patients (Ireland)	304
Defects of cutaneous sensibility	202
Definitions of insanity	1
— of medical attendant	287
Dejection	16
Delire des actes	158
Delirium	16
— acute	60
— — (diagnosis)	165
— — (etiology)	138
— — (pathology)	194
— — (prognosis)	175
— — (treatment)	263
— hallucinatory	56
— of fevers (diagnosis)	147
— of young children	131
— — — (diagnosis)	164, 165
— — — (etiology)	145
— — — (prognosis)	188
— with remissions	17
— with unconsciousness	133
— tremens	120
— — (treatment)	268
Delusion	132
Delusions, etc.	164
— genuine	17
— of ambition	156
— of persecution	156
— of pride	156
— of suspicion	155
— (pathology)	215
— spurious	18
— systematised	155
— unsystematised	156, 158

GENERAL INDEX.

Delusional insanity 61
— — (diagnosis) .. 154, 166
— — (etiology) 138
— — (pathology) 195
— — (prognosis) 175
— mania.. 92
— melancholia 97
Demeanour, change of .. 156
Dementia, acute 117
— — (etiology) 144
— — (prognosis) 186
— agitated terminal (diagnosis) 161
— alcoholic 126
— epileptic (diagnosis) .. 161
— organic .. vii, 54, 160
— paralytic 54
— paretic 71
— partial 100, 159
— puerperal 113
— senile 117, 160
— — advanced.. 166
— simple primary 99
— terminal 65
— — (diagnosis) 167
— — (etiology) 138
— — (pathology) 195
— — (prognosis) 176
Depression 18
— and exaltation, alternating 12, 21
— with change of habits .. 156
Deprivation, idiocy by.. .. 85
— of senses 66
— — insanity from (etiology) 138
Desire, sexual, loss of 216
Despondency, religious .. 19
Destructiveness.. .. 19, 245
Deterioration, mental 99
— moral 34
— primary mental 159
— — (etiology) 139
— — (prognosis) 176
Devils, animals, etc., seen by patient 164
Diabetic insanity 66
— — (diagnosis) 163
— — (etiology) 139
— — (prognosis) 176
Diagnosis, differential, of the forms of insanity .. 154
— of insanity 146
— references to paragraphs on 168
Diarrhœa 19
Diet in home treatment .. 251
Differential diagnosis of the forms of insanity .. 154

Diminutives, tendency to use 19
Dipsomania 30, 156
Dirty in habits 19
Discharge, form of notice of .. 283
Discontentedness 19
Disorders, abdominal 49
— of cutaneous sensibility (morbid anatomy) 202
Disposition and conduct, change in .. 15, 19, 156
— to connect all things with self 15
— to squander 43
— to wander 47
Distaste for work 48
Distortion of objects and persons 19
District lunatic asylums (Ireland) 303
— — — rules for (Ireland) .. 309
Disturbance, emotional .. 20
— of bodily functions 14
— auditory (morbid anatomy) 204
— gustatory 206
— intestinal 32
— olfactory (morbid anatomy) 206
— trophic 46
— visual (morbid anatomy) .. 205
Doubting insanity 70
— — (diagnosis) 158
Dread, unfounded 19
Dreaming, theory of 235
Dreaminess 19
Dress and undress, inability to 19
— suggestive of exalted ideas 19
Dressing fantastically, etc. .. 19
Drink, intense craving for .. 19
Drinking and eating, excess in 21
Drowsiness 20
Drugs and alcohol, increased reaction to 39
Dulness and indifference .. 20
Dura mater, the (macroscopical morbid anatomy) .. 189
Durhæmatomata 189
Duties of medical attendant .. 287
— of person having charge of single patient 280
Dynamograph, Morselli's .. 153
Dynamometric indications .. 20
Dysmenorrhœa and irregular menstruation .. 20, 251
Dyspepsia 20

ESQUIROL, pp. v., 2, 83
Ears, lesions of 218
Eating and drinking, excess in 21

GENERAL INDEX. 329

	PAGE
Eccentricity, insanity from (diagnosis)	146
Echolalia	20
Eclampsic idiocy	84
Egotism, increased	20
Electric currents, sensations of in head	27
— sensibility abolished, etc.	20
Electricity in home treatment	252
— in hysterical insanity	265
Emaciation	20
Embarrassment, gastric	23
Emergency, certificate of (Scotland)	300
Emissions, nocturnal (treatment of)	252
Emotions, blunted	21
Emotional disturbance	20
— exaltation	132
Energy, diminished	21
Enfeeblement, mental	33
England and Wales, causes of insanity in	134
— — certification of insane patients in	269
— — laws as to keeping single patients	279
Enjoyment of life, lost, etc.	21
Ephemeral mania	92
Epilepsy and epileptic insanity (treatment)	263
— general, pathology of	196, 239
Epileptic dementia (diagnosis)	161
— insanity (etiology)	139
— fits	27
— idiocy	84
— insanity	67
— — (prognosis)	177
— seizures	21
Epileptiform attacks	21
Epileptisches irresein	67
Erotic ideas or tendency	21
— tendency	216
Erotomania	64
— (diagnosis)	156
Esquirol's classification	2
Etiology	134
Evidence the, of the insane	319
Evil presentiments of	38
Exacerbations at menstrual periods	21
— nocturnal	35
Exaggeration, ideas of	21
— of trifles	21
— proneness to	21
Exaltation	21

	PAGE
Exaltation and depression, alternating	12, 21
— emotional	132
— partial	106
— — (etiology)	143
— — (prognosis)	184
— with change of habits	157
Examination of patient in asylum	279
— of sexual organs	153
Examination of single patient	279
Examining a patient, method of	150
Excess, sexual	21
— in eating and drinking	21
— alcoholic tendency to	12
Excited melancholia	97
Excitement, motor and mental	22
— sexual	42
Exciting causes of insanity	135
Excretions, deficient	22
Exercise in home treatment	251, 256
Exertion, mental and bodily, incapacity for	22
Exhaustion, nervous	22
Exophthalmic goître, insanity with (etiology)	69, 139
— — — of (treatment)	264
— — — with (diagnosis)	163
Expenditure, extravagance as to, etc.	22
Expression, facial	22
— vacant	47
Extravagance as to expenditure, etc.	22
Extremities, cold and bluish	22
— rigidity of	40
— sores on	43
Eyes, averted	22
— neuroses affecting	35
— downcast	22
— fixed	22
— glistening	22
— hollow	22
— injected	22
— shutting	42

FALRET, *p.* 51; Ferrier, 204, 205, 215, 229

	PAGE
Face, flushed	22
— haggard	22
— sudden redness of	22
Facial circulation defective	22
— expression	22
— — absence of	166
— — anxious or terrified	167
— — intelligent or ecstatic	167

Facial muscles, movements of	23
Faculties, intellectual, absent	32
— reproductive, absent	40
Faradism, reactions to..	152
Fatigued easily ..	23
Fatuity ..	23
Fears, vague	47
Features, contracted	23
Feeble grip	24
Feeding, forcible	23
Feigned insanity (diagnosis)..	146
Female patient, penalty for carnal knowledge	287
Females, frequency of insanity occurring in	169
Fevers, delirium of (diagnosis)	147
Feverishness	23
Fickleness	23
Fingers, picking	38
Fixed ideas	29, 132
Flesh, gaining	23
Fly or hide, tendency to	23
Folds, naso-labial, flattened, etc.	35
Folie à Deux	70
— — — (etiology)	139
— — double forme	51
— alternante	51, 158
— circulaire	51
— du Doute	70
— — — (etiology)	139
— — — (prognosis)	177
— — — (treatment)	266
— épileptique	67
— hystérique	81
— raisonnante	100, 113
— — (etiology)..	139
— — (prognosis)	177
— systématisé	61
Food, refusal of	23, 39, 247
Force, want of, in thought and action	23
Forcible feeding	254
Forensic psychiatry	269
Forgetfulness	23
Form for admission of paying patients (Ireland)	308
— for criminal lunatics	296
— for receiving pauper lunatic	292
— of certificate of death	290
— of medical certificate (Ireland)	302
— of medical statement	288, 290
— of medical visitation book..	282
— of notice of admission	280
— of notice of discharge	283
— of urgency order (England and Wales)	270
Forms for reception of patients (Scotland)	297
— of insanity	137
Formication	23
— sensation of ..	34
Frenosi sensoria acuta	56
Frenzy, transitory	92
Frequency of principal forms of insanity	169
Friends, antipathy to ..	39
Fulness in head	27, 38
Functions, bodily, disturbance of	14
GALEZOWSKI, *p.* 78; Gamgee, 221; Gowers, 194, 202, 204, 221, 225, 229; Grasset, 69; Griesinger, 1, 25, 28, 58, 59, 72, 75, 76, 82, 90, 93, 94, 95, 97, 99, 178, 189, 190, 191, 194, 195, 197, 199, 248; Guggenbühl, 86; Guy, 321	
Gaining flesh	23
Gait, unsteady, etc.	23
Galvanism in insanity, exophthalmic goître	264
— reactions to ..	152
Ganglia, central, cells of the	221
Gastric embarrassment	23
— hypochondriasis	95
Gastro-enteric insanity	96
— — (diagnosis)	157
Gemüthswahnsinn	100
General epilepsy, pathology of	vi, 239
— mental weakness	213
— paralysis of the insane	71
— — — (etiology)	139
— — — (pathology)	196
— — — prodromal stage	158
— — — (prognosis)	177
— — — (treatment)	265
— prognosis	171
Genetous idiocy	84
Genuine delusions	17
Gestational insanity	80
— — (diagnosis)	163
— — (etiology)..	140
— — (prognosis)	178
Gesture, constantly making the same	16
Giddiness	23
Giving away property ..	23
Glance, vivacious	24
Gland, thyroid, lesions of	217
Gloominess	24
Goître, exophthalmic, insanity with	69

GENERAL INDEX.

	PAGE
Gouty insanity	110
— — (diagnosis)	155, 157
— — (etiology)	143
— — (prognosis)	184
Grave delirium	60
Grayness, premature	38
Grinding teeth	24
Grip, feeble	24
Groaning	24
Grouping of forms of mental aberration	131
Gunjah, insanity from (diagnosis)	165
Gustatory disturbances (morbid anatomy)	206
— hallucinations	24, 25
Gyratory impulses	24
HACK TUKE, pp. 253, 263, 317; Hoffbaner, 85; Huchard, 67, 81	
Habits, change of	24, 156
— dirty in	19
— wet in	48
Hachish, insanity from (diagnosis)	165
Hæmatoma auris	24
Hæmic causes of insomnia	226
Hæmorrhages, mucous	24, 80
Hair, lesions of	218
— pulling out	38
Hallucinations	132, 164
— auditory	25
— compound	26
— gustatory	24, 25
— in chronic alcoholic insanity	123
— of smell	26, 42
— production of	239
— sexual	26
— simple	24
— tactual, etc.	26
— visceral	26
Hallucinatory delirium	56
— psychoneurosis	56
Hands, wringing of	48
Happiness, feeling of	26
Head, flashes of heat to	27
— large	27
— measurement of	152
— pains in	27
— sensation of electric currents in	27
— fulness in	27
— pressure in	27
— temperature of	45
— very small	27

	PAGE
Headache	26, 73
Hearing impaired	27
— voices	27
Heart, lesions of	216
— palpitation of	27
Heat, flashes of, to head	27
— sense abolished, etc.	27
Hebephrenia	111
Height and weight of patient	152
Hemianæsthesia	27
Hemiplegia	27
Henbane in insomnia	244
Home treatment	249
— — minor details in forcible feeding during	255
Homicide or suicide	248
Homicidal impulse	27
— mania	156
Hopelessness	27
Hospitals, lunatic	260
Husband and children injuring	31
— having charge of wife	286
— repugnance to	40
Hydrocephalic idiocy	84
Hygiene and therapeutics	242
Hyoscine hydrobromate	244
Hyoscyamine in insomnia	244
— motor excitement	246
Hyperacousia	27
Hyperæsthesia, cutaneous	27
Hyperkinesis	86
Hypnotism, hypothesis as to	vii, 235
— in home treatment	253
— in insomnia	245
Hypnotics	243
Hypochondriacal melancholia	94
Hypochondriasis	27, 81, 95
— sexual	96
Hyperbulia	27
Hypocrisy, self-accusations of	41
Hypomania	90
Hysteria	27
Hysterical convulsions	27
— insanity	81
— — (diagnosis)	166
— — (etiology)	140
— — (prognosis)	178
— — (treatment)	265
— — chronic (diagnosis)	162
Hystero-epileptic fits	27
IRELAND, pp. 70, 83, 84, 85, 149, 239	
Ideas about sin	42
— ambitious	12
— and words, automatic	14

GENERAL INDEX.

Ideas, concentration of intellectual operations round one set of 15
— confusion of 15
— erotic, or tendency 21
— exalted, dress suggestive of 19
— fixed 29, 132
— incorrect of place 38
— mobile and futile 27
— mobility of 34
— multiple, absurd, etc. .. 27
— of exaggeration 21
— of persecution 37
— of poisoning 38
— of satisfaction 41
— of time, incorrect 46
— of unworthiness 47
— of wealth 48
— painful 36
— paucity of 28
Ideation, slowness of 42
Ideenjagd vii
Identity, mistakes of 28
Idiocy 83, 166
— (pathology) 197
— (prognosis) 178
— (treatment) 265
— pronounced cases of .. 167
— and imbecility (diagnosis) 161
— — (etiology) 141
— by deprivation 85
— the brain in 237
Idiot, committal warrant for dangerous (Ireland) .. 306
Ill-being, sense of 28
Illinois, certification of the insane in the State of .. 316
Illness of patient 287
Ill-treatment of patient .. 286
Illusions 28
Imagination weakened .. 29
Imbeciles 157
Imbecility 85
— congenital 83
— (prognosis) 178
— (treatment) 265
— and idiocy (diagnosis) .. 161
— — (etiology) 141
— with epilepsy 166
Immediate relief of urgent symptoms 243
Immoral, grossly and openly .. 29
Impaired sight 42
Impatience 29
Imperative conceptions 29, 104
Impiety, self-accusations of .. 41

Improvidence of actions .. 30
Impulse, homicidal 27
— morbid .. 30, 132, 156
— cursatory 16
— gyratory 24
Impulsive insanity 86
— — (etiology) 141
— — (diagnosis) 156
— — (prognosis) 178
Inability of patient to feed himself 247
— to write 48
Incoherence 30
Inconsistencies in speech and acts 31
Indecision 31
Index to paragraphs on diagnosis 168
Indications, dynamometric .. 20
Indifference 31
— and dulness 20
Inertia 31
Infantile convulsions, succeeding 31
Infant, perceptions of an .. 221
Inflammations, delirium of (diagnosis) 147
Inflammatory idiocy 85
Influenza .. .. vi, 243, 252
Inhibitory insanity 86
Inhibition, loss of power of .. 31
Injuring husband and children 31
— self 31
Inquietude 31
Inquisition, persons found lunatic by 290
Insane, admission into State Hospital, New York .. 316
— certification of, in the State of Connecticut 316
— — — Illinois 316
— — — Massachusetts .. 316
— — — New York 314
— — — Pennsylvania .. 316
— commitment of (New York State) 314
— general paralysis of the .. 71
— — — (etiology) 139
— — — (prognosis) 177
— — — (treatment) 265
— in Ireland, certification of 301
— Scotland, certification of .. 296
— patients, certification of in England and Wales .. 269
— responsibility of the .. 320
— testamentary capacity of the 317

GENERAL INDEX.

	PAGE
Insane, the evidence of the	319
Insanities, primary, cases of	241
Insanity, acute alcoholic (prognosis)	187
— adolescent	49
— — (diagnosis)	157, 163
— — (etiology)	137
— — (prognosis)	173
— — (treatment)	263
— advanced chronic confusional (diagnosis)	162
— alcoholic	120
— — (pathology)	202
— amenorrhœal	131
— — (diagnosis)	163
— — (etiology)	145
— — (prognosis)	188
— anæmic	50
— — (diagnosis)	157
— — (etiology)	137
— — (prognosis)	173
— cataleptic	50
— — (diagnosis)	166
— — (etiology)	137
— — (prognosis)	173
— causes of	137
— choreic	50
— — (etiology)	137
— — (pathology)	193
— — (prognosis)	173
— — (treatment)	262
— chronic alcoholic	123
— — — (diagnosis)	159
— — confusional (diagnosis)	158
— — epileptic (diagnosis)	161
— — hysterical (diagnosis)	162
— circular	51
— — (etiology)	137
— — (prognosis)	174
— — maniacal phase (diagnosis)	163
— climacteric	53
— — (diagnosis)	164
— — (etiology)	137
— — (pathology)	193
— — (prognosis)	174
— — (treatment)	263
— confusional	56
— — (etiology)	138
— — (pathology)	194
— — chronic (prognosis)	175
— — primary (prognosis)	174
— consecutive	58
— — (diagnosis)	155, 163
— — (etiology)	138

	PAGE
Insanity, consecutive (pathology)	194
— — (prognosis)	175
— cyanosis from bronchitis, etc. (prognosis)	175
— definitions of	1
— delusional	61
— — (diagnosis)	154, 166
— — (etiology)	138
— — (pathology)	195
— — (prognosis)	175
— diabetic	66
— — (diagnosis)	163
— — (etiology)	139
— — (prognosis)	176
— diagnosis of	146
— differential diagnosis of the forms of	154
— doubting (diagnosis)	158
— epileptic	67
— — (etiology)	139
— — (prognosis)	177
— — (treatment)	263
— feigned (diagnosis)	146
— forms of	137
— in which crimes are committed	320
— met with in asylums	170
— frequency of forms of at death	170
— principal forms of	169
— from abdominal disorders (etiology)	137
— absinthe	126
— asthma (diagnosis)	165
— bronchitis (diagnosis)	165
— cannabis indica	127
— — (diagnosis)	165
— cardiac disease (diagnosis)	165
— chloral	127
— coarse brain disease (diagnosis)	160
— — — (etiology)	137
— — — (prognosis)	174
— cyanosis (diagnosis)	165
— deprivation of senses	66, 138
— eccentricity (diagnosis)	146
— lead	127
— mercury	129
— morphia (diagnosis)	164
— opium	126
— — (diagnosis)	164
— — (prognosis)	188
— gastro-enteric	96
— gestational	80
— — (diagnosis)	163

GENERAL INDEX.

	PAGE
Insanity, gestational (etiology)	140
— — (prognosis)	178
— gouty (diagnosis)	155
— (etiology)	143
— — (prognosis)	184
— hypochondriacal	94
— hysterical	81
— — (diagnosis)	166
— — (etiology)	140
— — (prognosis)	178
— — (treatment)	265
— impulsive	86
— — (diagnosis)	156
— — (etiology)	141
— — (prognosis)	178
katatonic	88
— — (etiology)	141
— lactational	89
— — (diagnosis)	164
— — (etiology)	141
— — (prognosis)	179
— legal tests of	320
— masturbational (diagnosis) 157, 161, 163	
— — (etiology)	141
— — (prognosis)	181
— — (treatment)	263
— menstrual	107
— metastatic	100
— — (diagnosis)	163
— — (etiology)	142
— — (prognosis)	183
— mild choreic (diagnosis)	162
— moral	100
— — (etiology)	142
— — (prognosis)	183
— — causes of	134, 135
— neurasthenic	101
— — (etiology)	142
— — (prognosis)	183
— no two cases alike	145
— of Bright's disease	50
— — — (etiology)	137
— — — (prognosis)	173
— of cardiac disease and asthma	59
— of cyanosis from bronchitis	59
— of masturbation	92
— of myxœdema	101
— — (diagnosis)	161
— — (etiology)	142
— — (prognosis)	183
— — (treatment)	266
— of oxaluria	105
— — (etiology)	143
— — (prognosis)	183

	PAGE
Insanity of paralysis agitans	105
— — — (etiology)	143
— — — (prognosis)	184
— of phosphaturia	105
— — (etiology)	143
— — (prognosis)	183
— of prisoners	295
— of puberty	111
— old maid's	104
— — (etiology)	142
— — (prognosis)	183
— ovarian	104
— — (diagnosis)	156
— — (etiology)	142
— — (prognosis)	183
Insanity, paralytic	54
— partial (diagnosis)	167
— pellagrous	107
— — (diagnosis)	163
— — (etiology)	143
— — (prognosis)	184
— — (treatment)	267
— periodical	107
— — (diagnosis)	158
— — (etiology)	143
— — (pathology)	200
— — (prognosis)	184
— phthisical	108
— — (diagnosis)	154, 155
— — (etiology)	143
— — (prognosis)	184
— physical causes of	134, 135
— podagrous	110
— — (etiology)	143
— — (prognosis)	184
— post-connubial	111
— — (diagnosis)	162
— — (etiology)	143
— — (prognosis)	184
— post-febrile	58
— predisposing causes of	135
— primary or acute confusional (diagnosis)	162
— principal forms of	2
— pseudo of somnambulism	117
— — — (prognosis)	186
— pubescent	111
— — (diagnosis)	163
— — (etiology)	143
— — (prognosis)	184
— — (treatment)	263
— puerperal	112
— — (etiology)	143
— — (pathology)	200
— — (prognosis)	185
— — (treatment)	267

	PAGE
Insanity, reasoning	113
— rheumatic	114
— — (diagnosis)	161
— — (etiology)	144
— — (prognosis)	185
— saturnine	127
— — (diagnosis)	161
— — (pathology)	202
— — (prognosis)	188
— — comatose form (diagnosis)	162
— senile	116
— — (etiology)	144
— — (pathology)	200
— — (prognosis)	186
— — (treatment)	263
— severe choreic (diagnosis)	162
— Sibbald's gastro-enteric (etiology)	137
— stuporous	117
— — (etiology)	144
— supervening suddenly	254
— symptoms of	240
— syphilitic	118
— — (diagnosis)	159
— — (etiology)	144
— — (pathology)	201
— — (prognosis)	186
— — (treatment)	267
— — expansive (diagnosis)	163
— toxic	120
— — (etiology)	144
— — (pathology)	202
— — (prognosis)	187
— — (treatment)	267
— transitory neurasthenic	103
— traumatic	129
— — (diagnosis)	164
— — (etiology)	145
— — (pathology)	202
— — (prognosis)	188
— treatment of special forms of	262
— uterine	131
— — (diagnosis)	163, 164
— — (etiology)	145
— — (prognosis)	188
- weight of brain in	192
— with Bright's disease (diagnosis)	163
— with exophthalmic goître	69
— — — (diagnosis)	163
— — — (etiology)	139
— with paralysis agitans (diagnosis)	155
Insomnia	31
— (therapeutics)	243

	PAGE
Insomnia, exciting causes of	226
— in home treatment	251
Instability	32
Instruction, incapable of	32
Interest, loss of in surroundings	32
International Congress of 1867	135
Interrogation of self	41
Intestinal and stomach lesions	217
Intestinal disturbances	32
Intoxication, alcoholic and other (diagnosis)	148
Intractability and wildness	48
Introspection, shallow, etc.	32
Irascibility	32
Ireland, certification of the insane in	301
Irritability	32
Irritation of skin	42
JACKSONIAN epilepsy (treatment)	264
Jaws, champing of	32
Jealousy, insane	32
Judgment, defective	32
— — morbid anatomy of	213
— — weighing own	48
KAHLBAUM, pp. 26, 88, 89; Kolk, 49; Krafft-Ebing, 4, 9, 26, 41, 42, 46, 56, 57, 61, 65, 71, 90, 96, 100, 101, 104, 129, 139, 142, 177, 242, 244, 246, 250, 253	
Katatonia or katatonic insanity	vi, 88, 161
— cataleptic phase of	166
— (etiology)	141
— (pathology)	199
— (prognosis	179
Kidneys, lesions of	217
Kill, impulse to	32
Kleptomania	156
Krafft-Ebing's classification	4
— — recent classification	v, 9
LEWIS (Bevan), pp. 39, 69, 185, 190, 191, 195, 196, 197, 199; Lombard, 45; Luciani, 208, 221; Luys, 239	
Lactational insanity	89
— — (diagnosis)	164
— — (etiology)	141
— — (prognosis)	179
Language, obscene, etc.	32

	PAGE		PAGE
Language, incoherent	165	Lunatics, legal tests of	320
Lassitude	32	— (not pauper) not under control	294
Laughter	32	— pauper	291
— motiveless	34	— responsibility of	320
Laws as to keeping single patients, England and Wales	279	— testamentary capacity of	317
		— the evidence of	319
— of 1889 (New York State)	315	— uncertified	291
Laziness	32	— wandering	294
Lead, insanity from	127	Lungs, lesions of	216
Legal regulations	269	Lycanthropia	156
— tests of insanity	320	Lypemania	93
Lesions of non-nervous organs	216	Lypothymia	93
Lethargy	33		
Life, enjoyment of lost	21		
Listlessness	33	MAGNAN, pp. 45, 62, 63, 69, 121, 122, 123, 124, 125, 126, 187, 188, 258; Marcé, 51, 87; Maudsley, 86, 88, 172, 181, 183, 188, 196, 220, 236, 238, 319, 320; Max Müller, 149; Mendel, 56, 90; Mercier, 320; Merson, 53; Meynert, 191, 196, 199, 201, 204, 218, 219, 221, 222, 223, 225, 227, 228, 231, 239, 252; Mickle, 72, 75, 193, 197, 199, 200, 210; Morel, 2, 5; Morselli, 56, 70, 71, 86, 88, 107, 129, 153, 236; Monti, 40; Münsterberg, 153	
Lip, swelling of upper	47		
Liver, lesions of	217		
Localisation, cerebral	vi, 202		
Localised paralysis	33		
Lochia altered, etc.	33		
Locomotion, disturbances of	33		
London college of physicians' classification	8		
Loquacity	33		
Loss of ability to write	33		
— of self-confidence	41		
— of self-control	41		
— of sexual desire	216		
Low company, disposition for	33		
Lunacy Act, 1890, prosecution under	278	Magistrates' certificate for paying patients (Ireland)	312
— report of the commissioners in, 1889	134	Making words	47
		Malaise	33
Lunatic Asylum, admission to (Ireland)	308	Males, frequency of insanity occurring in	169
— — rules for (Ireland)	309	Mania	89
— hospitals	260	— (etiology)	141
— order for transmission of (Scotland)	300	— (pathology)	200
		— (prognosis)	179
Lunatics, admission into State hospitals, New York	316	— acute (diagnosis)	163
— by inquisition	290	— chronic advanced (diagnosis)	162
— certification of, in the State of Connecticut	316	— delusional (diagnosis)	165
		— destructive (diagnosis)	156
— — — Illinois	316	— hallucinatoria	56
— — — Massachusetts	316	— in young children	238
— — — Pennsylvania	316	— periodical (diagnosis)	163
— commitment of (New York State)	314	— puerperal	112
		— — (diagnosis)	163
— committal warrant for dangerous (Ireland)	306	— senile	116
		— — (diagnosis)	161
— criminal	294	— simple	90
— cruelly treated	294	— transitory (diagnosis)	165
— dangerous	294	Manie sans délire	100
— — (Ireland)	303	Manipulations, awkwardness in	33

GENERAL INDEX.

	PAGE
Manner, fierce	33
— foolish	33
— jolly	33
Manual inability, etc.	33
Marasmus	33
Massachusetts, certification of insane in the State of	316
Massage in home treatment	253
Masturbation	33
— self-reproaches of	41
Masturbational insanity	92
— — (diagnosis)	157, 161, 163
— — (etiology)	141
— — (prognosis)	181
— — (treatment)	263
Masturbatory melancholia	165
Medical attendant, definition of	287
— — duties of	287
— certificate (Scotland)	299
— — for dangerous lunatic (Ireland)	308
— for paying patients (Ireland)	310
— certificates (Ireland)	302, 305
— — persons disqualified from signing	278, 280
— superintendents, duties of	287
— visitation book, form of	282
Medicines in general paralysis of the insane	265
Megalomania	63
Melancholia	93, 166
— (etiology)	141
— (pathology)	v, 200
— (prognosis)	181
— (treatment)	266
— acute (diagnosis)	163
— agitated	97
— attonita	99
— delusional	97, 164
— in childhood	238
— masturbatory	165
— organic	56, 160
— puerperal	113
— religious	98
— senile	116, 161
— simple	94
— stuporous	99, 167
— suicidal	98
Melancholic frenzy (diagnosis)	165
Melancholy folie raisonnante	114
Memory, defective	33
— — morbid anatomy of	213
— loss of	33
— of patient	155
Mendacity	33

	PAGE
Meningitis, cerebral (diagnosis)	148
Menstrual insanity	107
— periods, exacerbations at	21
Menstruation irregular, etc.	33
— — and dysmenorrhœa	20
Mental aberration, grouping of forms of	131
— deterioration	99
— disease, how developed	239
— enfeeblement	33
— pain	36, 131
— restlessness	40
— torpor	46
— treatment	256
— weakness	33
— — general	213
Mercury, insanity from	129
Metastasis	34
Metastatic insanity	100
— — (diagnosis)	163
— — (etiology)	142
— — (prognosis)	183
Method of examining a patient	150
Methylal in insomnia	245
Microcephalic idiocy	84
Migraine	34
Military patients (Ireland)	303
Minor details in home treatment	255
Misanthropic	34
Mischievousness, tendency to	34
Mistakes of identity	28
Mixed causes of insanity	136
Moaning	34
Mobility of character	34
— — ideas	34
Mockery, tendency to	34
Monomania	61
— — (diagnosis)	154
— — (pathology)	195
— — (prognosis)	175
Monomonie affective	100
Monoparaplegia	34
Monophobia	105
Monoplegia	34
Monopsychosis	61
Monotony of speech and action	34
— — thoughts and movements	34
Mood quickly changing	34
Moral causes of insanity	134, 135
— deterioration	34
— insanity	100
— (etiology)	142
— (prognosis)	183
— perverseness	34
— treatment	256

	PAGE		PAGE
Moralisches irresein	100	Nervousness	35
Morbid anatomy	189	Neuralgia, general, etc.	35
— — of symptoms	202	Neurasthenia	101
— impulse	132, 156	— (prognosis)	183
— self-feeling	41	— (treatment)	266
— sensations	34	— Weir-Mitchell's treatment of	267
— sensiblerie	214	Neurasthenic degenerative psychoses	104
Morel's classification	2	— insanity (etiology)	142
Moroseness	34	— psychoneuroses	103
Morphia in motor excitement	246	— psychoses	103
— — paranoia	267	Neuritis, optic	36
— — insanity (diagnosis)	164	Neuroses affecting eye	35
— — suicidal tendency	249	Never speaking	35
Morphinismus (treatment)	268	New York, certification of the insane in the state of	314
Morphismus, chronic	155	— — Pauper Asylum, statistics of	170
Morselli's dynamograph	153	
Motiveless actions	34	
— laughter	34	
Motor excitement, etc.	245	Nocturnal emissions (treatment)	252
— — remedies in	245, 246	
— restlessness	34	Nocturnal exacerbations	35
— symptoms (morbid anatomy of)	208	Noisiness	35, 245
		Non-medicinal treatment	256
Movement, aversion to	35	Nosophobia	35
— resistance to, etc.	40	Notice of desire to have a personal interview	281
— voluntary	47	
— and thought, monotony of	34	— — discharge form of	283
— choreic	15	— — right to personal interview	281
— rhythmical	40	
Mucous hæmorrhages	24, 80	
Muscles, lesions of	217	Nourishment in motor excitement	246
— rigidity of	40	
Muscular contractures	35, 79	Nymphomonia	30, 156
— relaxation	35	— (prognosis)	178
— resistance diminished	35	Nystagmus	36
— sense abolished, etc.	35	
— weakness, limited	35	Ord, p. 142
Mutilation of self	41	Objects and persons, distortion of	19
Mutism	35	Obscenity	36
Muttering of patient	166	Obstinacy	36
— isolated words	35	Occupation, change of	36
Mysophobia	29, 105, 158	Occupations etc. in home treatment	256
— (prognosis)	184	Oddness and peculiarity, from birth	36
Myxœdema, insanity of	101	
— — — (diagnosis)	161	Odour exhaled	36
— — — (etiology)	142	Œdema	
— — — (prognosis)	183	Old maid's insanity	104
— — — (treatment)	266	— — — (diagnosis)	155
		— — — (etiology)	142
Nasse, p. 58; Newington, 56		— — — (prognosis)	183
Naked, stripping in public	43	Olfactory disturbances (morbid anatomy)	206
Naso-labial folds flattened, etc.	35	
Natural affection, loss of	35	— hallucinations	26, 36
Necrophilism	30, 156	Onanism	41
Neglect of lunatics	294	Opium in motor excitement	246
Nervous exhaustion	22	— in insanity (diagnosis)	126, 164

GENERAL INDEX.

	PAGE
Opium in insanity (prognosis)	188
— in suicidal tendency	249
Optic nerve changes, etc.	36
Optic neuritis	36
Order for reception of private patients	276
— — transmission of lunatic (Scotland)	300
Organic dementia	vii, 54, 160
— melancholia	56
Organs, lesions of non-nervous	216
Originating power, want of	36
Othæmatoma	24, 36
Othæmatomata	80
Outbreak, abruptness of	11
Ovarian insanity	104
— — (diagnosis)	155, 156
— — (etiology)	142
— — (prognosis)	183
Ovaries and uterus, lesions of	217
Oxaluria, insanity of	105
— — — (etiology)	143
— — — (prognosis)	183

PILGRIM, *p.* 316; Pinel, 2, 190; Prichard, 2; Pritchard, 24

	PAGE
Painful ideas	36
— sensations, heat, etc.	36
Pain, mental	36, 131
Pains	27, 86
Pallor	37
Palpitation	27, 37
Paraldehyde in insomnia	243
Paralysis agitans	155
— — insanity of	105
— — (etiology)	143
— — (prognosis)	184
— — alcoholic, pseudo-general	125, 160
— and paresis	37
— confirmed general	160
— general (pathology)	196
— — prodromal stage	158
— localised	33
— of vocal cord or cords	47
— saturnine, pseudo-general	128, 160
— syphilitic, pseudo-general	119, 160
Paralytic dementia	54
— idiocy	84
Paranoia	v, 61
— (diagnosis)	154
— (pathology)	195
— (prognosis)	175

	PAGE
Paranoia (treatment)	267
— neurasthenic	104
— — masturbatory (treatment)	266
— psiconeurotica	56
— rudimentaria impulsiva	86
Paraplegia	37
Paresis and paralysis	37
— localised	33
— vaso-motor	47
Paretic dementia	71
Parsimony	37
Partial dementia	100, 159
— emotional aberration	105
— — — (etiology)	143
— — — (prognosis)	184
— — — (treatment)	266
— exaltation	106
— — (etiology)	143
— — (prognosis)	184
— insanity (diagnosis)	167
Passive suffering	37
Pathological anatomy	189
Pathology and pathogenesis	218
Pathophobia (prognosis)	184
Pathos	37
Patient, absence of facial expression in	166
— absolutely mute	166
— and business	287
— change of residence	286
— chancery	290
— commitment of (New York State)	314
— death of	289
— examination of, in asylum	279
— illness of	287
— ill-treatment of	286
— inability of, to feed himself	247
— incoherence of language of	165
— loquacity of	165
— making inarticulate noises	167
— military (Ireland)	303
— method of examining a	150
— muttering of	166
— non-paying, certificate for (Ireland)	312
— pauper	291
— penalty for carnal knowledge of female	287
— private, admission into State Hospital, New York	316
— private order for reception of	276
— — petition for	274
— refusal of food by	247

GENERAL INDEX.

Patient, refusal to speak .. 166
— removal of to another asylum 287
— single, duties of person having charge of .. 280
— suitable for private care .. 257
— unable to speak 167
— visitation of 287
Pauper asylums.. 259
— lunatics 291
Pazzia catatonica 88
— epilettica 67
— isterica 81
Peculiarity from birth 36
Peevishness 37
Pellagra (treatment) 267
Pellagrous insanity 107
— — (diagnosis) 163
— — (etiology).. 143
— — (prognosis) 184
Penalty for carnal knowledge of female patient .. 287
Pennsylvania, certification of the insane in the State of 316
Penuriousness 37
Perception and sensation .. vii
Perceptions of an infant .. 221
Peripheral causes of insomnia 226
Periodical insanity 107
— — (diagnosis) 158
— — (etiology).. 143
— — (pathology) 200
— — (prognosis) 184
Periodicity 37
Periods, menstrual, exacerbations at 21
Persecutional delusions .. 156
Persecution, ideas of 37
Personal interview, certificate as to 281
— — notice of desire to have 281
— — — right to 281
Persons and objects, distortion of 19
— deemed lunatics 294
— disqualified from signing medical certificate .. 278
— found lunatic by inquisition 290
Perspiration, profuse 38
Perverseness, moral 34
Perversion, sexual .. .., 107
Petition for private patient .. 274
— to the Sheriff (Scotland) .. 296
Phosphaturia, insanity of .. 105
— — (etiology).. 143
— — (prognosis) 183

Photopsia 38
Phrase, repetition of 39
Phthisical insanity 108
— — (diagnosis) .. 154, 155
— — (etiology) 143
— — (prognosis) 184
Physical causes of insanity 134, 135
Pia mater, the (macroscopical morbid anatomy) .. 190
Picking fingers 38
Pilocarpine in threatening mania 247
Piscidia erythrina in insanity 263
— — insomnia 245
Place, desire for change of .. 14
— incorrect ideas of 38
Planomania 156
Pneumonia, delirium of (diagnosis) 147
Podagrous insanity 110
— — (diagnosis) 157
— — (etiology).. 143
— — (prognosis) 184
Poisoning, ideas of 38
Post-connubial insanity .. 111
Post-connubial insanity (diagnosis) 162
— — — (etiology) 143
— — — (prognosis) 184
Post-febrile insanity 58
Pot. bromide in motor excitement 246
Power, originating, want of .. 36
Predisposing causes of insanity 135
Pregnancy, insanity of .. 80
Premature grayness 38
Pre-occupation with ideas about sin.. 42
Presentiments of evil 38
Pressure or fulness in head, sensation of 38
— sensation of, in head .. 27
Pride, delusions of 156
Primary confusional insanity 56
— insanities, cases of 241
Prisoners found insane .. 294
Private asylums, admission into (Ireland) 301
— — (licensed houses) .. 261
— care (single patients) .. 257
— patient, order for reception of 276
— — petition for 274
— — admission into State Hospital, New York .. 314

Privy Council regulations (Ireland)	303
— — — 1876 (Ireland)	309
Prognosis, general	171
— special	173
Proneness to exaggeration	21
Propensities, morbid	38
Property, giving away	23
Prophylaxis, when to commence	242
Prosecution under Lunacy Act, 1890	278
Prostration	38
Pseudo-insanity of somnambulism	117
— — — (etiology)	144
Psychiatry, forensic	269
Psychical troubles	79
Psychlampsia	89, 93
Psychokinesia	86
Psychoneurosis, hallucinatory	56
— transitory neurasthenic	162, 167
Psychoses, neurasthenic	103
Ptosis	38
Puberty, insanity of	111
Pubescent insanity	111
— — (diagnosis)	157
— — (etiology)	143
— — (prognosis)	184
— — (treatment)	263
Puerperal dementia	113
— insanity	89, 112
— — (etiology)	143
— — (pathology)	200
— — (prognosis)	185
— — (treatment)	267
— mania	112
— — (diagnosis)	163
— melancholia	113
Pulling out hair	38
Pulse	38
— slow	42
Punctiliousness	38
Pupils contracted, etc.	38
— dilated	38
— the, to examine	152
Pyromania	30, 156
QUARRELSOMENESS	39
Questions answered, but conversation not sustained	163
— answered irrelevantly	12, 162
— answering of, by patient	154
— not answered by patient	165
Quietude	39
ROBERTSON, p.106; Ross, 140, 173, 230, 234	

Ramollissement	54
Reaction time	39
— — patient's, ascertain	152
— to alcohol or drugs increased	39
— to Faradism	152
Reasoning insanity	113
References to paragraphs on diagnosis	168
Reflexes exaggerated	39
— investigation of	152
Refusal of food	23, 39, 247
Regulations, legal	269
— of Privy Council, 1876 (Ireland)	309
Relatives and friends, antipathy to	39
Relaxation, muscular	35
Religious despondency	19
— melancholia	98, 164
— tinge	39
Remedies in insomnia	243
— in motor excitement	245, 246
Remedial treatment	243
Remissions with delirium	17
Remorse	39
Removal of patient to another asylum	287
Repetition of words to self	39
Report of the Commissioners in Lunacy, 1889	134
Reproductive faculties absent	40
Repugnance to husband	40
Re-representative cognitions and feelings	214
Residence, change of, with patient	286
Resistance, muscular, diminished	35
— to movement, etc.	40
Respirations, affected	40
Responsibility of the insane	319
Restlessness, mental	40
— motor	35, 40
Restraint, certificate as to mechanical means of	285
Retention of urine	40
Retina, changes in	40
Rheumatic insanity	114
— — (diagnosis)	161
— — (etiology)	144
— — (prognosis)	185
Rheumatism, cerebral	114
Rhyming speech	40
Rhythmical movements	40
Rigidity, cataleptic	40
— of muscles, etc.	40

GENERAL INDEX.

SANDER, *p.* 61; Sankey, 110; Saulle, 70; Savage, 1, 8, 25, 26, 27, 58, 59, 65, 67, 69, 73, 82, 89, 97, 98, 99, 112, 113, 116, 127, 140, 144, 183, 185; Schüle, 61; Sepilli, 208; Sibbald, 49, 96, 120, 137, 184, 187; Skae, 4, 64; Spencer (Herbert), vii, 220, 223, 227; Spitzka, 4, 6, 11, 14, 15, 17, 18, 20, 21, 25, 26, 27, 28, 29, 30, 33, 34, 40, 42, 46, 47, 50, 52, 56, 57, 58, 60, 61, 62, 64, 65, 66, 68, 71, 82, 87, 88, 93, 99, 100, 108, 111, 117, 118, 124, 129, 130, 131, 138, 141, 144, 145, 170, 175, 176, 179, 182, 184, 194, 195, 219, 245; Stretch-Dowse, 201, 243; Sully, 28

Salix Nigra in erotism, emissions, etc. .. 252, 258
Sallowness 40
Satisfaction, ideas of 41
Saturnine insanity 127
— — (diagnosis) 161
— — (pathology) 202
— — (prognosis) 188
— — comatose form (diagnosis) 162
— pseudo-general paralysis 128, 160
Satyriasis 30, 41, 156
Savage's classification 8
Scotland, certification of the insane in.. 296
Screaming 41
Seasonal and barometric conditions 14
Secondary dementia (prognosis) 176
Secretions, diminished .. 41
Seizures, epileptic 21
Self-abasement 41
— absorption 41
— abuse 41
— accusations of hypocrisy, etc. 41
— confidence, loss of 41
— control, loss of 41
— disposition to connect everything with 15
— esteem, exaggerated or diminished 41
— feeling, morbid 41
— injuring 31
— injury 41
— interrogation 41
— mutilation 41

Self, repetition of words to .. 89
— reproaches of masturbation 41
— talking to 44
Senile dementia .. 117, 160
— insanity 116
— — (etiology).. 144
— — (pathology) 200
— — (prognosis) 186
— — (treatment) 263
— mania 116
— — (diagnosis) 161
— melancholia 116, 161
— speech 43
Sensation and perception .. vii
Sensations, initiatory, of an infant 223
— morbid 34
— of electric currents in head 27
— of fulness in head 27
— of pressure in head.. 27, 38
— painful, heat, etc. 36
Sense of being controlled, etc. 16
— of heat abolished 27
— of ill-being 28
Senses, absence of two or more 41
— acute 41
— deprivation of .. 66, 157
— — insanity from (etiology) 188
— enfeebled 41
— muscular, abolished, etc. .. 35
— the various, to investigate the acuteness of.. vi, 152
Sensibility, disorders of .. 202
— electric, abolished, etc. .. 20
— enfeebled, etc. 41
— extreme 41
— perverted, etc. 11
Sensiblerie, morbid 214
Sentences uncompleted .. 41
Sentiments, alteration of .. 41
Seriousness increased 41
Sexual appetite diminished .. 41
— — lost 41
— desire, loss of 216
— excess 21
— excitement .. 42, 252, 257
— hallucination 26
— hypochondriasis 96
— neurasthenia 102
— organs, examination of .. 153
— perversion 42, 107
Shutting eyes 42
Sibbald's gastro-enteric insanity (etiology) 187
Sight, impaired.. 42

GENERAL INDEX. 343

	PAGE
Similarity of repeated attacks in same patient	42
Simple hallucinations	24
— mania	90
— melancholia	94
— primary dementia	99
Single patient, duties of person having charge of	280
— — examination of	279
— — (private care)	257
— — laws as to keeping, England and Wales	279
Sin, ideas about	42
Sitophobia	42
— and anorexia	215
Size and form of cerebral substance	191
Skae's classification	4
Skin, changes in	42
— irritation	42
— lesions of	217
Sleep, how produced	224
— theories of	vii, 224, 225
— walking in	235
Sleeplessness	42
Slovenly and untidy	42
Slowness of ideation	42
Slow pulse	42
Small head	27
Smell, hallucinations of	42
Sodii salicyl. as a bactericide	262
Solitariness	42
Somatic stigmata	42
Sombreness	43
Somnambulism, pseudo-insanity of	117
— — — (etiology)	144
— — — (prognosis)	186
Somnolence, or coma	15
Sores on extremities, tendency to	43
Speaking, never	35
Special prognosis	173
Specific gravity of cerebral substance	192
Speech, abnormalities of	43
— and action, monotony of	34
— — inconsistencies in	31
— congenitally absent, etc.	43
— or action limited	33
— rhyming	40
— senile	43
Spermatorrhœa	43
Spinal cord in general paralysis	197
— neurasthenia	102
Spitzka's classification	6

	PAGE
Spitzka's statistics of New York pauper asylum	170
Spleen, lesions of	217
Spontaneity impaired	43
Spurious delusions	18
Squander, disposition to	43
State of New York, certification of the insane in	314
— restless, etc.	43
Statement accompanying urgency order	273
— form of medical	288, 290
Statements inconsistent, etc.	43
Statistics of frequency of principal forms of insanity	169
Status epilepticus (treatment)	264
Stigmata, somatic	42
Stomach and intestinal lesions	217
Stripping naked in public	43
Stunnings	43
Stupidity	43
Stupor	43, 133
— anergic	117, 167
— — (etiology)	144
— — (prognosis)	186
— delusional	56
— pathology of	225
Stuporous insanity	117
— melancholia	99
Subsultus tendinum	44
Suffering, passive	37
Suicide or homicide	248
— threatening to commit	45
Suicidal mania	156
— melancholia	98
— tendency	44, 215
Sulphonal in insomnia	243
Superintendents, medical, duties of	287
Suppression of catamenia	44
Suspicious delusions	155
Suspiciousness	44
Swelling of upper lip	47
Symmetry, bodily, want of	14
Symptoms influenced by physiological periods	44
— morbid anatomy of	202
— motor (morbid anatomy)	208
— of insanity	240
— of mental aberration	131
— treatment of at home	251
— urgent, immediate relief of	243
Syphilitic insanity	118
— — (diagnosis)	159
— — (etiology)	144
— — (pathology)	201

GENERAL INDEX.

	PAGE
Syphilitic insanity (prognosis)	186
— — (treatment)	267
— pseudo-general paralysis	119, 160
Systematised delusions	155
THUDICHUM, p. 221; Tuke (Batty), 136, 191, 219; Tuke (Hack), 253, 268, 316	
Taciturnity	44
Tactile hallucinations	26, 44
Talkativeness	44
Talking to self	44
— with uncompleted sentences	41
Taste, hallucinations of	44
Teeth, grinding	24
Temper and character, change of	74
— change of	45
— unequal	45
Temperature	44
— of head	45
Tendency to mischievousness	34
— to suicide	44
Terminal dementia	65
— — (diagnosis)	167
— — (etiology)	138
— — (pathology)	195
— — (prognosis)	176
Terminaler blödsinn	65
Terror	45
Testamentary capacity of the insane	317
Tests of insanity, legal	320, 321
Theomania	64
Theories of sleep	224
Therapeutics and hygiene	242
Thought and action, want of continuity in	16
— — — force in	23
Thoughts and feelings, concentration of, on health, etc.	15
— — movements, monotony of	34
Threatening to commit suicide	45
— violence to others	45
Thyroid gland affected	45
— — lesions of	217
Time, ideas of incorrect	46
— reaction	39
Timidity	46
Tinnitus aurium	46
Tissues, lesions of	216
Tobsucht	89
Tongue	46
Torpidity	46
Torpor, mental	46
Toxic insanity	120

	PAGE
Toxic insanity (etiology)	144
— — (pathology)	202
— — (prognosis)	187
— — (treatment)	267
Transitory mania	92
— — (diagnosis)	165
— neurasthenic insanity	103
Traumatic idiocy	84
— insanity	129
— — (diagnosis)	164
— — (etiology)	145
— — (pathology)	202
— — (prognosis)	188
Treatment at home	249
— of special forms of insanity	262
— symptoms at home	251
— remedial	243
Tremor	46
Trifles, annoyance at	12
— exaggeration of	21
Trophic disturbances	46
Trunk, rigidity of	40
Tumours in brain	55
Typhomania	60
Typhus fever, delirium of (diagnosis)	147
ULTIMATE care and treatment	249
Uncertified lunatics	291
Unconsciousness, apparent	13, 46
Unsociability	46
Unsystematised delusions	156, 158
Untidy and slovenly	42
Unworthiness, ideas of	47
Upper lip, swelling of	47
Urgency order, form of (England and Wales)	270
— — statement accompanying	278
Urine	47
— retention of	40
Uterine insanity	131
— — (diagnosis)	163, 164, 165, 167
— — (etiology)	145
— — (prognosis)	188
Uterus and ovaries, lesions of	217
VERITY, pp. 199, 200; Virchow, 199; Voisin, 36, 39, 40, 43, 72, 73, 74, 75, 76, 77, 78, 79, 80, 140, 177, 197	
Vacant expression	47
Vacillation	47
Vague fears	47
Vanity	47
Vascular changes	47

GENERAL INDEX.

	PAGE
Vaso-motor paresis	47
Ventricles, the (macroscopical morbid anatomy)	193
Veratrum viride in motor excitement	246
Verbigeration	vi., xiii., 47
Verrücktheit, acute primäre	56
Vertigo	23, 47
Violence	47
— threatening to others	45
Visceral hallucinations	26
— neurastheniæ	102
Visitation book, form of medical	282
— of patients	287
Visual disturbances (morbid anatomy of)	205
— hallucinations	47
Vocal cord or cords paralysed	47
Vociferation, abusive	47
Voices, hearing	27, 47
Voluntary boarders	261, 279
— movement	47
Voracity	47

WATTEVILLE, *pp.* 252, 265; Weatherly, 257; Westphal, 56; W i g l e s w o r t h, 36; Wilson (Erasmus), 143, 184; Wolfenden, 69; Wundt, 82

Wahnsinn	56
Wander abroad, disposition to	47
Wandering lunatics	294
Weakness	48
— congenital, mental or moral	133

	PAGE
Weakness, general mental	33, 213
— in back	14
— muscular, limited	35
— of will	158
Wealth, ideas of	48
Weeping	48
Weighing own judgments	48
Weight and height of patient	152
— of brain in insanity	192
Weir-Mitchell's treatment of neurasthenia	267
Wet in habits	48
Wife in charge of husband	286
Wildness and intractability	48
Will, affections of the	48
— weakness of	158
Word-making	47
Words and acts, paying attention to	36
— — — own	48
— ideas, automatic	14
— muttering isolated	35
— repetition of, to self	39
Work, distaste for	48
Wringing hands	48
Wrists flexed	48
Write, inability to	48
Writing altered	48

YOUNG children, delirium of	131
— — — (etiology)	145
— — — (prognosis)	188
ZWANGSVORSTELLUNGEN	29

INSANITY.

Its Classification, Diagnosis and Treatment;

A Manual for Students and Practitioners of Medicine.

BY

E. C. SPITZKA, M. D.,

Professor of Medical Jurisprudence and of the Anatomy and Physiology of the Nervous System, at the New York Post-Graduate School of Medicine, President of the New York Neurological Society, etc.

In this, the first systematic treatise on Insanity published in America since the days of the immortal Rush, the author has made its definitions, classifications, diagnosis and treatment plain and practical; and has laid particular stress upon points comparatively new and has succeeded in presenting the subject in such a manner that the rudiments of this difficult and intricate branch of medicine may be easily acquired and understood.

☞ This important work has already been adopted as the *Standard Text-Book* in the College of Physicians and Surgeons of New York, the College of Physicians and Surgeons of Baltimore, the Rush Medical College of Chicago, the College of Physicians and Surgeons of St. Louis, and the Medical-Chirurgical College of Philadelphia.

The Boston Medical and Surgical Journal says: "Conservative and in accordance with the highest principle of scientific investigation, which accepts no half-truth, but proven facts alone. . . . Its chief merit consists in its effort to present the subject in a clear, accurate, and scientific manner."

The Louisville Medical News says: "The book is written in a clear and forcible style, and while the practical side of the question is kept constantly in the foreground, it abounds in incidents, historical and modern, which admirably illustrate the points made by the author, and contribute largely to the entertainment of the reader."

The Weekly Medical Review says: "It cannot be neglected by any one desiring a clear and comprehensive review of the whole subject of insanity."

The New York Medical Record says: "The accomplished author displays throughout a masterly grasp of his intricate subject, and a familiarity with its bibliography which is in the highest degree commendable. . . The presentation of his arguments is direct and decided, his illustrations usually apt and well put, and his expositions of the most important points forcible."

The Cincinnati Lancet and Clinic says: "A great variety of useful information and an intelligent discussion."

The American Medical Weekly says: "It is clear, it is up to the times, and last but not least, it is practical."

The New England Medical Monthly says: "By far the best book that has appeared in English in this department of Science."

In One Large Octavo Volume, 424 pages. Illustrated. $2.75.

E. B. TREAT, Publisher, 5 Cooper Union, New York.